T3-BOI-748

FROM FUSION
TO LIGHT SURFING

FROM FUSION TO LIGHT SURFING

Lectures on Plasma Physics
Honoring John M. Dawson

Edited by

Thomas Katsouleas
University of Southern California

ADDISON-WESLEY PUBLISHING COMPANY
The Advanced Book Program
Redwood City, California • Menlo Park, California
Reading, Massachusetts • New York • Don Mills, Ontario
Workingham, United Kingdom • Amsterdam • Bonn
Sydney • Singapore • Tokyo • Madrid • San Juan

Publisher: *Allan M. Wylde*
Physics Editor: *Barbara Holland*
Marketing Manager: *Laura Likely*
Production Manager: *Jan V. Benes*
Production Assistant: *Karl Matsumoto*
Cover Design: *Irene Imfeld*

Library of Congress Cataloging-in-Publication Data

From fusion to light surfing: lectures on plasma physics
honoring John M. Dawson/edited by Thomas Katsouleas.
p. cm.
Includes bibliographical references.
1. Plasma (ionized gases)–Congresses. 2. Dawson, J. M. (John Myrick), 1930-.
I. Katsouleas, Thomas. II. Dawson, J. M. (John Myrick), 1930-.
QC717.6.F75 1991 530.4'4–dc20 91-11995
ISBN 0-201-55444-5

This book was prepared by the editor, using the TEX typesetting language.

1 2 3 4 5 6 7 8 9 10-MA-95 94 93 92 91

Contents

Foreword

Over the years, Professor John Dawson and I have developed a very special relationship. In fact, I think of him as my mentor, and I think he thinks of me as his tormentor. But I am not alone. Over the course of his career, John Dawson has mentored more than thirty other graduate students and generously shared his inspiration and physical insight with numerous colleagues. So it was with great enthusiasm that many of these colleagues, students, and former students from around the world gathered on Catalina Island on September 24th and 25th, 1990, to attend a symposium on the Physics of Plasmas and to honor Professor John Dawson on his sixtieth birthday.

According to Webster's Unabridged Dictionary, the word "symposium" traces its etymology to the ancient Greek. It arises from "sym" which means together plus "pinein" which means to drink. Thus, a symposium was originally a gathering at which to drink together. A more current definition is a meeting, especially one at which discussion of ideas takes place. By either definition the "International 'Dawson' Symposium on the Physics of Plasmas" was a resounding success.

This book contains the invited lectures that were presented at the Dawson Symposium and some of the introductions to those lectures. Included at the end of the text is a listing of the 283 publications John Dawson has published to date. The title of the Symposium as well as this book, *The Physics of Plasmas*, reflects the breadth of John Dawson's contributions to the field of plasma physics. So, too, do the lectures in this volume span the spectrum of topics and capture the excitement of research in plasma physics today. The lecture topics range from magnetic and inertial fusion to novel accelerators and light sources, computer simulation, space plasmas and basic plasma physics.

At this point, a brief history of Professor John Dawson is appropriate.

John was born on September 30, 1930, in Champaign, Illinois. Although he grew up on a farm, John's interests lay more in the stars than in the fields. In high school, he ground his own lenses to build a telescope. Not satisfied with just a large telescope, he wanted a motor-driven mount for it in order to be able to track and photograph stars. He was unable to afford a standard off-the-shelf mount, so instead he used the only thing he could find: a surplus Japanese machine gun mount! That was John's first contact with the Japanese.

It would turn out to be only the first of many encounters with the Japanese. Over the course of his career, John would make more than 20 trips to Japan. Through these visits and his special personality, John developed US/Japan collaboration in fusion energy and plasma physics.

John's interest in science led him to the University of Maryland, where he received his Bachelor's, Master's, and Ph.D. degrees in physics. After graduating from Maryland, he took his first job at Princeton University, working in fusion on Project Matterhorn. He spent 17 years at Princeton University. In 1960 he started doing computer modeling and was among the first to open the field of computer simulation of plasmas. In 1966, John became head of the theoretical group at Princeton University and from 1968-70, John set up the plasma simulation group at Naval Research Lab.

In 1973, John came to UCLA. John's history at UCLA and some of his many accomplishments were described at the Symposium by Dr. Raymond Orbach, Professor of Physics and Provost of the University. The text of his speech is reprinted here:

John Dawson, as Tom has indicated, has been at UCLA since 1973. In his career, he has written over 200 papers on the fourth state of matter, contributed to twelve books, sixty internal reports, and forty-three conference proceedings. But I think the most important statistic is the one that Tom said; namely, 30 Ph.D. students who have worked under his tutelage.

For those who may be familiar with one part of his works and not another, let me just state by title those areas in which John has pioneered. Obviously, computer simulation of plasmas is what he is most famous for. It is not a narrow area at all. One of the things that makes John's work so exciting is that he uses simulations as a complement to analytic theory and to experiments, and his insight generally makes him regarded as the father of particle simulation of plasmas.

He has used computer simulation to investigate nonlinear damping of large amplitude waves, trapped electron oscillations and instabilities, diffusion across magnetic fields, nonlinear behavior of beam and laser driven instabilities, particle acceleration by plasma waves. In addition, his more recent work has focused on interaction of plasmas with intense radiation.

He was one of the first to propose and champion the use of intense lasers to heat small pellets and to mechanically confine gaseous plasmas to thermonuclear temperatures. He has worked on the high frequency resistivity of plasmas and plasma heating by laser driven instabilities. As you will hear tonight, he invented cyclotron resonance techniques for economically separating isotopes of any element, and the consequences of that to the medical field are to be discussed this evening and are quite exciting; he must feel very proud about it.

He has proposed a laser-driven particle accelerator, and very apropos to our symposium site, the surfatron. That can be observed in practice out the window and has high theoretical energy gain and small synchrotron losses, though I doubt that the surfers are aware of that. He has worked in thermonuclear fusion – and my favorite one was the wet wood burner. I did not even know that wet wood burned until I read about it in his research.

And then, of course, there is his simulation of space plasmas and plasma lenses for focusing particle beams and frequency up-conversion of electromagnetic radiation with the use of an over-dense plasma. These are just by title, and many of these topics will be touched upon during this Symposium.

John received a Fullbright scholarship in 1964, and spent the year at the Institute of Plasma Physics in Nagoya, Japan. That began a long and fruitful collaboration with Nagoya and with a large number of Japanese scientists.

He was Chairman of the Division of Plasma Physics of the American Physical Society. He is a fellow of the American Physical Society and the American Association for the Advancement of Science. He has received the Research and Publication Award of the Naval Research Laboratory. He was elected to the National Academy of Sciences in 1977. He received from TRW Systems the Exceptional Scientific Achievement Award in 1977. He has received the James Clerk Maxwell Prize in Plasma Physics from the American Physical Society and was named California Scientist of the Year in 1978. *Science Digest* lists his work in the year's "Top 100 Innovations and the Men and Women Behind Them" in the 1985 issue.

If that were all the work of John, we would be congratulating him for his accomplishments, but John offers much more. His sense of service and his guidance to students is what makes him so special to us all. In terms of service, he was Director for the Center of Plasma Physics and Fusion Engineering at UCLA from 1972 to 1987, and currently serves as Associate Director of the Institute of Plasma and Fusion Research.

He is a member of a number of committees, and I will just list a few: The MIT Plasma Fusion Center Visiting Committee; the National Research Council Physics Survey Series Committee; the Advisory Board for the Institute of Theoretical Physics at Santa Barbara; the National Research Council Committee for the Review of DOE and Inertial Confinement Fusion Program; the American Physical Society Subcommittee on International Scientific Affairs; the Joint Steering Committee of the US-Japan Joint Institute for Fusion Theory; the Board on Physics and Astronomy of the National Research Council; and the Plasma Science Committee of the National Research Council.

I would like to read from John's own words because I think they say better than anything else what John's commitment is to the field:

> "One outgrowth from the field of computer modeling is that it has given us wonderful tools to explore the richness of plasma physics. With it, we have been able to look at nonlinear processes and processes where there are many competing phenomena. We are able to sort out which processes are important ones. We are also able to explore with machines what would be difficult or impossible to explore experimentally. Such studies lead us into modeling space plasma phenomena, and a large number of basic plasma processes.
>
> The study of intense waves in plasmas leads directly to the ideas for the use of plasmas in accelerators and lenses for high energy particles. This in turn leads to ideas for using plasmas to generate light, to upshift the frequency of light, and to make various types of unique light sources. These studies reveal a richness of physical phenomena which was not really anticipated. I believe that this richness is not fully recognized in the plasma community even now.
>
> All of the above lead to a great wealth of interesting, intriguing, and important problems to attack. These make good thesis problems, and working them through gives students experience with computing, analysis, plasma physics, data analysis, and general methods for going about attacking problems in the real world.
>
> The approach also develops physical intuition. Of course, some students are very good at this type of work while others find it difficult. They expect the computer to give them the answer they are looking for in a straightforward way, which it rarely does, and they do not know what to do when they get unexpected results.
>
> I use a type of apprentice method with my students. I team up a senior student with a starting student. This method seems to work very well. The students are more on top of the mechanics of computing and also many advanced computing techniques than I am. I work with the students in understanding the physics, in advising them in what approach should be

taken, and what should be done next. I spend quite a lot of time doing this."

Now I want to read just two very short quotes from students about John. (These are anonymous):

"I am currently engaged in several projects with John and his group, despite the fact that my focus is experimental and his is theory and simulation. We share a concrete sense for plasmas and an appreciation for new methods for data documentation. He is one of the bright stars of my graduate education. I will owe him a considerable debt."

Finally, there is this quote:

"As for being a thesis adviser, from what I have heard and seen, he would be considered among the best. From my experience, he has a natural ability to guide student's research skillfully and efficiently. It is clear that spending time with his graduate students is a high priority, and he always makes time to see us. In fact, sometimes I get the feeling that he prefers it over many other duties."

This is the man that we have come to honor on his 60th birthday. Thank you all for joining us in this celebration.

To Professor Orbach's thank-you, I would add my deepest appreciation to the speakers, the members of the International Steering Committee, and the Internal Organizing Committee. The Symposium and these proceedings were made possible by the generous support of the UCLA Institute for Plasma and Fusion Research, the UCLA Physics Department, TRW, and the Lawrence Livermore National Laboratory-Plasma Physics Research Institute. Special thanks also to a talented and motivated staff for contributing generously with their time and energy: Charlotte Carter, Chris Chang, Maria Guerrero, Annie Love, James Rose, Karen Shapiro, and Sophie Spurrier.

Finally, the expert assistance and support of Allan Wylde, Vice President for Advanced Books, Jan Benes, Karl Matsumoto, Barbara Holland, and their colleagues at Addison-Wesley is greatly appreciated.

Happy Birthday, John!

Tom Katsouleas

Department of Electrical Engineering/Electrophysics
University of Southern California
Los Angeles, CA 90089

B. Coppi
Massachusetts Institute of Technology

The Gift of Prophecy Introductory Remarks To Session I

There is hardly any area of plasma physics that has not been lit at some stage by the creativity of John Dawson; and since I am a long-time friend of his I am grateful for the respect and admiration that U.C.L.A. has surrounded him with since we lost him from the East Coast. In fact, one of my most cherished experiences when coming to U.C.L.A. is to sit around with John as we used to do at Princeton years ago and daydream of various ideas that we would like to see pursued. Of course, scientific daydreaming is always looked upon with some suspicion. Therefore, I shall not dwell on all the subjects of our private conversations; I should mention three that come to mind, and that are no longer controversial.

In 1967, we were both with Princeton University and we were debating about possible novel directions that the Plasma Physics Laboratory could take. In one of those conversations John suggested that we join forces and propose experiments on plasma physics in high magnetic fields. In those days he was thinking of linear experiments and I was concerned with toroidal configurations.

Years later, that is, today, it has become clear that the most promising and expeditious experiments to demonstrate ignition by fusion are by high magnetic field machines. In fact, the Alcator experimental program that I developed at MIT, immediately after leaving Princeton for Cambridge (MA) in 1969, has uncovered a series of new physical regimes of plasmas confined by high magnetic fields, confirming beyond expectations how farsighted John's intuition was. Moreover, it is commonly and even officially (see the recent report by the Fusion Policy Advisory Committee of the U.S. Department of Energy) recognized in the U.S. and in Japan that an economical and efficient power producing fusion reactor will have high magnetic fields. It is a fortunate and befitting circumstance that the ARIES fusion reactor studies pursued at U.C.L.A. have been the first ones to reach this conclusion.

John and I have always wondered why fusion reactors should be viewed as useful only as prospective power stations. He has in fact proposed a number of applications of fusion reactors as radiation sources, producing, for instance, large amounts of cyclotron emission. A subject connected with this that until recently was looked upon with great suspicion – and indeed it was barred from funding by the European agencies until the I.A.E.A. Conference held in Nice in 1988 – is that of deuterium-helium3 fusion reactors. Almost all agencies in charge of fusion research have viewed the thinking of reactors that are not based on the familiar deuterium-tritium reaction as a dangerous distraction. In fact, 80% deuterium-tritium reaction is in the form of energetic neutrons, while considerations of various factors indicate that a desirable type of fusion reactor should minimize the fraction of energy produced in the form of neutrons. This is possible in principle by burning deuterium-helium3 plasma mixtures.

Well, during the late seventies John did not indulge in lengthy debates about the physics limitations that would confine us to foresee the use of the deuterium-tritium reaction only. He went ahead, in his own style, proposing the investigation of deuterium-helium3 reactors with good scientific and general policy arguments, while I waited for at least a year after him until I could prove the existence of a new stability regime that I called the "second stability region" of parameter space, where plasmas, relative to the pressure of the confining magnetic field, are found to be stable. On this basis I could find the courage to propose the first deuterium-helium3 burn experiments that could be built with present technologies, utilizing the knowledge gained on the properties of high density plasmas produced and confined in high magnetic field experiments. I do not know whether I would have come out in the open with this line of work, without John's support and understanding. It is sad that a decade has gone by, but no substantial program of research has been allowed to grow in this general area.

Before I leave the floor to Harold Furth to tell us about the circumstances that had John involved in the formulation of the "wet wood burner" concept, I would like to mention a week-long meeting that we had at the Institute for Advanced Study at Princeton, as guests of M. Rosenbluth, to analyze the two-component-tokamak experiment which had developed out of the wet wood burner idea. Specifically, we were considering a thermal plasma in which a high energy particle population was being injected and maintained. This is clearly a situation far away from thermal equilibrium. We were, of course, afraid that a legion of micro-instabilities could be unleashed, at least theoretically, to degrade irreparably the confinement of both the thermal and the high energy particle population, and that the concept may prove unfeasible. We found instead that no serious micro-instability could be excited.

Years later the experimental observations not only have confirmed this finding, but they have gone beyond them. In fact, the injection of a high energy particle population has been discovered to be capable of suppressing, in a controlled way, an important class of macroscopic instabilities (so-called $m = 1$ modes), that lead to "sawtooth" oscillations of the central part of the plasma column. The agreement with the theory that we have developed to explain this phenomenon seems to be excellent[1]. These oscillations are a serious threat to the ability of a magnetically confined plasma to attain fusion ignition conditions, and their suppression has been a long sought objective.

Thus we have another clear demonstration of how worthy John's instincts are, as well as a humbling lesson for not having devoted sufficient effort to study the properties of two-component plasmas much earlier. After all, the two-component condition is the natural one for fusion burning plasma regimes where the temperature of the thermal component, in the range 10 to 100 keV, is maintained by the multi-MeV charged particles that are produced in it by the fusion reactions.

References

1. B. Coppi, P. Detragiache, S. Migliuolo, F. Pegoraro, and F. Porcelli, *Phys. Rev. Letters*, **63**, 7733 (1989).

H.P. Furth
Princeton Plasma Physics Laboratory

Non-Maxwellian Fusion Plasmas and Other Curiosities

Figure 1 is a historic photograph. This was the situation at Princeton Plasma Physics Laboratory just two years before John Dawson got there, and they were thinking of rather large fusion reactors back then. Somehow these football colleges seemed to size things in 100 yard increments; as long as it does not exceed the 100 yard specs too much, it is okay. This is a good size fusion reactor. Once John got into the act, things got a little bit better.

Now, Fig. 2 is a collection of photographs of long-time members of the Princeton Theory Division.

This was war time, and so whenever anybody arrived at Plasma Theory Group, they had a mug shot. So here we have these candid mug shots of various people in the audience. One thing you will notice about John Dawson's mug shot is that it is practically an invariant. This was taken a long time ago, and if he were to submit a candid mug shot of himself today, he would look exactly the same ... like me. For those in the back row, Fig. 3 is an enlargement.

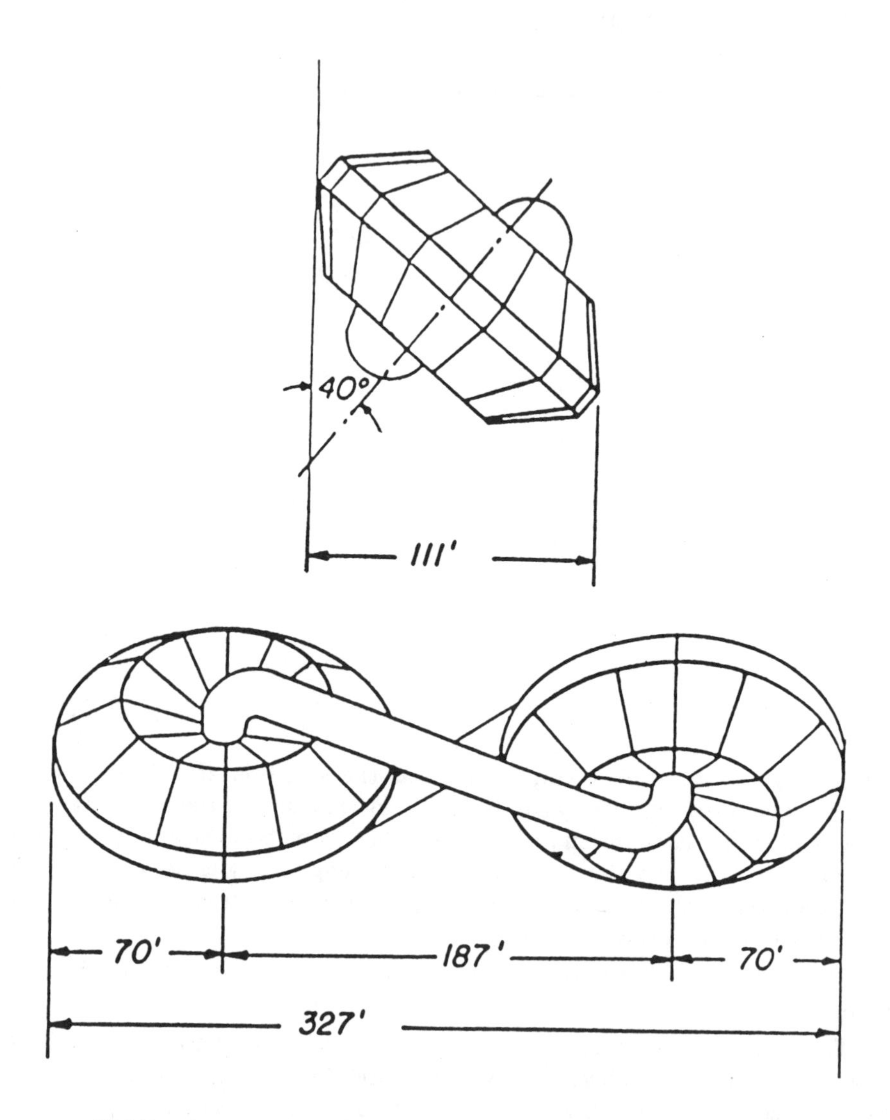

FIGURE 1 Project Matterhorn's vision of a steady-state magnetic fusion reactor using the stellarator principle of plasma confinement. The Model-D reactor study was prepared by Project Matterhorn as a classified report, two years before John Dawson's advent at Princeton.

FIGURE 2 A photograph of John Dawson, among other distinguished long-time members of the Princeton Theory Division.

~1956
(26 AD)

FIGURE 3 An enlargement taken from Figure 2. The notation "26 AD", accompanying the ordinary calendar year "1956", refers to years elapsed since John's birth, in 1930.

I have put down here the actual calendar date and a transcribed date in units of Dawson days. By my calculations, you see that he was born in 1930 or thereabouts, so you will track that as we go along. The AD stands for "After Dawson".

At the time that John Dawson came into this business, things did not really look too wonderful, and ten years later they looked absolutely terrible. Not that it was John's fault. Not a great deal was happening, then, as you can see in Fig. 4. I laid out here on the Dawson time scale where we would be, given the same rate of logarithmic progress in the Lawson diagram. Poor John would be about 100 before one ever got to "break-even".

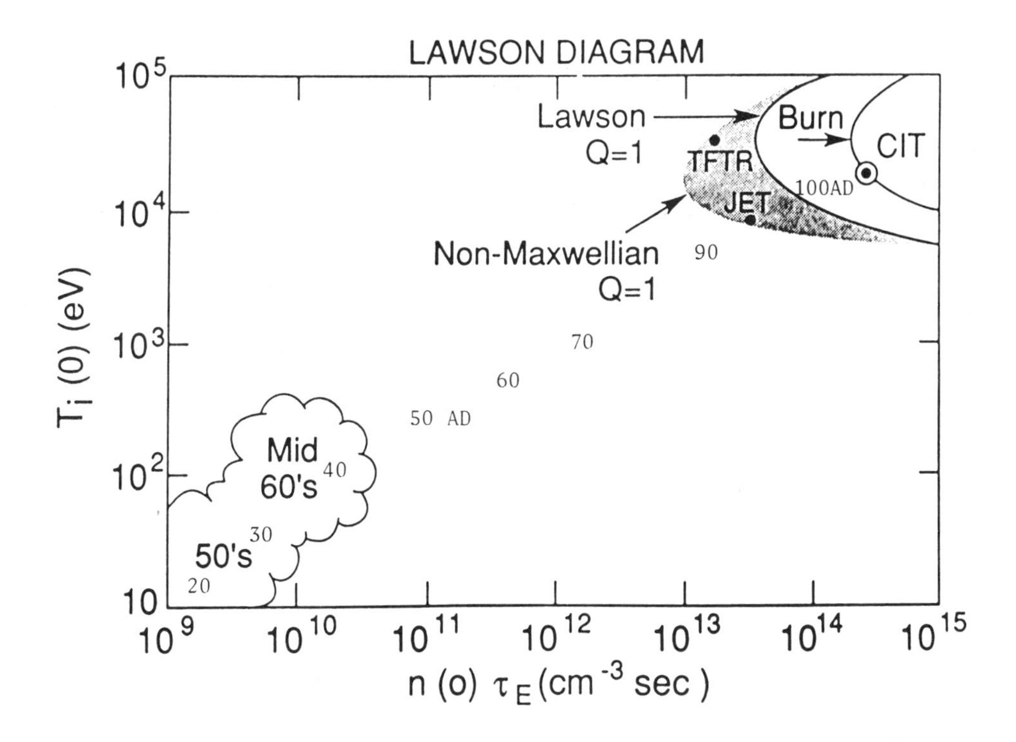

FIGURE 4 The standard Lawson Diagram for DT fusion fuel, along with supplementary data.

Several favorable trends have taken place. One thing is Fig. 4 itself; after all, it was done at the last IAEA Conference, 1988, and that was "After Dawson 58". There are points in the upper right of the figure that should not have happened until Dawson was 94 or so. Real acceleration took place; this acceleration was largely due to the tokamak. Another development that accelerated progress was something called Non-Maxwellian Plasmas. This pulled forward the break-even curve somewhat from the Lawson Curve, which after all calls for electrons and ions to be the same temperature and well thermalized and thus makes it needlessly tough to reach break-even.

So, between the two of those events the main question one asks is "Why don't these machines achieve break-even yet?" Practically, they are within striking distance, I think. I think maybe at the 1990 IAEA meeting that we are going to attend in a few weeks, people will stimulate themselves and will at least mentally do it. It is sort of a mental process to calculate in the DT case where you are. DD is well defined experimentally, but with DT there is an element of the spirit that enters.

John contributed mightily to the idea of the two-component tokamak where particles are not constrained to the usual thermal distribution. The general idea came about during a lunchtime conversation where I said that our new tokamak compression experiment was a wonderful thing that worked in all respects. But it bent the needle in the neutron counter when you compressed. First you heated it with neutral beams, then you compressed it and it blew the lid off. I was complaining that this was a terrible thing, and John said that this was not all together a bad thing. From there the idea took off.

The idea is if you take deuterons in not too terribly high an energy range and shoot them into a plasma, then there is a period of great productivity during the slowing-down time, where you have a lot of fusion reactions (see Fig. 5). Then the question was, "Is that good for anything useful?" This question is addressed more clearly in a paper by J.M. Dawson, H.P. Furth, and F.H. Tenney in *Phys. Rev. Lett.*, **26**, 1156 (1971).

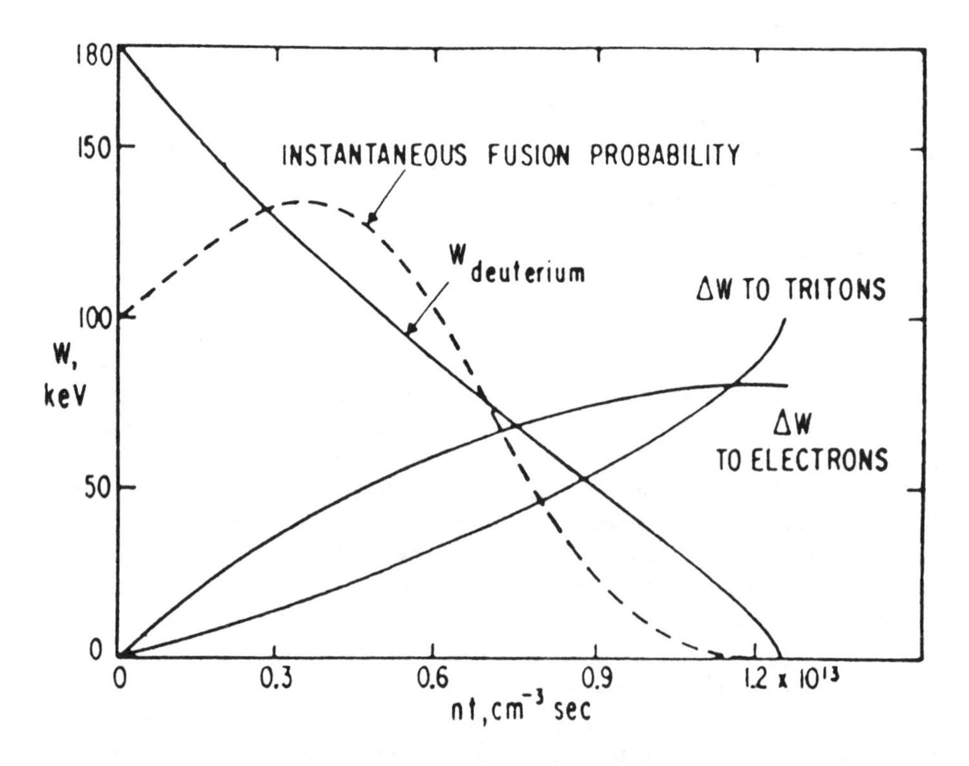

FIGURE 5 This figure points to the physical reason for the enhancement of Q-values in Non-Maxwellian DT plasmas: the DT reaction rate is strongly peaked in the energy range $W \geq 100$ keV.

In Fig. 6 we have outlined the conditions for getting break-even when shooting deuterium into tritium, and they are fairly lenient. You have to get the electron temperature to 4 or 5 keV, which is not too demanding. I should say parenthetically these days in large tokamaks, like TFTR and JET and others, these temperatures are exceeded by a factor of two.

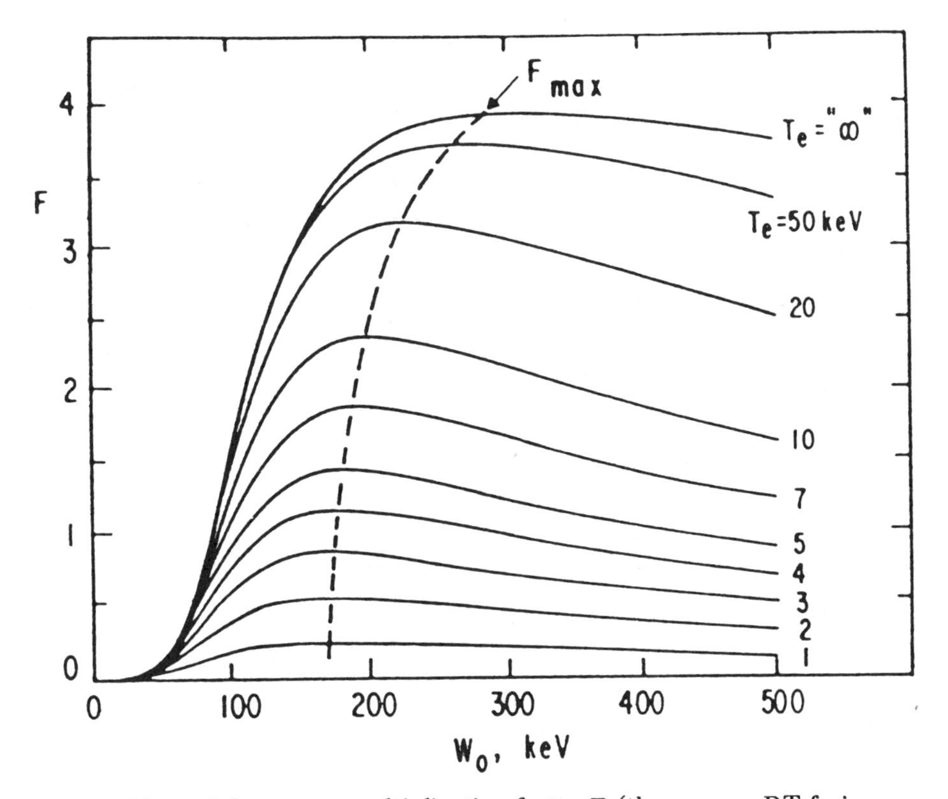

FIGURE 6 Plots of the energy multiplication factor F (the average DT fusion yield during the slowing-down of energetic deuterions, divided by the initial investment in deuterion energy W_o). The horizontal line at $F = 1$ corresponds to the usual break-even condition. The attainment of $Q = 1$ calls mainly for the ability to sustain 4-keV-range electron temperatures (T_e) by the thermalization of 100-keV range deuterons.

Also, a good energy to use for the energetic deuterium components is between 100 and 120 keV. By coincidence, this is an energy range which is virtually dictated by technological considerations. If the energy is too low, then you end up not getting the "oomph" that you should, and you cannot penetrate with the neutral beam. If the energy is too high, then it shoots right through, and that is not good. It so happens that the usual beam energy which makes good sense to use is something like 100 - 120 keV.

These are the two conditions which you should use to feed all the energy to the plasma: electron temperatures exceeding 5 keV and injection energies exceeding 100 keV. both of these are sort of a natural, so much so that everybody now uses them without really embracing them intellectually. They just use them and make energy.

It was not always so. In 1975, there was a meeting of the so-called four big tokamaks in Dubna. There were a variety of aims and dates (see Fig. 7). In the case of Japan's JT60, they were not going to head directly toward DT; they were going to head for new physics (which was a pretty good idea) and also long pulse and current drive (also a good idea). In this they succeeded very well. In the case of T20, they thought they could make pretty much a fusion reactor immediately, which indeed, given large amounts of money, they could have done. Incidentally, they did not involve the whole world. International involvement is advantageous from two points of view. One is that there is more money, and the other is it is a nice integrated international activity, and an opportunity for us all to work together. Essentially the T20 was not abandoned, but has served as an inspirational model for ITER which we are all working on today.

TABLE I. COMBINED LIST OF LARGE-TOKAMAK PARAMETERS (July 1975)

No.	Parameter	Symbol	Unit	JET	TCT (TFTR)		T-20	JT-60	No.
1.	Major Radius	R	m	2.96	2.48		5	3	1
2.	Plasma Minor Radius	a	m	1.25	0.85		2	1	2
3.	Plasma Half Height	b	m	2.10	0.85		2	1	3
4.	Aspect Ratio	R/a		2.37	2.9		2.5	3	4
5.	Elongation Ratio	b/a		1.68	1.0		1	1	5
6.	Magnetic Field on Axis	B_o	T	3.4*	5.2		3.5	5	6
7.	Maximum Field	B_{max}	T	6.9*	9.5		7.8	11	7
8.	Plasma Current	I	NA	4.8*	2.5	1.0	6	3.3	8
9.	Safety Factor at Surface	q_a		6	3.0	7.5	2.3	3.5	9
10.	Safety Factor at Axis	q_o		1	1.2	1.2	1.1	1	10
11.	Current Rise Time		sec	1.0	0.12	0.05	1.1	0.1-1	11
12.	Mean Ion Temperature	T_i	keV	5	6.0	6.0	7-10	5-10	12
13.	Mean Ion Density	n	m^{-3}	$5x10^{19}$	$4x10^{19}$	$8x10^{19}$*	$(0.5-5)x10^{19}$	$(2-10)x10^{19}$	13
14.	Mean Electron Temperature	Te	keV	5	6.0	6.0	7-10	5-10	14
15.	Plasma Pressure Ratio (Poloidal Beta)	β_p		1	1.0	2.5	1	1	15
16.	Plasma Pressure Ratio (Toroidal Beta)	B_t		0.03	0.007	0.015*	0.03	0.02	16
17.	Energy Confinement Time	τ_E	sec	1	?	0.2	2	0.2-1	17
18.	$n\tau_E$		m^{-3} sec	$0.5x10^{20}$	?	$1.5x10^{19}$*	10^{20}	$(2-6)x10^{19}$	18
↓ 64				τ_E and β	Non-Maxw. $Q \gtrsim 1$		↓ ITER	Long Pulse	

FIGURE 7 Taken from a summary of the IAEA-sponsored conference on "Large Tokamaks', which took place in Dubna (USSR) in July 1975. The summary of this conference (as published in Nuclear Fusion, **15** (1975) p. 309), provided a listing of the proposed Large-Tokamak parameters, part of which is shown enlarged.

In the case of the United States, the machine that was initially built was called the TCT for two component tokamak. This was because it was noted that a good approach to break-even was to shoot in energetic deuterons to bombard a tritium plasma. This had a fairly reasonable goal of achieving $Q = 1$. Actually, the Department of Energy, led by various prudent souls, prohibited us from making that claim. This was sort of a comfortable place to start from but increasingly it has become obvious that this machine *can* hit $Q = 1$. I hope this will have become more obvious by the time of the 1990 IAEA meeting.

The JET device was aimed at goals of large energy confinement. This goal used to be set to be one second, which they have now done. They are also beginning to reach their goals of substantial "nt" data values, so this also has been a substantial success. They were able to do this with the same two-component technique. So at the moment, for the people capable of getting to break-even, the two-component approach is essentially the only game in town. That is illustrated somewhat by Fig. 8.

In Fig. 8 we have the fraction of particles that are made by Maxwellian processes and by non-Maxwellian processes. This viewgraph was made for TFTR. Starting with a neutral beam power of about 10 megawatts going up to 30 or so, the fraction of thermal reactions increases and it reaches about 20 percent. This is in deuterium, but it will be about the same in DT. This means essentially that the whole other fraction, the other 4/5th of the fusion reactions, has been made by non-thermal two component processes. In Fig. 8 I have also plotted the JET point on this. JET has also gotten the same result (the same fraction of thermal heating). That is how the matter stands at the moment[1].

The grand summary of these results is plotted in Fig. 9. This is from the DOE's Fusion Policy Advisory Committee (FPAC) report. What is plotted there as a token of progress is the real fusion power that is being generated versus years. A very good upswing is noted. The leading edge is all non-thermal (Non-Maxwellian). The other points representing honorable methods, like thermalized plasmas, are struggling with a disadvantage factor. The other points (Maxwellian plasmas) are headed toward the ignition regime, where the plasmas will have to be more or less thermalized because of the high "nt".

---

[1]At the 1990 IAEA meeting, TFTR and JET reported experimental DD results projecting to Q_{DT}-values in the range of initial reactor-physics interest. In both cases, the non-Maxwellian reactions were expected to provide more than half the DT fusion power.

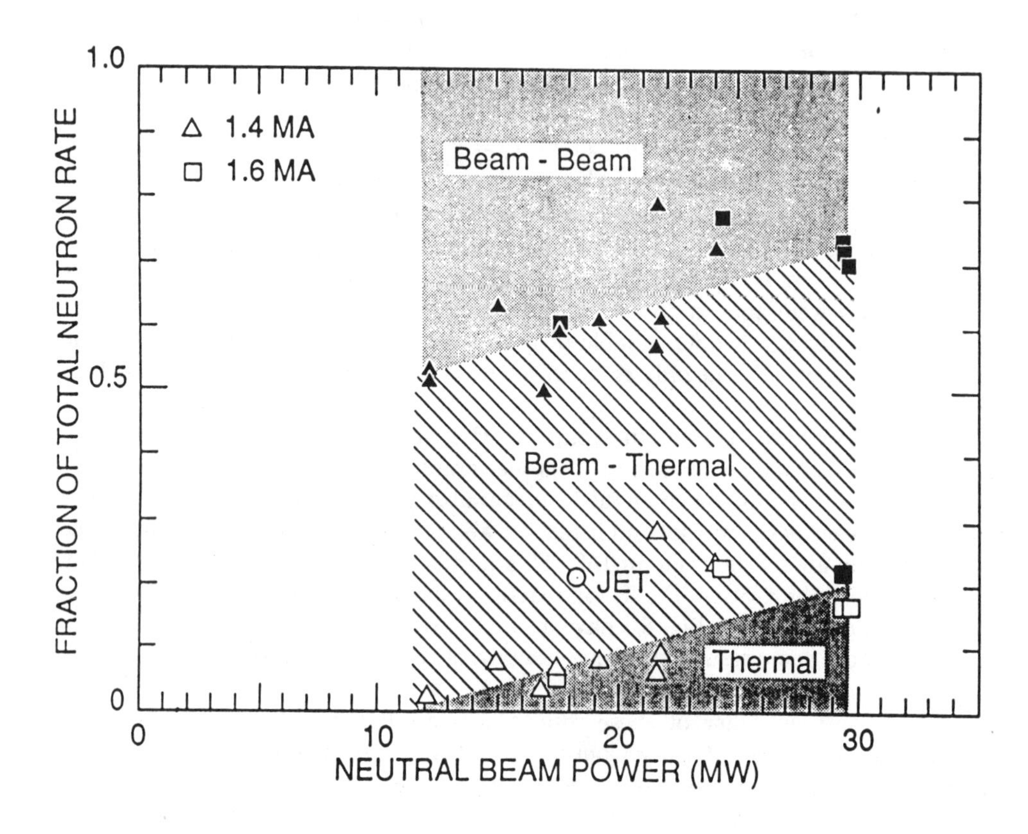

FIGURE 8 An illustration of the relative contributions to the total fusion power that arise from various DD reaction processes in TFTR and JET: In both these experiments, the non-Maxwellian contribution has been dominant. Present-day projections for reaching the "break-even regime" in DT operation are all based on utilizing the two-component approach outlined in Figs. 5 and 6.

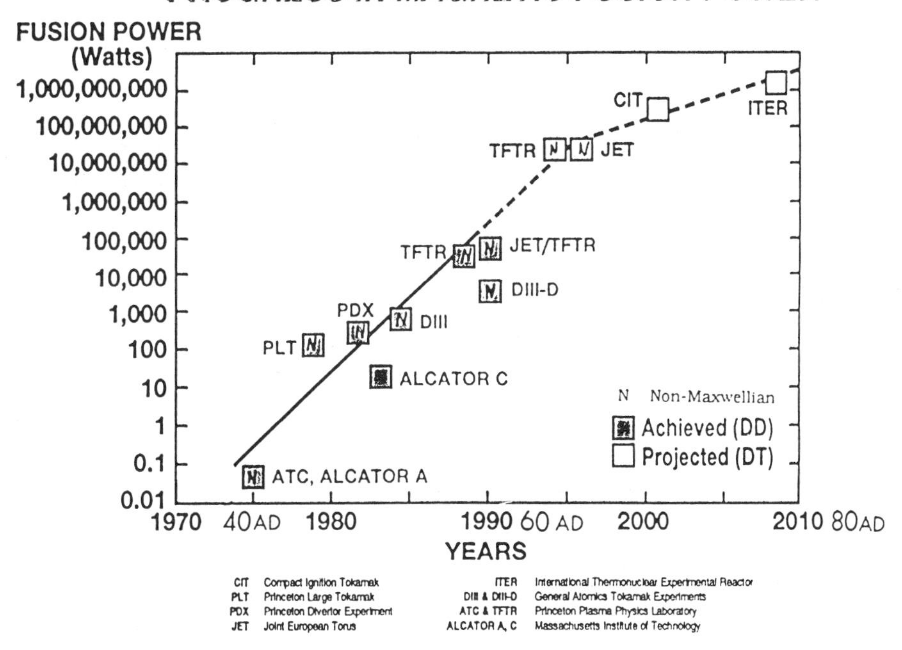

FIGURE 9 This was included in the 1990 report of the DoE's Fusion Policy Advisory Committee (FPAC). In DD operation, actual fusion power levels up to 10^5 watts have been reached with non-Maxwellian plasmas – from which DT fusion power levels of several tens of megawatts are projected.

When one puts in tritium, it is non-unique how much of a multiplication factor one will get from the two component approach. For instance, if you put in half deuterium and half tritium, then you get only half of the multiplication factor that you would if you shoot deuterium at tritium. So what you will get is a little bit ambiguous, depending on just exactly how you run the experiment. I would say the outlook is pretty good that by the two component method we will get to break even before long. This on the real time scale, or on the Dawson scale, is not 100 years, it is something considerably less.

In the end, we must all come around to good honorable thermalized plasmas in order to get ignition. In that case, what we would normally think of as big machines, such as ITER, are not the football field size (see Fig. 10). ITER is only about 1/4 of that, or about the distance to the 25 yard line. So things have shrunk a bit. Of course the very nice thing to do is to hang on to the high field technique and in this case, one can also project ignition.

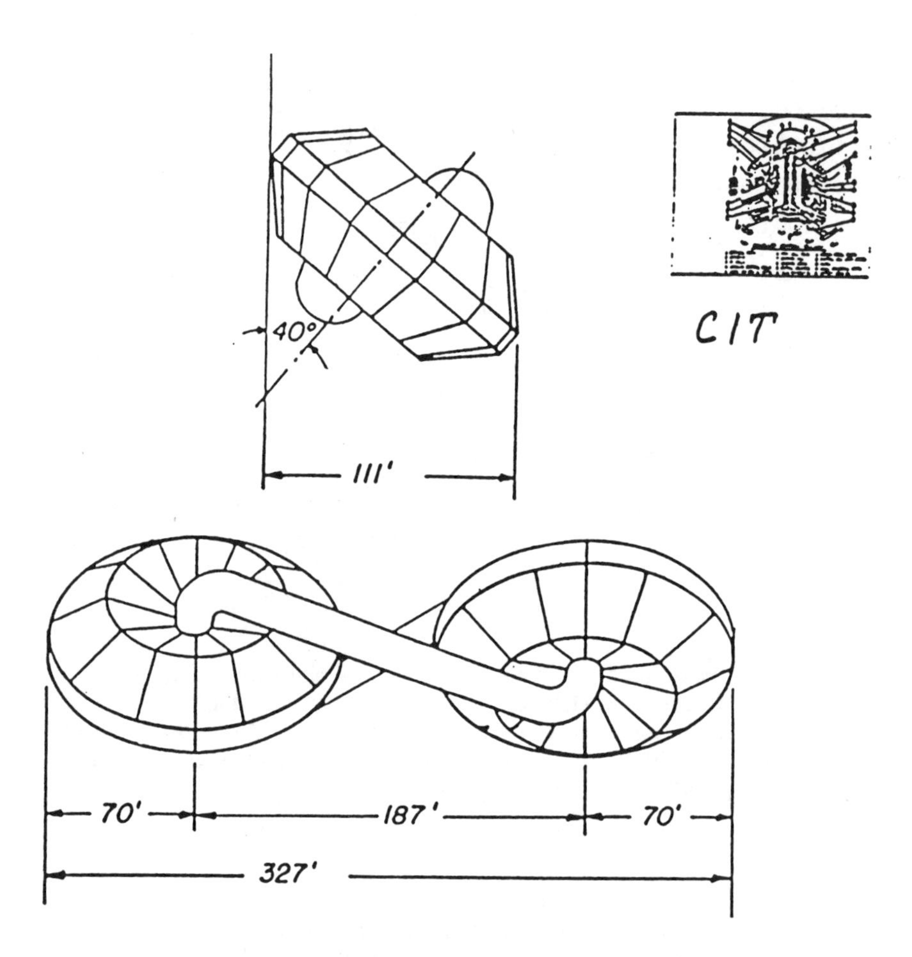

FIGURE 10 Illustrates the relative sizes of the Model-D reactor concept (shown in Fig. 1) and the proposed Compact Ignition Tokamak (CIT). While the non-Maxwellian approach cannot be used to minimize the size of an ignition-level DT burner, high-field tokamak confinement provides the basis in CIT for some other types of economy.

That it can even be believed that a machine of that size can get ignition represents considerable progress, I think.

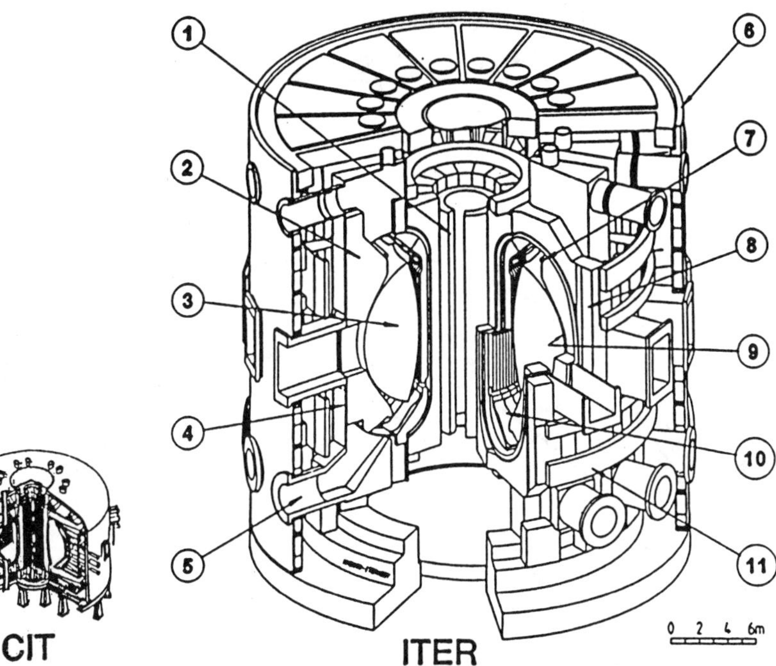

FIGURE 11 A comparison of the sizes of the short-burn CIT experiment and the more fully reactor-like ITER project, which is designed to produce long-pulse burning plasmas in superconducting steady-state magnetic to fields. Relative the football-field-sized Model-C Stellarator, of Fig. 1, even the ITER Tokamak would extend only to the 25-yard line.

Finally, let me say that there is a lot more to John Dawson's participation in fusion with original ideas than just the two-component business. Figure 12 is symbolic of this. A long time ago – a remarkably long time ago (back in 1981) – John, without asking permission from NASA or anybody, was already onto how interesting the business of ^{3}He was. And he did not shy from considering ^{3}He-^{3}He and other mad things which are interesting to explore.

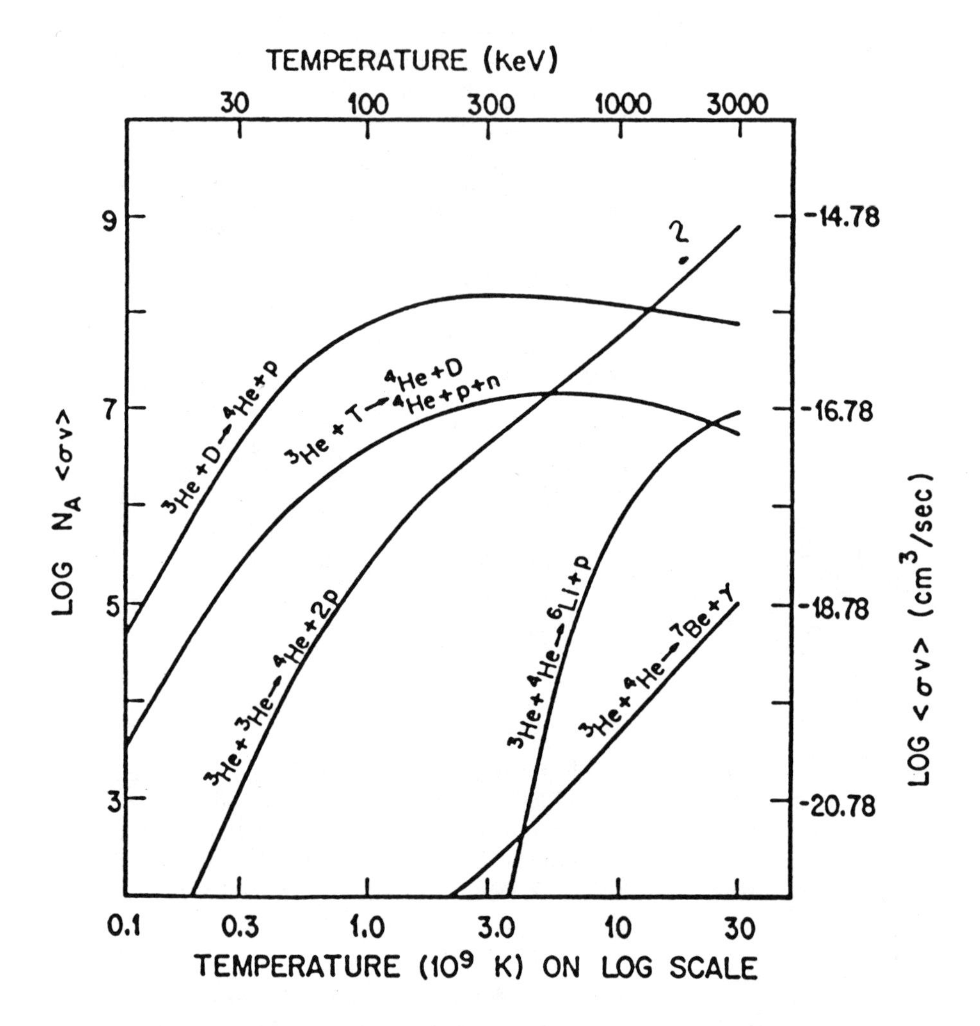

FIGURE 12 Taken from *Fusion*, Volume 1, Part 2, edited by Edward Teller and published by Academic Press in 1981. Beyond the challenge of the DD and DT fusion experiments, new horizons might be made accessible by inventive ideas.

So with that I would like to conclude. John, it has been a real pleasure to be here to honor you.

P.K. Kaw
Institute for Plasma Research
Bhat, Gandhinagar 382 424 India

AC Helicity Injection: Current Drive by Nonresonant RF Forces

Abstract

The intimate relationship between AC helicity injection and current drive by parallel RF forces on non-resonant particles in a magnetized plasma is examined. It is demonstrated that in a collisionless plasma, parallel non-resonant RF forces can drive no steady currents in a periodic system like a torus, because these forces have zero circulation. The current drive is thus shown to be fundamentally tied to the dissipation of high-frequency waves. An example of parallel current drive by dissipative surface modes around the ion cyclotron range of frequencies is then given using the language of AC helicity injection, and its efficiency estimated.

Introduction

About twenty-three years ago, I first met John Dawson as a fresh postdoc at Princeton. He quickly initiated me into the physics of laser plasma interactions and we had a great time exploring effects due to nonresonant radiation forces like ponderomotive forces in unmagnetized plasmas. Then about eight years back, he briefly visited Princeton while I was still there, and we collaborated again – this time on synchrotron current drive in tokamaks. The topic I have chosen for discussion today, viz., current drive by parallel nonresonant RF forces (or what is more fashionably called AC helicity injection) brings back memories of the above-mentioned collaborations with John Dawson. John is a great teacher, and I learned a lot from him. The memories I cherish most are those where he would share a little cartoon picture which usually contained a profound physical insight, or showed how to get most of the results (within a factor of 2 or π) on the back of a napkin at lunch. I dedicate this paper to him.

Physics of AC Helicity Injection

The concept of magnetic helicity and helicity injection was popularized in the physics of toroidal discharges by the papers of Taylor[1] and Jensen and Chu[2]. If $\mathbf{A}$ and $\mathbf{B}$ respectively denote the vector potential and the magnetic field vector, we may define the helicity K associated with each magnetic surface as

$$K = \int_{\Psi} \mathbf{A} \cdot \mathbf{B}\, d\tau \qquad (1)$$

K is an invariant of ideal MHD equations and is a measure of the linkage between toroidal and poloidal fluxes. A given profile of current in a toroidal discharge like a tokamak corresponds to a $K(\Psi)$. We know that non-ideal effects like resistive dissipation lead to a decay of plasma current and hence of K. Thus, to get a steady state discharge we must find methods of overcoming the resistive decay of K.

If we define K_T as the helicity for the whole plasma volume, we may use the non-ideal Ohm's law to derive the following evolution equation[2,3]:

$$\begin{aligned}\frac{d}{dt}K_T = & \int_V d\tau \left[-2\eta\, \mathbf{J} \cdot \mathbf{B} + 2\frac{T_{e1}}{e} (B \cdot \nabla)\, \ell n \frac{n_e}{n_o} \right] \\ & - \int dS\, \hat{n} \cdot (\mathbf{E} \times \mathbf{A}) - \int dS\, \hat{n} \cdot \mathbf{B} \left[\phi - 1.71 \frac{T_e}{e} - \frac{T_{eO}}{e} \ell n \frac{n_e}{n_o} \right] \\ & + \int dS\, \hat{n} \cdot \mathbf{v}(\mathbf{A} \cdot \mathbf{B}) \end{aligned} \qquad (2)$$

where $\mathbf{E}$, ϕ, $\mathbf{J}$, e, n_e, η have usual meanings, $\hat{n}$ is the unit normal to the surface S and we have written $T_e = T_{eo} + T_{e1}$ such that $(\mathbf{B}\cdot\nabla)T_{eo} = 0$. The first integral on the right corresponds to the volume sinks (due to resistivity) and sources (due to $\nabla n \times \nabla T$ type field generation terms) of helicity. The second integral corresponds to the AC helicity injection at the plasma surface. By oscillating $\mathbf{E}$ and $\mathbf{A}$ at the plasma surface with the right phase relationship, one can hope to overcome the volume sink of helicity; these terms are finite even when the plasma has closed magnetic surfaces (for a more accurate description of how helicity injection is finite in a plasma bounded by a conducting wall with a cut or with a vacuum region between the outermost surface and the wall, please see refs. 4,5). The third integral corresponds to the injection of dc helicity along open lines at the surface using electrostatic and thermostatic forces. The last term corresponds to effects due to moving boundaries (first introduced in the context of helicity injection by Bellan[6]). The importance of the pressure gradient terms in the above equation derived from the non-ideal Ohm's law was first pointed out in ref. 3.

We now concentrate our attention on the AC helicity injection, viz. effects associated with the second integral on the right side of eqn. (2). Bevir and Grey[7] first used this term to introduce the idea of F-Θ pumping in reversed field pinches, viz. the idea of oscillating toroidal and poloidal fluxes to overcome the resistive decay of plasma currents. Experiments by Shoenbeg et al[8] and Bellan[9], however, gave only negative results. More recently, Ohkawa et al.[10] have introduced the idea of helicity injection using waves with helicity in the ion-cyclotron range of frequencies. The basic idea of AC helicity injection schemes is to generate a *parallel non-resonant dc force* by oscillating fields to compensate for the frictional drag on electrons in the plasma volume, i.e. it is a parallel ponderomotive force type effect. Note that whatever helicity is injected at the surface through the second integral is available for use throughout the plasma volume, either because the waves with finite helicity penetrate into the interior and dissipate there (equivalent to replacing the surface integral by a volume integral using the divergence theorem), or because whatever currents are generated at the plasma surface are assumed to find their way to plasma center by relaxation processes.

The idea of using nonresonant radiation forces to drive a plasma current is very distinct from the usual RF current drive schemes which rely on *resonant* wave-particle interactions. In the conventional RF schemes, one injects an *anisotropic* spectrum of waves into the plasma and lets resonant electrons absorb them. Unfortunately, these current drive schemes are inefficient and one has not been able to find any parameter space with acceptable efficiency. The first question we would like to address is that, since the non-resonant radiation forces act on all the particles, can their use have some key merits similar to Ohmic current drive which has excellent efficiency[11]? In Ohmic current drive, the power absorption is weak ($P \sim J^2$ in contrast to $P \sim J$ of resonant RF current drive schemes) because there is a lowest order cancellation between particles with $\mathbf{F}\cdot\mathbf{v} > 0$ and those with $\mathbf{F}\cdot\mathbf{v} < 0$. Since nonresonant RF forces also involve all particles, one may ask "Can nonresonant RF current drive give an Ohmic type performance?" The answer is unfortunately negative, as the arguments[11] in the following section demonstrate.

Current Drive by Nonresonant RF Forces

We first demonstrate that nonresonant RF forces in a collisionless plasma *cannot* drive steady parallel currents due to very fundamental considerations[11]. Consider a cold plasma with electromagnetic fields $\mathbf{E}_1, \mathbf{B}_1 = \mathbf{E}_1(r), \mathbf{B}_1(r)\ \exp(-i\omega t) + \text{c.c.}$, and let us write the electron fluid equations of motion to various orders in the field amplitudes. To the first order, we have

$$-i\omega \mathbf{v}_1 + \mathbf{v}_1 \times \mathbf{\Omega} = -\frac{e}{m}\mathbf{E}_1 \tag{3}$$

where Ω is the cyclotron frequency due to background d.c. magnetic field. To second order, the d.c. force on the electron fluid is given by[12]

$$\mathbf{F}_p = -n\,\nabla\Psi \;+\; \mathbf{B}_o \times (\nabla \times \mathbf{M}) \tag{4}$$

where

$$\mathbf{M} = -(ien/4\omega)\,(\mathbf{v}_1 \times \mathbf{v}_1^*) \tag{5a}$$

$$\Psi = (m/4)\,\left[\,|\mathbf{v}_1|^2 + (i/\omega)\mathbf{\Omega}\cdot(\mathbf{v}_1 \times \mathbf{v}_1^*)\,\right] \tag{5b}$$

We note from eqn. (4) that the parallel nonresonant RF force

$$F_{P\parallel} = \hat{b}\cdot\mathbf{F}_p = -n\nabla_\parallel\Psi = -\nabla_\parallel(n\Psi) \tag{6}$$

where we have assumed that n is constant on a line (similar conclusions follow if $n = n(\Psi)$). In a periodic system such as a torus, this force can drive no current, since it is gradient of a scalar:

$$\oint v_\phi d\phi = \oint F_{p\phi}/m\nu = 0 \tag{7}$$

Thus parallel nonresonant RF forces in a collisionless cold plasma can drive no currents in a periodic system because they have zero circulation.

We now demonstrate that the more complete Hamiltonian theory[13] of nonresonant RF forces in warm plasmas essentially supports the above theorem. Using Lie transform methods and non-canonical variables, it has been shown that the parallel motion of "oscillation center" of a non-resonant particle in given RF and magnetostatic fields is given by

$$\frac{d}{dt}\vec{X} = \hat{b}\left(P + \frac{\partial K_2}{\partial P}\right) \tag{8a}$$

$$\frac{d}{dt}\vec{P} = -\nabla_{\parallel}\left(\mu\Omega + \phi + K_2\right) \tag{8b}$$

where we have taken $e = m = c = 1$, $\vec{X}$, $\vec{P}$ denote the parallel position and momentum coordinates of the oscillation center, μ is the magnetic moment, and K_2 is the ponderomotive Hamiltonian given by

$$K_2 = \frac{|E|^2}{\omega^2} + \sum_{\ell=-\infty}^{\infty}\left(\ell\frac{\partial}{\partial\mu} + k_{\parallel}\frac{\partial}{\partial P}\right)\frac{|H_\ell|^2}{\omega - \ell\Omega - k_{\parallel}P} \tag{9a}$$

$$H_\ell = \frac{i}{\omega}\left[\frac{\ell\Omega}{k_\perp}J_\ell\left(k_\perp\sqrt{\frac{2\mu}{\omega}}\right)\hat{k} + 2i\frac{\Omega\mu}{k_\perp}\frac{\partial J_\ell}{\partial\mu}\hat{b}\times\hat{k} + PJ_\ell\hat{b}\right]\cdot \mathbf{E}e^{-i\omega t+\ldots} \tag{9b}$$

Note from (8a) that the parallel velocity of the oscillation center differs from the parallel momentum P by the quantity $\partial K_2/\partial P$. This quantity, when averaged over a distribution of particles, is finite for a finite temperature plasma. Using (8b) and averaging its toroidal component over the periodic ϕ direction, we may now write

$$\frac{d}{dt}\oint d\phi\left(v_\phi - \frac{B_\phi}{B}\frac{\partial K_2}{\partial P}\right) = 0 \tag{10}$$

The $(\partial K_2/\partial P)$ term therefore acts like a pseudo-vector potential arising because of RF fields. If it is *monotonically* varied in time, it may be used in a manner similar to ohmic electric field to sustain parallel plasma currents against the collisional drag on electrons. However, by its very nature this can only lead to a pulsed system with the same defects as Ohmic current drive. Thus, if we ignore these pseudovector potential effects, we confirm the basic conclusion of cold plasma theory that *parallel nonresonant RF forces in a collisionless plasma are unable to drive steady parallel currents in a periodic system because they have zero circulation.*

The above discussion ignored the collisional absorption of high-frequency waves. It is well known that if high-frequency dissipation is retained, new current drive terms arise because of the RF force on non-resonant particles. For unmagnetized plasmas, it was demonstrated by Bezzerides et al.[14] in the context of magnetic field generation in laser pellet interactions that one drives a d.c. current

$$\mathbf{J}_o = -i\frac{n_o e^3}{m^2\omega^3}\nabla\times\left[\mathbf{E}_1^*\times\mathbf{E}_1\right] \tag{11}$$

Note that the collision frequency has cancelled from the two sides of (11) because the resistive decay of the dc current on the left side is balanced by the collisional nonresonant RF force on the right. We demonstrate by a specific calculation in the next section that even in a magnetized plasma we may use nonresonant RF forces to drive parallel currents *provided effects due to dissipation of high-frequency waves are included.*

We conclude this section by reiterating that AC helicity injection is nothing but current drive using parallel nonresonant forces due to waves in a plasma. Unfortunately, these forces can drive parallel currents only if significant high-frequency dissipation of the waves occurs. The efficiency of such current drive schemes can therefore not be very high. The overall attractiveness of AC helicity injection is thus limited by fundamental considerations.

Surface Helicity Injection Using ICRF[15]

In this section we give an example of parallel current drive in a magnetized plasma using nonresonant RF forces due to dissipative surface waves in the ion cyclotron range of frequencies. For our discussion, however, we shall use the language of AC helicity injection introduced in Sec. 2. Using the simplest non-ideal Ohm's Law $\mathbf{E}+\mathbf{v}\times\mathbf{B}=\eta\mathbf{J}$ and ignoring the terms due to dc helicity injection on open lines, we may rewrite eqn. (2) in the form

$$\frac{d}{dt}K_T = -2\int_V \eta\mathbf{J}\cdot\mathbf{B}\,d\tau + \int \eta\hat{n}\cdot(\mathbf{A}\times\mathbf{J})\,dS + \int \hat{n}\cdot[\mathbf{A}\times(\mathbf{E}+\mathbf{v}\times\mathbf{B})]\,dS \tag{12}$$

The last term is the AC helicity injection term. Early calculations[2,7] had ignored the effects due to $\mathbf{v}\neq 0$ at the plasma boundary and had concluded that significant current drive by F-Θ pumping may result. However, after the experiments of Shoenberg et al[8] and Bellan[9] gave negative results, it was pointed out by Bellan and his colleagues[6] that the culprit is the $\mathbf{v}\times\mathbf{B}$ contribution in the last term. The conclusion was that as the plasma and magnetic field move together at the boundary (except for resistive slippage) when low-frequency oscillating fields are employed ($\omega \ll \Omega_i$), the net AC helicity injection is reduced to a rather weak resistive effect.

Recently, it has been pointed out[15] that the above problem can be remedied by working with frequencies close to ion cyclotron frequency ($\leq \Omega_i$) in a plasma where the ion waves are significantly damped, by viscosity for example. The appropriate Ohm's law to use now is the one with the Hall term included, i.e.,

$$\mathbf{E}+\mathbf{v}\times\mathbf{B}=\eta\mathbf{J}+(\mathbf{J}\times\mathbf{B})/en \tag{13}$$

Note that $\mathbf{E}\cdot\mathbf{B}$ still depends on η and the basic helicity conservation physics is unaffected. However, since the magnetic field and plasma no longer move together due to finite ion inertia effects, the last term in eqn. (12) may now become significant. We may now write the steady-state helicity drive equation as

$$2\int \eta \mathbf{J}\cdot\mathbf{B}\, d\tau = \frac{1}{en_o}\int \left\langle \hat{n}\cdot\mathbf{J}_1\,(\mathbf{A}_0\cdot\mathbf{B}_1 + \mathbf{A}_1\cdot\mathbf{B}_0)\right\rangle dS \tag{14}$$

where $\langle\rangle$ denotes averaging over the fast time scale and we have only retained the largest (non-resistive) contributions to the right side. This equation basically shows that in a steady state the volume sinks of helicity are balanced by helicity injection terms at the surface. A more complete analysis involving an overall conservation law for the kinetic helicity defined as $H = \int \mathbf{P}\cdot\nabla\times\mathbf{P}\, d\tau$ where $\mathbf{P} = (m_i\mathbf{v}/e) + \mathbf{A}$ shows that the surface term on the right side of eqn. (14) may be interpreted as injection of cross-helicity (terms associated with $\mathbf{v}\cdot\nabla\times\mathbf{A}$) into the plasma. It is interesting to note that the new helicity injection terms are proportional to $\hat{n}\cdot\mathbf{J}_1$ and hence to oscillating surface charge produced by ion inertia. This suggests that coupling to surface waves plays a rather important role in this helicity injection scheme. Furthermore, as demonstrated below, the right side is non-zero only if the surface modes are damped modes, i.e., dissipation of high-frequency waves is crucial to this current drive scheme. This is consistent with our discussion of current drive due to parallel nonresonant forces in Sec. 3. In our subsequent discussion we *mock up* the damping of high-frequency waves by inclusion of an ion viscosity term $\overline{\mu}\nabla^2\mathbf{v}$ in the ion equation of motion and find as expected that $J_{\parallel} \propto \overline{\mu}$. In reality, wave damping may arise even in a collisionless plasma through minority cyclotron damping, and we shall assume that the $\overline{\mu}\nabla^2$ term mocks up the effect in an approximate fashion.

To make an estimate of the right side of (14) we need to consider dispersion and damping characteristics of surface waves near the ion cyclotron frequency. The basic equations are

$$\nabla(\nabla\cdot\mathbf{E}_1) - \nabla^2\mathbf{E}_1 = 4\pi\, i\omega\mathbf{J}_1 \tag{15a}$$

$$\mathbf{E}_1 + \mathbf{v}_1\times\mathbf{B}_0 = \frac{\mathbf{J}_1\times\mathbf{B}_0}{en_o} + \eta\mathbf{J}_1 \tag{15b}$$

We treat a slab geometry with $y-z$ plane as the vacuum-plasma interface (plasma occupying $x>0$ region), magnetic field in the z-direction and plasma and vacuum field solutions of the form

$$\mathbf{E}_1^P \propto \exp\left[-\alpha x + i\,k_y\, y + i\,k_z z - i\omega t\right] \qquad x>0$$

$$\mathbf{E}_1^V \propto \exp\left[(\omega^2-k^2)^{1/2}x + ik_y y + ik_z z - i\omega\, t\right] \qquad x<0$$

where (note: c = 1)

$$\alpha^2 = k^2 - A - \overline{\omega}^2 A^2/\Omega_i^2\,(k_z^2 - A)\,, \quad \overline{\omega} = \omega + ik_y^2\overline{\mu}$$

and

$$A = -\omega\,\overline{\omega}\,\omega_{pi}^2/(\overline{\omega}^2 - \Omega_i^2)\ .$$

α^{-1} is the radial decay length of the modes. The modes in general have elliptical polarization with

$$E_{x1}^P/E_{y1}^P = i(\overline{\omega}A/\Omega_i + \alpha k_y)/(k^2 - A) \tag{16a}$$

Matching solutions at the boundary gives the dispersion relation, which for $k_z \ll k_y$ is given by

$$\omega\overline{\omega} = 2k_z^2 V_A^2 \left(1 - \frac{\overline{\omega}}{\Omega_i}\right) \Big/ \left[1 - \frac{k_z^2}{2k_y^2}\left(1 + \frac{\overline{\omega}^2}{\Omega_i^2}\right)^{-1}\right] \tag{16b}$$

We may use equation (16b) to obtain the damping characteristics of steady-state periodic waves with real ω, k_z.

Substituting for the properties of the mode on the right side of (14), we get an estimate for the steady state d.c. current J_{zo} as

$$J_{zo} \simeq 4\frac{R}{a}\,\frac{ek_z|E|^2}{\eta m_i(\Omega_i^2 - \omega^2)}\left(\frac{\overline{\mu}k_y^2}{\omega}\right)\ . \tag{17}$$

We have assumed that this current is uniformly distributed in the plasma interior due to some anomalous process of current penetration. Note that current drive is proportional to the damping of the ion cyclotron wave and is significant only when the damping is large. Unlike eqn. (11), there is no cancellation of ν here because the resistive drag on electrons and the high-frequency dissipation arise due to separate irreversible effects. The power absorption per unit volume may be written as

$$p = \frac{\omega_{pi}^2\omega}{(\Omega_i^2 - \omega^2)}\,\frac{|E|^2}{4\pi}\,\frac{\overline{\mu}k_y^2}{\omega} \tag{18}$$

From eqn. (17) we note that the current drive effect vanishes for $m_i \to \infty$, since it is critically dependent on ion-inertia. We may now combine eqs. (17) and (18) to write the usual figure of merit for current drive as $(\omega \sim \Omega_i)$:

$$IR/P \simeq 2^{\frac{3}{2}}\left(\frac{R}{\Delta a}\right)\left(\frac{1}{\eta c B_o}\,\frac{c}{\omega_p}\right) \tag{19}$$

where I is the total current driven, P the total power absorbed, and Δa the width of surface wave region. Note the unfavorable $n^{-\frac{1}{2}}$ dependence on plasma density if Δa is taken to be independent of density. One could argue with some justification that $(\Delta a)_{\min} \sim \alpha^{-1} \sim k^{-1} \sim c/\omega_{pi}$ so that the density dependence cancels. However, for practical reasons $(\Delta a) > (\Delta a)_{\min}$ may have to be used; hence we retain the density dependence and treat Δa as an external parameter. In terms of practical parameters, we may rewrite (19) as

$$I/P = T_{ev}^{\frac{3}{2}} \Big/ \left(B_o \, n_{13}^{\frac{1}{2}} \ell n \wedge \Delta a_m \right) \quad \frac{\text{Amps}}{\text{Watt}}$$

where Δa_m is measured in meters and n_{13} in units of 10^{13} cm^{-3}. For $T \sim 10$Kev, $B_o \sim 5T$, $n \sim 2.10^{14}$cm^{-3}, $\Delta a \sim 10$cms, we find

$$I/P \sim 1 \, \text{Amp/Watt}$$

which is a rather interesting range of current drive efficiencies. Note, however, that all the current is driven on the surface and depends on pouring about 10 MW of power in the outer 10 cms at the plasma edge, in contrast to Ohmic drive, where the power is benignly distributed over the whole volume.

Conclusion

We have demonstrated that AC helicity injection in magnetized plasmas is the same as current drive due to parallel nonresonant RF forces. We find that this current drive mechanism is fundamentally dependent on dissipation of high-frequency waves and therefore is unlikely to have a very good efficiency (such as Ohmic, for example). We have also shown that the negative effect of cancellations due to moving plasma boundaries observed in conventional F-Θ pumping can be overcome by the use of surface waves in the ion cyclotron range of frequencies. Ion inertia effects are shown to lead to an injection of cross-helicity through coupling to oscillating surface charges associated with surface wave modes. This then results in significant current drive magnitudes with interesting efficiencies. However, most of the current drive takes place at the surface and one has to assume that it finds its way into the plasma core by anomalous current penetration processes, hopefully without damaging the good confinement properties of the discharge. Obviously, experiments designed to investigate these questions are needed to resolve such important issues.

References

1. Taylor, J.B., *Phys. Rev. Letters* **33**, 1139 (1974).
2. Jensen, T.H. and Chu, M.S., *Phys. Fluids* **27**, 2881 (1984).
3. Fisch, N. and Kaw, P.K., *Bull. Am. Phys. Soc.* **33**, (1988).
4. Taylor, J.B., *Revs. Modern Physics* **58**, 741 (1986).

5. Finn, J.M. and Antonsen, T.M., *Comments Plasma Phys. and Contr. Fusion* **9**, 111 (1985).
6. Bellan, P., *Phys. Rev. Letters* **57**, 2383 (1986).
7. Bevir, M.K. and Grey, J.W., *Proc. of RFP Workshop* (LASL, New Mexico 1981, eds. H.P. Lewis and P.A. Gerwin) Session III, Paper A-3.
8. Shoenberg, K.F. et al., *Phys. Fluids* **27**, 548 (1984).
9. Bellan, P., *Nucl. Fusion* **29**, 78 (1989).
10. Chan, V.S., Miller, R.L. and Ohkawa, T., *Phys. Fluids* **B2**, 944, 1441 (1990).
11. Kaw, P.K., Invited Talk, Symp. Plasma Sci. and Techn., Plasma Sci. Soc. India, Jaipur (1985).
12. Klima, R. and Petrzilka, *J. Phys. A: Gen. Phys.* **11**, 1687 (1978).
13. Grebogi, C., Kaufman, A.N. and Littlejohn, R.G., *Phys. Rev. Letters* **43**, 1668 (1979).
14. Bezzerides, B., Dubois, D.F., Forslund, D.W. and Lindman, E.L., *Phys. Rev. Letters* **38**, 495 (1977).
15. Avinash, K. and Kaw, P.K., 23rd IAEA Intl. Conf. Plasma Phys. and Contr. Fusion, Washington, D.C., 1990, paper CN-53/E-3-15.

R.N. Sudan
Laboratory of Plasma Studies
Cornell University Ithaca, NY 14853

SubGrid Modeling in Reduced MHD Turbulence

Introduction

I am deeply honored to make this presentation at the celebration of John Dawson's 60th birthday. I have known John for a very long time – almost from B.C. (*B*efore *C*oppi arrived at Princeton!) I have participated in two investigations[1,2] with him and thus observed at first-hand his creativity in action. Not only is John a scientist of great originality, but he has always been supportive of new ideas also. When I began daydreaming of how to develop the technology of intense ion beams, in 1970-1971, John's encouragement was invaluable to me.

John has made a wide spectrum of contributions to plasma physics and technology, but to my mind his greatest achievement is the unraveling of plasma behavior through particle computations. Whenever a new technology arises, it is grafted onto a previous way of thinking. When the first automobile was invented, it took the shape of a horseless carriage. A few decades elapsed before it acquired its modern characteristic appearance. This is no less true of computers. Most early applications of computers extended the domain of the solution of differential equations beyond those obtained by analytical means. It took John Dawson[3] and Oscar Buneman[4] to appreciate the true potential of computers and develop the new technique of "particle simulation" wholly defined by the inherent capability of the computer and owing very little to analytical solutions!

But alas, even a modern supercomputer can handle only a system with large but finite degrees of freedom. The representation of a system with impossibly large numbers of degrees of freedom by a finite set is one of the central issues in computational mathematics. By various clever artifacts, Dawson and others[5] have attempted successfully to solve this problem for particle codes. In hydrodynamics, the technique of "large eddy simulation" is popular; where the large scale eddies are computed, whereas the action of the small eddies is represented by a so-called empirical "eddy diffusion" term in the differential equations[6] . More recently, Yakhot and Orszag[7] have obtained this "eddy diffusion" term by a systematic technique for averaging the small-scales first introduced by Wilson[8], and known as the Renormalization Group Theory (RNG). This "eddy diffusion" term represents the action of all scales smaller than the grid size Δ on scales larger than Δ which are actually being computed. Hence the name "subgrid modeling" for this topic of research. In the subsequent sections, I wish to summarize a calculation by Longcope and myself[9] in which the RNG method is applied to the reduced MHD equations[10,11] in order to obtain renormalized operators for the viscous and resistive dissipation that effectively represent all the small scale Alfven wave turbulence less than the computational grid size.

Mathematical Model of a Solar Coronal Magnetic Loop

It has long been recognized that the ambient solar corona is heated by the mechanical energy of the turbulent convective motions of the photosphere transmitted to the corona through the medium of the solar magnetic field especially in the closed loops[12]. Ignoring magnetic curvature, gravity and density stratification, the loop can be modeled as a uniform magnetic field $\mathbf{B}_0 = B_0\hat{\mathbf{z}}$ anchored in the turbulently convective photosphere at $z = 0$ and $z = L$. The problem being addressed is the dissipation of the energy in the random motions of the feet of the field lines into heat in the volume of the flux tube. The coronal temperature is maintained at $1 \sim 2 \times 10^6$ °K from the power input of ~ 1 Watt/cm^2 from the photosphere[13] . But because the Lundquist number $\mathrm{Rm} = 4\pi\sigma v_A a/c^2 \sim 10^{10}$ is very large for the scale size of 10^4km for the loop width a, it is not immediately clear that the energy will actually get dissipated in the flux tube volume or get reabsorbed in the photosphere and chromosphere at the ends of the loop. In order for it to be absorbed in the volume of the loop, the effective Lundquist number $\mathrm{Rm}_* = 4\pi\sigma_* v_A a_*/c^2$ must be orders of magnitude smaller. This means that the turbulent motions generate a transverse fine structure which drastically reduces the Lundquist number or, alternatively, the effective resistivity and viscosity are increased by orders of magnitude in the presence of this fine structure[14] . In other words, the energy injected at large scales cascades to very fine scales where Rm_* is order unity and effectively absorbed by the molecular dissipative processes. This picture has emerged from treatments of hydrodynamic turbulence, but what we need to show is that this same physical picture is true for MHD and to provide quantitative expressions for the effective viscosity and resistivity.

Because the loops scale as $a \ll L$, i.e., width much smaller than the length with $B_0 \gg B_\perp$, we may employ the equations of reduced MHD[10,11] to describe the dynamics of the coronal loop:

$$\frac{\partial}{\partial t}\nabla_\perp^2\psi + \mathbf{v}_\perp \cdot \nabla_\perp \nabla_\perp^2 \psi = \frac{1}{4\pi\rho}\left[B_0 \frac{\partial}{\partial z}\nabla_\perp^2 A + \mathbf{B}_\perp \cdot \nabla_\perp \nabla_\perp^2 A\right] + \nu_0 \nabla_\perp^4 \psi \quad , \tag{1}$$

$$\frac{\partial}{\partial t}A + \mathbf{v}_\perp \cdot \nabla_\perp A = B_0 \frac{\partial}{\partial z}\psi + \eta_0 \nabla_\perp^2 A \quad , \tag{2}$$

where ψ and A are the stream function and the z-component of the vector potential respectively; defined by the relations

$$\mathbf{v}_\perp = \nabla_\perp \psi \times \hat{\mathbf{z}} \, , \qquad \mathbf{B}_\perp = \nabla_\perp A \times \hat{\mathbf{z}} \, , \tag{3}$$

ν and η are the molecular values of the perpendicular viscosity and resistivity of the coronal plasma. Equation (1) expresses transport of parallel vorticity and (2) comes from Ohm's law to lowest order in $a/L \ll 1$. The plasma velocity obeys $\mathbf{E} \times \mathbf{B}$ drift and hence ψ is proportional to the electrostatic potential. Typically the velocity is specified as a random function on $z = 0$ and $z = L$ planes to simulate the photospheric motions. We wish to compute the set of equations (1) to (3) on a three-dimensional grid. Since the number of degrees of freedom are approximately $\mathrm{Rm}^{9/4}$, clearly for $\mathrm{Rm} \sim 10^{10}$ this is an impossible task on any present-day computer and even those we may envisage for the future. Alternatively, we aim to solve equations (1) to (3) limited to a small sub-range (say two decades in any one dimension) of wave numbers where $0 < k < K$ and $K = \pi/\Delta$ is the grid size but with renormalized dissipation coefficients $\nu(K)$ and $\eta(K)$ in place of molecular ones. These new dissipation coefficients represent the transfer of energy to scales smaller than Δ, which eventually at the very smallest scales is converted to heat by η_0 and ν_0. It is the objective of the theory of subgrid modeling to obtain expressions for $\nu(K)$ and $\eta(K)$ in terms of $\nabla_\perp^2\psi$ and J that are being computed in the range $0 < k < K$.

Renormalized Group Analysis of SubGrid Modeling

If the spectrum of small scale fluctuations in the steady state is self similar, then one can argue that the subgrid scales may be represented by such a spectrum. The rate of cascade of energy to small scales will then lead to the effective dissipation coefficients we seek. For such a self similar spectrum to exist, the basic equations

(1) to (3) must in a sense be scale invariant. The technique of dynamical renormalization group[15,16,7] is the most effective way of establishing whenever a system of equations representing a physical model possesses scale invariance.

The procedure assumes that the boundary and initial conditions are replaced by external forcing functions F_Ω and F_J in equations (1) and (2) respectively, with Gaussian statistics and known variance. The forcing functions are taken to have power law dependence in Fourier space up to an ultraviolet cutoff Λ. Equations (1) to (3) are transformed from $(\mathbf{x}, t)$ to $(\mathbf{k}, \omega)$ space. The first step is to separate field quantities and the forcing functions into two parts: (a) region $e^{-r}\Lambda < |\mathbf{k}| < \Lambda$ and (b) region $|\mathbf{k}| < e^{-r}\Lambda$, where r is taken to be infinitesimally small. The quantities in region (a) are solved formally in terms of (b) perturbatively to low order in the nonlinearity. The perturbation series for the quantities in region (b) are then averaged over the statistics of the forcing functions in the narrow band (a). The resulting averaged equations are rescaled in such a manner as to revert the system back to the same ultraviolet cutoff Λ. This procedure results in equations self similar to the original set except that the dissipative parameters are altered by an infinitesimal amount. It is then possible to write differential equations for the dissipative parameters as a function of r. Under certain conditions these differential equations have fixed points of solution which represent a model system with self similar scaling. The mathematical details of this procedure are too lengthy to be incorporated here. The interested reader is referred to the calculation by Longcope and Sudan[9] and earlier references. The principal results of this analysis are[9]:

1. In the long wavelength limit $|\mathbf{k}| \to 0$ the Prandtl number $\mathrm{Pr}_* = \nu(k)/\eta(k)$ varies between $1 \le \mathrm{Pr}_* \le 1.7321$ and depends upon $w = (\langle F_\Omega^2\rangle - \langle F_J^2\rangle)/(\langle F_\Omega^2\rangle + \langle F_J^2\rangle)$. The range of Pr_* corresponds to $0.44415 = w_{\mathrm{cr}} < w < 1$.
2. The turbulent β_T defined to be the ratio of spectral kinetic and magnetic energy densities is given by $\beta_T = \mathrm{Pr}_*/(\mathrm{Pr}_*^2 - 1)$. From (1), it is evident that $0.8660 \le \beta_T \le \infty$.
3. The energy spectrum
$$E(k_\perp) \propto k_\perp^{1-m/2} \ , \tag{4a}$$
and
$$\mu(k_\perp) = (\nu + \eta) \propto k_\perp^{-m/4} \ , \tag{4b}$$
where the forcing function $\langle FF\rangle \propto k_\perp^{-m}$
4. Applying Kolmogorov's cascade principle for the energy density fixes $m = 16/3$ so that $E(k_\perp) \propto k_\perp^{-5/3}$ and $\mu(k_\perp) \propto k_\perp^{-4/3}$.
5. The renormalized Reynolds and Lundquist numbers are,

$$\mathrm{Re}_* = \frac{1}{2}\Big(\frac{\beta_T}{1+\beta_T}\Big)^{1/2}(1 + \mathrm{Pr}_*^{-1})\mathcal{R} \ , \tag{5a}$$

$$\mathrm{Rm}_* = \frac{1}{2}\Big(\frac{\beta_T}{1+\beta_T}\Big)^{1/2}(1 + \mathrm{Pr}_*)\mathcal{R} \ , \tag{5b}$$

where $\mathcal{R}$ is function of w whose magnitude is of order unity. By equating the work done by the external forcing functions to the energy dissipated in the entire spectrum by the molecular dissipation, it is possible to derive the following expressions for $\nu(K)$ and $\eta(K)$:

$$\nu(K) = (1 + \rho_*)\mu(K) \ , \qquad \eta(K) = (1 - \rho_*)\mu(K) \ , \tag{6a}$$

$$\mu(K) = \sqrt{\alpha(w)}\, K^{-2}\Big[(1 + \rho_*)\Omega^2 + (1 - \rho_*)J^2\Big]^{1/2} \ , \tag{6b}$$

where $K = \pi/\Delta$, Δ is the perpendicular grid spacing, $\alpha(w)$ is a known function of w, $\Omega = -\nabla_\perp^2\psi$ and $J = -\nabla_\perp^2 A$ and $\rho_* = [1 + 4\,|\nabla_\perp\psi|^2/|\nabla_\perp A|^2]^{-1/2}$. These expressions for ν and η are similar to the well known Smagorinsky[6] model in hydrodynamics.

Discussion

The equations of reduced MHD represent the interaction of very long, very thin structures of Alfven pulses along a strong uniform magnetic field. The interaction of counter propagating waves dominates the dynamics. We have shown that stationary, fully developed, self-similar turbulence applies to the smallest scales just as in Navier Stokes turbulence. However, no self similar RMHD turbulence occurs for $\beta_T < 0.8660$. Even if this condition is violated at the longest scales for the force free solar corona, it is likely that at the smallest scales $\beta_T \to 1$. Thus the grid size in simulations should be such that the simulations show $\beta_T \sim 1$ for $k_\perp \leq \pi/\Delta$. Then smaller scales will obey the Kolmogorov turbulent spectra and the effect of energy transfer may be represented by the sub grid model with $\nu(K)$ and $\eta(K)$ given by equation (6) .

Acknowledgements

I am indebted to D.W. Longcope for many discussions and calculations on this topic. Work supported by NASA Grant #NAGW-1006 and NSF Grant #86-00308.

References

1. W.L. Kruer, J.M. Dawson and R.N. Sudan, *Phys. Rev. Lett.*, 23, 238 (1969).
2. J. Stamper, K. Papadapoulos, R. N. Sudan, E. McLean, S. Dean and J. Dawson, *Phys. Rev. Lett.*, 2b, 1012 (1971).
3. J. Dawson, *Phys. Rev.* 113, 383 (1959); *Phys. Rev.*, 118, 381 (1960).
4. O. Buneman, *Phys. Rev.*, 115, 503 (1959).
5. See for example: C.K. Birdsal and A.B. Langdon, *Plasma Physics via Computer Simulation*, McGraw Hill: New York, NY (1985).
6. J. Smagorinsky, *Mon. Weather Rev.*, 91, 99 (1963).
7. V. Yakhot and S.A. Orszag, *Phys. Rev. Lett.*, 57, 1772 (1986).
8. K.G. Wilson and J. Kogut, *Phys. Rep.*, 12, 7 (1974).
9. D.W. Longcope and R.N. Sudan, "Renormalized Group Analysis of Alfvén Turbulence in Reduced MHD", Report No. LPS90-10, Laboratory of Plasma Studies, Cornell University, Ithaca, N.Y.
10. M.N. Rosenbluth, D.A. Monticello, H.R. Strauss and R.B. White, *Phys. Fluids*, 19, 1987 (1976).
11. H.R. Strauss, *Phys. Fluids*, 19, 136 (1976).
12. E.N. Parker, *Ap. J.*, 174, 499 (1972).
13. E.N. Parker, *Phys. Today*, July 1987, p. 36.
14. R.N. Sudan, "Heating in Stochastic Magnetic Field", Conference on Chromospheric and Coronal Heating Mechanisms, Heidelberg, Germany, June 1990.
15. S.K. Ma and G. Mazenko, *Phys. Rev.*, B, 11, 4077 (1975).
16. D. Forster, D.R. Nelson and J.M. Stephen, *Phys. Rev.*, A 16, 732 (1977).

Carl Oberman
Princeton Plasma Physics Laboratory
Princeton, NJ 08543

Atoms in Strong Electromagnetic Fields

Introduction

Because of John's longstanding interest in laser-matter interaction, I will take this opportunity to report on some fundamental calculations of atomic physics, necessitated by having to take into account the fact that some current laser field strengths in the focal region are comparable to or even greater than atomic fields. The time-honored technique of perturbation theory becomes prohibitively difficult and, for sufficiently large intensities, inadequate, and new techniques must be found to describe the laser-plasma interaction. Transitions involving the absorption or emission of many field quanta can occur under these circumstances.

However, since the laser frequency, ω, is in general much less than atomic frequencies, we are able to remain within the classical description of the electromagnetic field and are led to develop a description asymptotic in ω/Δ where $\Delta \equiv (E_2 - E_1)/2$ and E_1 and E_2 are two characteristic atomic frequencies.

I shall comment briefly on three pieces of work: multiphoton excitation with a simple atomic model; multiphoton ionization, again with a simple model; and finally a treatment of multiphoton ionization of a complex high Z atom in which we argue for a modified Thomas-Fermi description.

Multiphoton Excitation – Two Level System

Most experimental studies involving multiphoton processes have investigated multiphoton ionization. There is the suggestion, however, based on Rhodes' observation[1] in Krypton, that excitation by a many photon process has been observed experimentally. In light of these results, it has been proposed that selective excitation by low-frequency electromagnetic fields may be feasible and desirable.

The simplest model that addresses this problem is the two-level system given in Figure 1. There, $E_2 > E_1$, $\omega > 0$, $F > 0$, $\dot{F}/F \ll \omega$.

- **The two-level atom**

- R. Duvall's thesis

- R. E. Duvall, E. J. Valeo and C. R. Oberman, Phys. Rev. A **37**, 4685 (1988).

- Simple hamiltonian

$$i\frac{d\mathbf{a}}{dt} = \mathbf{H}(t) \cdot \mathbf{a} \quad ,$$

of the explicit form

$$i\frac{d}{dt}\begin{pmatrix} a_1 \\ a_2 \end{pmatrix} = \begin{pmatrix} E_1 & F\sin\omega t \\ F\sin\omega t & E_2 \end{pmatrix}\begin{pmatrix} a_1 \\ a_2 \end{pmatrix} \quad .$$

- J.H.Shirley, Phys. Rev. **138**, B979 (1965).

- N. B. Delone and V. P. Krainov, *Atoms in Strong Light Fields* (Springer, Berlin, 1985).

FIGURE 1

Here F is proportional to the laser field strength and the dipole moment connecting the two states. The field is treated semi-classically and in the dipole approximation – laser wavelength much greater than atomic size. (This model describes other physical problems such as a spin $-\frac{1}{2}$ particle in crossed static and oscillating magnetic fields.) Solutions of this problem have been obtained from perturbation theory (F_{small}) by Shirley[2] using a Floquet analysis and perturbation theory. We develop a theory in ω/Δ small but spanning the domain $F \sim \Delta$ and constrained only by $F \ll \Delta^2/\omega$. The identification with the quantized treatment is that $2\Delta/\omega$ is taken equal to the number of photons involved in the transition.

We first discuss the case where F is constant. In order to find the time evolution of the system, we write the first order system in second order form (Figure 2), and introduce the WKB representation.

- Can be written equivalently as a second order system

$$\frac{d^2y}{d\tau^2} + Qy = 0,$$

where

$$Q \equiv \left(\frac{\Delta}{\omega}\right)^2\left[1 + \left(\frac{F}{\Delta}\right)^2 \sin^2\tau + i\frac{\omega F}{\Delta^2}\cos\tau\right],$$

with

$$\Delta \equiv \frac{(E_2 - E_1)}{2}.$$

- Our solution is asymptotic in

$$\frac{\Delta}{\omega} >> 1 \quad ,$$

rather than perturbative in F/Δ.

- Because $\Delta/\omega >> 1$, $Q >> 1$ on the real time axis)

FIGURE 2

Solving the problem then amounts to finding the small changes in the amplitudes of this representation over each period of the applied field. Exactly as in the problem of calculating the classically forbidden above barrier transmission coefficient[3] or, isomorphically, the change in the adiabatic invariant, the small changes necessitate moving our WKB path for the phase integral into the complex-time plane to meet, in this case, the zeroes of Q, where the WKB breaks down. By asymptotic matching to the solution in the vicinity of $Q = 0$, or, equivalently, using the established connection formulas for the jumps in the solution on crossing a Stokes line in one of these regions (Fig. 3), we determine a transfer matrix which advances the system over one period of the applied field. The Nth power of this matrix then determines the state of the system after N periods. The method is to be contrasted with Shirley's approach using Floquet theory, where it is necessary to find the eigenvalues of very large matrices.

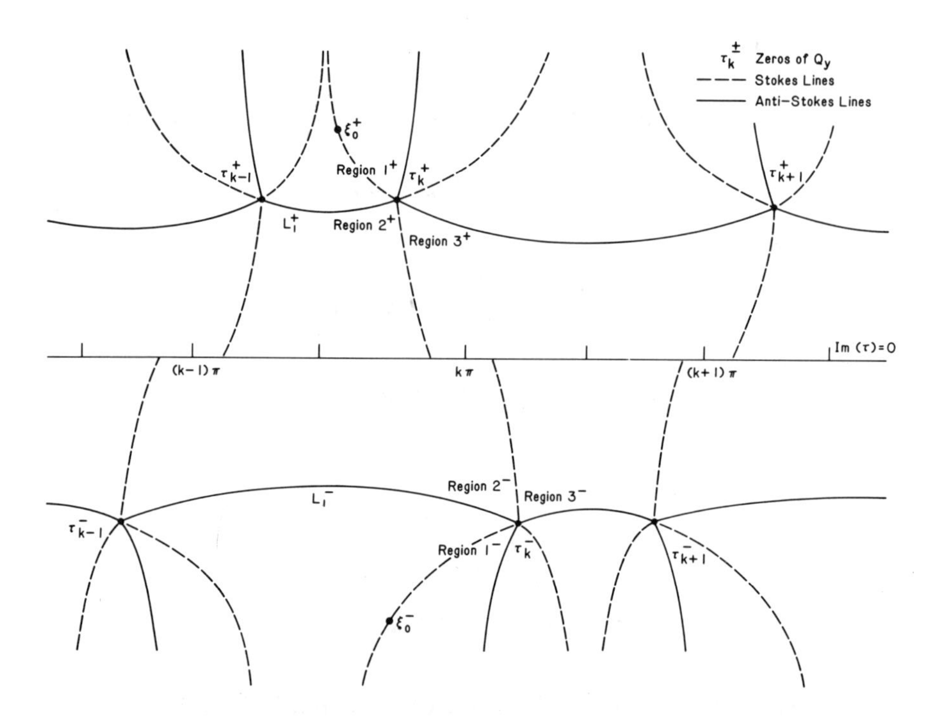

FIGURE 3 Stokes and anti-Stokes line structure for Q_y in the complex τ plane for an interval surrounding $\tau = k\pi$, where k is taken even. The structure repeats as τ changes by 2π.

We are thus able to find the excitation probability P of finding a particle in state 2, given that initially it was in state 1 in terms of the real phase advance per half period θ_0 and the phase integral $\varepsilon \equiv (\tau_k^-, \tau_k^+)$ between conjugate zeroes of Q (see Figs. 4, 5, and 6), which quantities are functions of field frequency and intensity. For small F/Δ, we recover the resonance shifts and $(F^2)^{\Delta/\omega}$ power for the generalized Rabi frequency[4] as found by a perturbative treatment of Shirley's

Floquet Hamiltonian. Simplified expressions for E and the resonant frequency $\omega_{\rm res}$ can also be given for $F/\Delta \gg 1$. We note the exponential smallness of u, since $\omega F/\Delta^2 \ll 1$. Because of the smallness of u, the resonance widths for P are very small. Further, to the extent our model is applicable we see variations in $\omega_{\rm res}$ that are large compared to u will occur in practice because of intensity variation due to finite pulse length. These considerations place very stringent criteria on the constancy of the intensity, if any particular resonance is to be maintained for as long as one Rabi period. In fact, because of the intensity dependence of $\omega_{\rm res}$, the condition $\omega_{\rm res} - (2p+1)\omega = 0$ for a given order p is satisfied precisely at only a single time t_p. If $p \gg 1$, then many resonances are passed over as the applied pulse evolves (see Fig. 7). In the short autocorrelation time limit, in which successive resonances contribute with random phases, the average excitation rate is given by $R = \pi u^2/4\omega$.

- We represent y asymptotically in the WKB form

$$y \sim c_k^+(\tau_k^+, \tau) + c_k^-(\tau, \tau_k^-) \quad , \qquad |\tau - k\pi| < \pi .$$

where

$$(a, b) \equiv Q^{-1/4}(\tau) \exp(+i\textstyle\int_a^b Q^{1/2} d\tau) \quad ,$$

- We compute the change in c_k over one period as we cross the Stokes' lines.

- Denoting $\mathbf{c}_k = \begin{pmatrix} c_k^+ \\ c_k^- \end{pmatrix}$, we eventually obtain a matrix relating $\mathbf{c}_{k+2}$ to $\mathbf{c}_k$,

$$\mathbf{c}_{k+2} = \mathbf{M} \cdot \mathbf{c}_k \quad .$$

where

$$\mathbf{M} = \begin{pmatrix} \exp(i2\theta_0) & 2\varepsilon \sin\theta_0 \exp(i\theta_0) \\ 2\varepsilon \sin\theta_0 \exp(-i\theta_0) & \exp(-i2\theta_0) \end{pmatrix} \quad .$$

is expressed in terms of

$$\varepsilon \equiv (\tau_k^-, \tau_k^+) ,$$

FIGURE 4

and of the (evidently real) phase advance per half period of the applied field

$$\theta_0 \equiv (0,\pi) = \frac{\Delta}{\omega}\int_0^{\pi}[1 + (\frac{F}{\Delta})^2 \sin^2\tau + i\frac{\omega F}{\Delta^2}\cos\tau]^{\frac{1}{2}} d\tau \quad ,$$

• ε has the asymptotic forms

$$\begin{aligned} |\varepsilon| &\sim (\frac{e\,F}{4\,\Delta})^{2\frac{\Delta}{\omega}} \quad , && F/\Delta << 1 \quad , \\ &\sim \exp(-\frac{\pi}{2}\frac{\Delta}{\omega}\frac{\Delta}{F}) \quad , && F/\Delta >> 1 \quad , \end{aligned}$$

• θ_0 has the asymptotic forms

$$\begin{aligned} \theta_0 &\sim \pi\frac{\Delta}{\omega}[1 + \frac{1}{4}(\frac{F}{\Delta})^2] \quad , && F/\Delta << 1 \quad , \\ &\sim 2\frac{F}{\omega} \quad , && F/\Delta >> 1 \quad . \end{aligned}$$

• The matrix which evolves $\mathbf{a}_k$ over an integral number N of periods is, finally,

$$\mathbf{a}_{k+2N} = \mathbf{M}^N \cdot \mathbf{a}_k \quad .$$

FIGURE 5

- Assuming $a^-(0) = 0$, $a^+(0) = 1$, setting $P \equiv |a^-_{k+2N}|^2$ and making the replacement $N \to \omega t/2\pi$, we obtain the result, for $\omega t \sim \varepsilon^{-1}$,

$$P \simeq \frac{u^2}{(\omega_{\rm res} - (2p+1)\omega)^2 + u^2} \sin^2\{\frac{1}{2}[(\omega_{\rm res} - (2p+1)\omega)^2 + u^2]^{\frac{1}{2}} t\}$$

where we have defined $u \equiv (2\omega/\pi)|\varepsilon|$ and $\omega_{\rm res} \equiv (2\omega/\pi)\theta_0$ where $l = 2p+1$, and p is an integer.

- The resonance condition is

$$\omega_{\rm res} - (2p+1)\omega = 0 \quad .$$

- In the small field limit (F/Δ small), we have

$$u \sim \frac{2\omega}{\pi}(\frac{eF}{4\Delta})^{2\frac{\Delta}{\omega}} \quad ,$$

$$\omega_{\rm res} \sim 2\Delta(1 + \frac{1}{4}(\frac{F}{\Delta})^2) \quad .$$

(These results are equivalent to those obtained perturbatively by Shirley.)

- For the case of large fields (F/Δ large) we have

$$u \sim \frac{2\omega}{\pi} \exp[-\frac{\pi}{2}\frac{\Delta}{\omega}\frac{\Delta}{F}] \quad ,$$

$$\omega_{\rm res} \sim \frac{4F}{\pi} \quad .$$

FIGURE 6

- **An Effect of Finite Pulse Length on high order resonances**

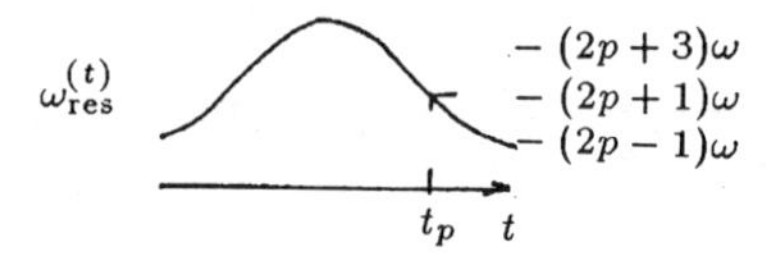

- Because of the intensity dependence of ω_{res} the condition

$$\omega_{res} - (2p+1)\,\omega = 0\,,$$

for resonance of a given order p is satisfied precisely at only a single time t_p.

- If $p >> 1$, then it can be that many resonances are passed through as the applied pulse evolves.

- In the limit in which successive resonances contribute with random phases one obtains an average rate of excitation

$$R = \frac{\pi u^2}{4\omega}\quad .$$

FIGURE 7

To test the relevance of this result to proposed realistic experimental conditions[5], this average excitation rate was evaluated for parameters relevant to the case of multiphoton excitation in Cd^{12+}, and was compared to the rate of multiphoton ionization from the upper excited state using the formula of Keldysh[6] ($\gamma \ll 1$). The ionization from the excited state was found to dominate the excitation for all intensities of interest.

Multiphoton Ionization in a Short Range Potential

Before attacking the difficult problem of a real atom subject to a very strong field, one would like to develop an intuition about the new effects caused by such a field by studying model potentials. Even the problem of an electron bound by a model potential and subject to a strong em field has not been solved analytically for any potential. The numerical solution is not an easy task either, the main reason being that strong fields cause ionization and hence any calculation would have to include the continuum spectrum.

Recent experiments on the ionization of atoms have analyzed the outcoming electron energy[5,7]. Several features, such as ponderomotive corrections to the ionization potential and above-threshold ionization, where the electron absorbs more than the minimum number of photons necessary for ionization, were observed. These features were already qualitatively predicted in earlier work[6,8,9]. In the present work, we will comment on some critical aspects of the strong field-many photon limit. In particular, we solve in this limit a model which represents the atomic potential as a δ-function. The analytical results obtained are asymptotically correct as the number of photons goes to infinity. The problem was also solved numerically, and very good agreement with the analytical results was obtained for even a few photons.

We approach the ionization problem first in a qualitative fashion. This will allow us to draw a number of conclusions without the necessity of going into complicated equations.

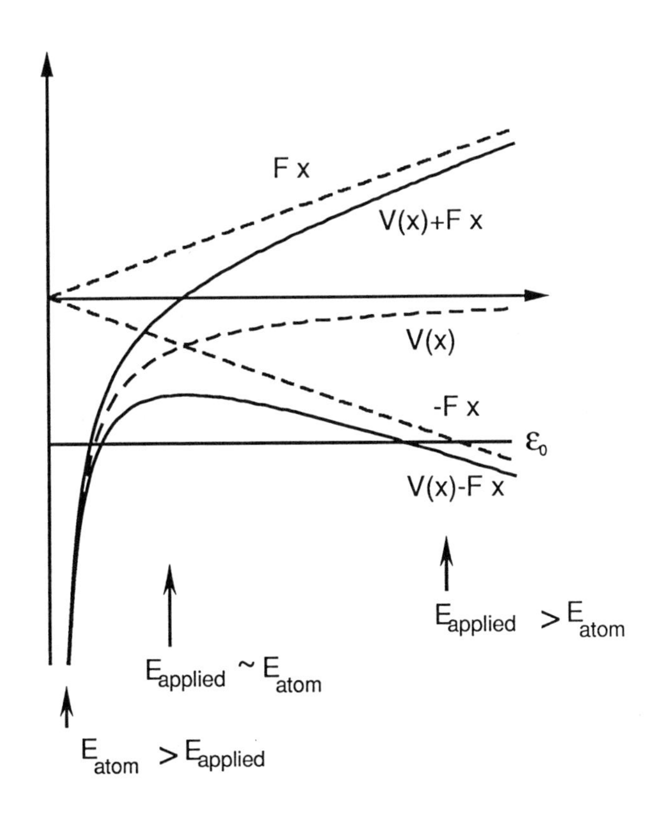

FIGURE 8

In Fig. 8, we have plotted a typical binding potential $V(x)$ and the potential due to an external, linearly polarized, monochromatic field, $V_{\rm ext} = Fx \cos(\omega t)$ (atomic units are adopted throughout), for each half a period of the field. It is clear that the barrier formed by these two potentials will be up and down in intervals of half a period. When the barrier is down, the electron in the bound state with energy ε_0 must traverse a barrier of width, roughly $\ell \sim \varepsilon_0/F$. The time $\tau_{\rm tunn}$ to tunnel through that barrier at (imaginary) velocity $v \sim \sqrt{2\varepsilon_0}$ is $\tau_{\rm tunn} \sim \sqrt{2\varepsilon_0}/F$. If $\tau_{\rm tunn}$ is much smaller than a period of the field, the electron has time to tunnel before the barrier goes up again, and the problem is basically static. This is called the adiabatic regime. If, on the other hand, $\tau_{\rm tunn}$ is larger than the period of the field, the electron cannot get out of the well before the barrier goes up and we should expect a frequency dependence of the ionization rate. This case is called the multiphoton case. A natural dimensionless parameter is then Keldysh's parameter (Fig. 9), which is defined as the ratio of the tunneling time to the period of the field, $\gamma = \tau_{\rm tunn}/\tau = \sqrt{2\varepsilon_0}\,\omega/F$.

- The ratio of the tunneling time to the period of the field

$$\gamma \equiv \tau_{\rm tunn}/\tau = \sqrt{2\epsilon_0}\omega/F.$$

is called the **Keldysh parameter**.

  - from the original work of L.V. Keldysh, Zh. Eksp. Teor. Fiz. **47**, 1945 (1964) [Sov. Phys. JETP **20**, 1307 (1965)].
  - cf, also A.M. Perelomov, V.S. Popov and M.V. Terentev, Zh. Eksp. Teor. Fiz. **50**, 1393 (1966) [Sov. Phys. JETP **23**, 924 (1966)].
  - If $\gamma << 1$ one has the **tunneling** or adiabatic regime.
  - If $\gamma >> 1$ one has the (frequency dependent) **multiphoton** limit.

FIGURE 9

The case $\gamma \gg 1$ is the multiphoton case, and $\gamma \ll 1$ corresponds to the adiabatic case. Besides γ, there are basically two more dimensionless parameters (Fig. 10) that characterize the ionization from a general binding potential $V(x)$: the ratio between the frequency of the applied field and a typical atomic frequency, $\omega/\omega_{\text{atom}}$, whose inverse is the number of photons involved in the process; and the ratio of applied electric field amplitude to a typical atomic field strength, F/F_{atom}.

- **Ionization in a constant Electric field**

- In a constant electric field F, the classical position of an electron is ($q = -1$ a.u.)

$$x_0 = -\frac{F}{2}t^2 .$$

- Then, simply,

$$S(t, t-T) = -\frac{F^2T^3}{24},$$

and

$$\chi(t) = \chi_0(t) + \frac{i^{1/2}B}{\sqrt{2\pi}} \int_0^t \frac{dT}{\sqrt{T}} \exp\left(-\frac{iF^2}{24}T^3\right) \chi(t-T) .$$

- Setting

$$\tilde{\chi}(p) \equiv \int_0^\infty dt \exp(-pT)\, \chi(t) ,$$

the transformed integral equation becomes

$$\tilde{\chi}(p) = \frac{\tilde{\chi}_0(p)}{[1 - C(p)]},$$

where

$$C(p) = \frac{i^{1/2}B}{\sqrt{2\pi}} \int_0^\infty \frac{dt}{\sqrt{t}} \exp\left(-pt - \frac{iF^2}{24}t^3\right) .$$

FIGURE 10

If we consider the relative magnitude of the applied field to the atomic field, we see also from Fig. 8 that we can distinguish three regions. The first one, which we might call "inside the atom", has $E_{\text{atom}} > E_{\text{applied}}$. If this relationship between field strengths did not hold inside the atom, the electron would be ripped off the atom in a few atomic periods. On the other hand, since in a real laser pulse there is always a finite rise time, with such a strong amplitude the atom will ionize before it reaches the peak. The second region is an intermediate one, where $E_{\text{applied}} \sim E_{\text{atom}}$ and, finally, we have the "outside of the atom", where $E_{\text{applied}} > E_{\text{atom}}$. These three regions exist for any typical binding potential subject to a linearly polarized field, and consequently any theory that uses a perturbation approach in $E_{\text{applied}}/E_{\text{atom}}$ or in $E_{\text{atom}}/E_{\text{applied}}$ cannot be justified, because neither assumption holds throughout all space.

In light of the previous discussion, it is worth pointing out here also the differences between the ionization problem and the scattering problem of an electron in the presence of a strong field. In the scattering problem, the electron can be considered always outside of the atom (external problem), and, under certain conditions, one can assume $E_{\text{applied}} > E_{\text{atom}}$ uniformly throughout all space. However, the ionization problem presents the extra difficulty that no matter how strong the external field is, there is a region where the electron is initially "inside the atom", where the binding field is larger and where the strong external field approximation breaks down. This difference makes the ionization problem much harder to solve than the scattering problem.

We will now turn to the language of Green's functions, or evolution operators, to put the intuitive description of the previous section into a mathematical form. This allows us to describe our approach and will also enable us later to relate it in a simple way to other works in the literature.

Schrödinger's differential equation for the wave function of an electron subject to a binding potential $V(x)$ and an external field V_{ext} is

$$i\frac{\partial|\psi>}{\partial t} = (T + V + V_{\text{ext}})\,|\psi> \equiv H|\psi>$$

where T is the kinetic energy operator. Similarly, the equation for the Green's function $G(t,t')$ of our problem is

$$\left(i\frac{\partial}{\partial t} - H\right) G(t,t') = i\delta(t-t')\ . \tag{1}$$

We can also formally write the equations for the Green's functions corresponding to the Hamiltonians $H_0 = T + V$ for an electron in the binding potential V, and to $H_{\text{ext}} = T + V_{\text{ext}}$ for an electron subject to V_{ext} only:

$$\left(i\frac{\partial}{\partial t} - H_0\right) G_0(t,t') = i\delta(t-t')\ , \tag{1}$$

$$\left(i\frac{\partial}{\partial t} - H_{\text{ext}}\right) G_{\text{ext}}(t,t') = i\delta(t-t')\ . \tag{1}$$

There is clearly a relationship between G and G_{ext} and between G and G_0. These relationships are given by the following expressions:

$$G(t,t') = G_0(t,t') - i\int_{t'}^{t} dt''\, G_0(t,t'')\, V_{\text{ext}}(t'')\, G(t'',t') \tag{2}$$

and

$$G(t,t') = G_{\text{ext}}(t,t') - i\int_{t'}^{t} dt''\, G_{\text{ext}}(t,t'')\, V(t'')\, G(t'',t')\ . \tag{3}$$

These integral equations for $G(t,t')$ are equivalent to Schrödinger's differential equation (1) because they satisfy this equation identically. $G(t,t')$ is the evolution operator of the system. That means that if $|\psi(t_0)\rangle$ is the wave function representing the system at time t_0, the wave function at time t is given by

$$|\psi(t)\rangle = G(t,t_0)|\,\psi(t_0)\rangle\ . \tag{4}$$

In particular, one is interested in the transition amplitude for going from an initial state $|\psi_i(t_0)\rangle$ at time t_0 to a final state $|\psi_f(t)\rangle$ at time t. This transition amplitude can be calculated in terms of G by

$$T_{fi} = \langle\,\psi_f(t)\,|G(t,t_0)|\,\psi_i(t_0)\,\rangle\ . \tag{5}$$

Unfortunately, the problem of calculating $G(t,t')$ for an electron subject simultaneously to a binding potential V and the potential V_{ext} due to a plane monochromatic wave has not been solved analytically for any V. For calculating the transition amplitudes, one has then to rely on numerical calculations or on some kind of approximation scheme.

The usual way of proceeding is to assume that $V_{\text{ext}} \ll V$ uniformly in space, and calculate G from an iterative procedure starting from Eq. (2). That is, one develops a perturbation series in the applied field strength.

The values of field strength that are now being used in the laboratory preclude this kind of approach. On the other hand, it is not possible to assume that $V_{\text{ext}} \gg V$ uniformly throughout all space, either, because inside the atom – as we saw qualitatively in the previous section – the atomic field is stronger than the applied field.

An alternative way of proceeding is to use either Eq. (2) or Eq. (3) as a starting point, but with the assumption that $\omega_{\text{atom}}/\omega \to \infty$; that is, we assume that a large number of photons is involved in the transition.

It proves to be more convenient to use Eq. (3). The reason is that the expression for G_{ext} is simpler, in general, than G_0. A typical binding potential V has both a bound and a discrete spectrum, making the expression for G_0 very involved. Instead, the Green's function, $G_{\text{ext}}(t,t')$, for an electron in the potential (we choose to work in the Fx gauge):

$$V_{\text{ext}} = Fx\,\cos(\omega t) \tag{6}$$

where F is the amplitude of the applied field, is well known. Furthermore, because of a result due to Feynman, we can write $G_{\text{ext}}(t,t')$, in the x representation $G(x,t;\,x',t') = \langle\,x\,|G(t,t')|\,x'\,\rangle$, simply as

$$G_{\text{ext}}(x,t;\,x',t') = \frac{\exp[iS_{cl}(x,t;\,x',t')]}{\sqrt{2\pi i(t-t')}} \tag{7}$$

Here $S_{cl}(x,t;\,x',t')$ is the classical action of an electron, subject to the potential (6), which starts at x',t' and ends at x,t. It is given by

$$S_{cl}(x,t;\,x',t') = \frac{1}{2(t-t')}\left\{(x-x') - [x_0(t) - x_0(t')]\right\}^2$$
$$-\frac{1}{2}\int_{t'}^{t} d\tau \dot{x}_0^2(\tau) + x\dot{x}_0(t) - x'\dot{x}_0(t') .$$

expressed in terms of the classical position $x_0(t)$ of an electron in the external field only; that is, $x_0(t)$ satisfies $\ddot{x}_0 = -F(t) \quad (q = -1\ a.u.)$.

It proves convenient to work with the following integral equation, derived from eqs. (3) and (4) in the x representation, making use of Eq. (7),

$$\psi(x,t) = \psi_0(x,t) - \frac{i^{1/2}}{\sqrt{2\pi}}\int_0^t dt' \frac{1}{\sqrt{t-t'}}\int_{-\infty}^{\infty} dx' \exp\left[iS_{cl}(x,t;\,x',t')\right] V(x')\,\psi(x't') . \tag{8}$$

The inhomogeneous term in this integral equation for $\psi(x,t)$ is the wave function $\psi_0(x,t) \equiv \langle x|\psi_0(t)\rangle$, which coincides with the bound state at $t=0$ and which at later times evolves with the Hamiltonian of the electron in the external field only.

We now specialize to the case of an electron bound by the potential

$$V(x) = -B\delta(x) . \tag{9}$$

This potential has only one bound state of energy, $B^2/2$ (the ionization potential is then $I_p = B^2/2$), in which the electron is going to be at $t=0$. If we specialize Eq. (8) for this potential, the x' integral is straightforward. After setting $x = 0$ in the same equation, we obtain an integral equation, now in time only, for $\chi(\tau) \equiv \psi(0,t)$:

$$\chi(t) = \chi_0(t) + \frac{i^{1/2}B}{\sqrt{2\pi}}\int_0^t dT \frac{1}{\sqrt{T}} \exp[is(t,t-T)]\,\chi(t-T) . \tag{10}$$

We have made the change of time variables $T = t - t'$ and we have set $S(t,t-T) \equiv S_{cl}(0,t;\,0,t-t)$, i.e.,

$$S(t,t-T) = \frac{1}{2T}\left[x_0(t) - x_0(t-T)\right]^2 - \frac{1}{2}\int_{t-T}^{t} d\tau\, \dot{x}_0^2(\tau) . \tag{11}$$

Note that we have eliminated in this way the spatial dependence from the problem. From the solution $\chi(t)$ of Eq. (10), we can reconstruct $\psi(x,t)$ by means of Eq. (8). However, as we will see in the next two sections, only χ is necessary for computing the ionization rate.

Consider now the ionization of an electron from the δ-function potential in a constant (dc) field. The method we use to analyze this problem is somewhat different from the conventional approach, and we present it here because it introduces some of the techniques we adopt to analyze the time-dependent field case.

In a constant electric field F, the classical position of an electron is ($q = 1\,a.u.$) $x_0 = -F^2/2$. In the potential, $V = -B\delta(x)$, the wave function at the site of the potential $\chi(t) = \psi(0,t)$ evolved according to Eq. (10), with $S(t, t-T)$ given by Eq. (11). Then, simply, $S(t, t-T) = F^2T^3/24$ and

$$\chi(t) = \chi_0(t) + \frac{i^{1/2}B}{\sqrt{2\pi}} \int_0^t dT \frac{1}{\sqrt{T}} \exp\left(-\frac{iF^2}{24}T^3\right) \chi(t-T) \,. \tag{12}$$

Equation (12) is in convolution form. This simply reflects the fact that we are dealing with a time-independent Hamiltonian. The problem is invariant under translations in time, and that means that Schrödinger's differential equation can be reduced to an eigenvalue equation and its corresponding integral equation is in convolution form. We can formally solve Eq. (12) by using Laplace transforms. Let us define

$$\tilde{\chi}(p) \equiv \int_0^\infty dt \, \exp(-pT)\, \chi(t) \,, \tag{13}$$

with a similar definition holding for $\tilde{\chi}_0(p)$. Then Eq. (12) becomes

$$\tilde{\chi}(p) = \frac{\tilde{\chi}_0(p)}{1 - C(p)} \,,$$

with $C(p)$ given by

$$C(p) = \frac{i^{1/2}B}{\sqrt{2\pi}} \int_0^\infty dt \frac{1}{\sqrt{t}} \exp\left(-pt - \frac{iF^2}{24}t^3\right) \,. \tag{14}$$

The inverse transform is defined by

$$\chi(t) \equiv \frac{1}{2\pi} \int_{-i\infty+\sigma}^{i\infty+\sigma} dp \, \exp(pt)\, \tilde{\chi}(p) \,, \tag{15}$$

where $\sigma > \sigma_0$. The Laplace transform is defined only for $Re(p) > \sigma_0$ where σ_0 is chosen so that the integral in Eq. (13) exists. $\tilde{\chi}(p)$ is analytically continued from $Re(p) > \sigma_0$ to $Re(p) < \sigma_0$. The inverse transform may then be evaluated by closing the p contour in the left half of the p plane. $\chi(t)$ therefore has contributions from the poles (and branch cuts) of $\tilde{\chi}(p)$. Specifically, the contribution to $\chi(t)$ from a pole of $\tilde{\chi}(p)$ at p_0 is

$$\chi(t) \cong A \exp(p_0 t) + \ldots \,.$$

Note that, by definition, all poles of $\tilde{\chi}(p)$, p_j have a real part less than σ_0. We may therefore find the ionization rate by finding the positions of the poles of $\tilde{\chi}(p)$. (Clearly, the ionization rate ω equals the negative real part of $2p_j$). Contributions from $\chi(t)$ from any branch cuts in the p plane are ignored for two reasons: first, they produce a decay of χ like t^{β}, where $\beta \geq 1/2$, so that for $\omega \ll 1$ and $t \equiv 0(\omega^{-1})$ the exponentially decaying terms dominate χ; and, second, because these algebraic terms represent contributions to χ from free wave packets that are spreading and propagating away from $x = 0$ and not contributions from a "quasibound" wave function. With "smooth" initial conditions, $\tilde{\chi}_0(p)$ has no poles, so that we may write

$$1 - C(p_j) = 0 \tag{16}$$

Our treatment is analogous to Landau's[11] treatment of the decay of plasma oscillations, where Eq. (16) takes the role of the "dispersion relation". By assuming that the initial conditions are smooth, we obtain a decay rate that does not depend on them. If we take the initial condition to be the bound state of the δ-function model, $\tilde{\chi}_0(p)$ has no poles.

Equation (16) is a transcendental equation for the p_js and, unfortunately, the integral $C(p)$ is not available in a closed form. We are mainly interested in the limit when $F \ll B^3$ (or, in physical terms, the applied electric field is smaller than the atomic field when the electron is in the bound state); in this limit $C(p)$ can be evaluated asymptotically near p_j via the method of steepest descents. When $F = 0$, there is only one pole ($p_0 = iB^2/2$). Consider the integral for $C(p)$, Eq. (14), in the complex t plane. The steepest descent path for the exponential argument $(-i\varepsilon t - iF^2t^3/24 - (1/2)\ell n(t))$ from $t = 0$ is (for $\varepsilon > 0$) along the negative imaginary axis to the saddle point at $t_0 = -i[8(\varepsilon + i\omega/2)/F^2]^{1/2}$, and then in the direction of the increasing $Re\, t$. This path is illustrated in Fig. 11 where the scaled variable $z \equiv tF/\varepsilon_0^{1/2}$ has been introduced in order to generate a universal curve. The path of integration consists of two parts, 0 to t_0, the dominant part C_D, and t_0 to ∞, the subdominant exponentially small part, C_S. If we set $t = -is$ along the first path,

$$\begin{aligned} C(p) &= \frac{B}{\sqrt{2\pi}} \int_0^{s_0} ds \frac{1}{\sqrt{s}} \exp\left(-\left(\varepsilon + i\frac{\omega}{2}\right) s + \frac{F^2 s^3}{24}\right) \\ &\quad + \frac{i^{1/2}B}{2\pi} \int_{t_0}^{\infty} dt \frac{1}{\sqrt{t}} \exp\left(-i\varepsilon t + \frac{\omega}{2}t - \frac{iF^2}{24}t^3\right) \\ &= C_D\left(\varepsilon + i\frac{\omega}{2}\right) + C_S\left(\varepsilon + i\frac{\omega}{2}\right) \end{aligned}$$

where $s_0 = it_0 \cong \sqrt{8\varepsilon 0}/F$. We note that ($\omega \ll \varepsilon$ and $\varepsilon \cong \varepsilon_0$)

$$C_S(\varepsilon_0) \cong 0\left(\exp\left(-i\varepsilon_0 t_0 - \frac{iF^2}{24} t_0^3\right)\right)$$
$$\equiv 0\left(\exp\left(-\frac{2}{3}\frac{B^3}{f}\right)\right) \ll 1 \,,$$

where we have used $\varepsilon_0 = B^2/2$. We therefore look for solutions of the equation

$$1 - C_D\left(\varepsilon + i\frac{\omega}{2}\right) = 0 \tag{17}$$

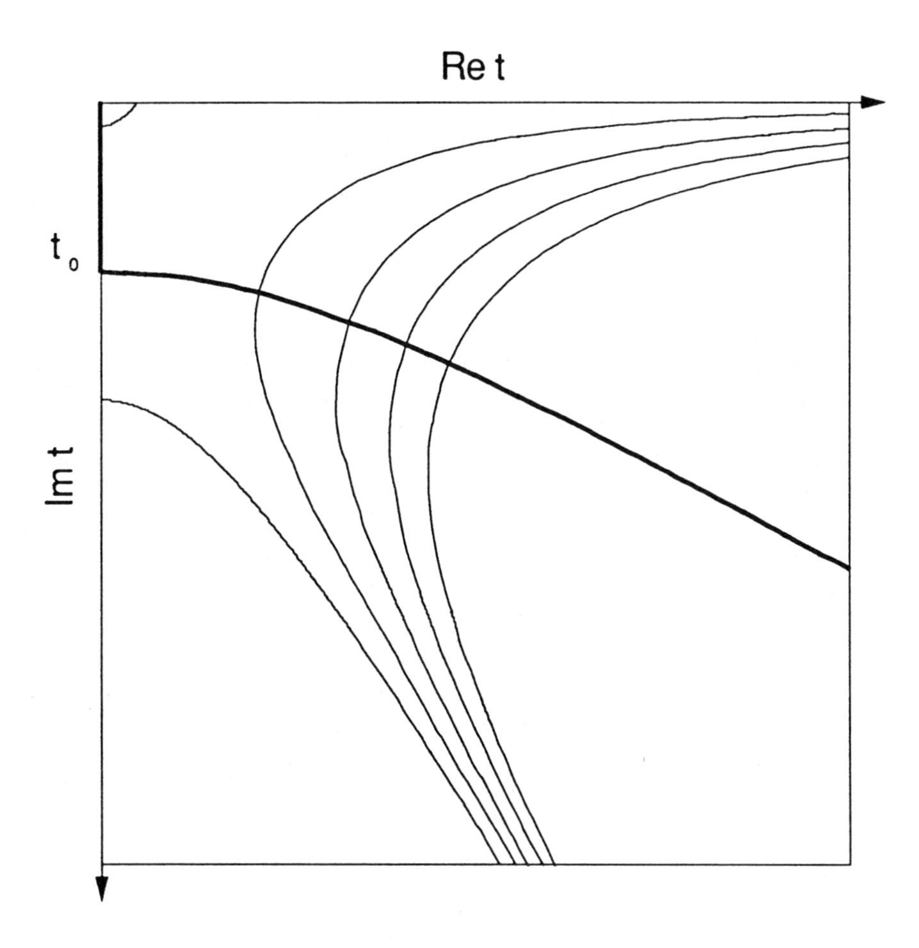

FIGURE 11

Equation (17) can be solved by expansion of the term $\exp(-F^2s^3/24)$ in the integrand to obtain an asymptotic series for ε in F/B^3. From Eq. (16), we obtain for ε and ω

$$\varepsilon = \frac{B^2}{2}\left(1 + \frac{5}{4}\frac{F^2}{B^6} + \ldots\right)$$
$$\omega = 0 \tag{18}$$

One may easily verify that ω is identically zero in Eq. (17). Therefore ω is determined by including C_S; specifically

$$\frac{\omega}{2}\frac{\partial C_D}{\partial \varepsilon} = -Im\, C_S(\varepsilon) \tag{19}$$

$C_S(\varepsilon)$ can be determined by noting that the dominant contribution to the integral comes from the integration over half the saddle point at t_0. Evaluating C_S and $\partial C_D/\partial\varepsilon$ to lowest order, we obtain the ionization rate

$$\omega = B^2 \exp\left(-\frac{2B^3}{3F}\right) . \tag{20}$$

In this evaluation it is sufficient to use $\varepsilon \equiv B^2/2$ in Eq. (20).

The separation of the integral into C_D and C_S is a crucial part of our analysis. The real frequency ε (the energy eigenvalue plus its shift) comes from C_D and it is given by an asymptotic series in F/B^3. The imaginary frequency $\omega/2$ (half of the ionization rate) comes solely from including C_S and is exponentially small in the parameter F/B^3. Since the exponentially small terms are in the asymptotic sense "beyond all orders" in F/B^3, it is essential that the solution of Eq. (16) (including only C_D) gives an ω that is identically zero. The separation of C into C_D and C_S enables us to separate the contributions to ω from the corrections to ε, which, although larger in the asymptotic sense, are irrelevant. A similar separation of the problem occurs in the many-photon limit treated in the next section.

We will study the problem of the ionization of an electron bound by the same single δ-function potential treated in the last section, but now subject to an external, monochromatic electric field, as represented by the potential, Eq. (6). The integral equation for χ, the wave function at the site of the potential $V(x) = -B\delta(x)$, is, again, given by Eq. (10). Now, however, due to the time dependence of V_{ext}, this equation is no longer in convolution form, a reflection of the fact that the corresponding Schrödinger differential equation cannot be reduced to an eigenvalue equation.

We look for a solution of the integral equation (10) in the eikonal form

$$\chi(t) = \exp\left(i\int_0^t d\tau\varepsilon(\tau)\right) . \tag{21}$$

with $\varepsilon(\tau)$ complex. Intuitively, one can think of Eq. (21) as a "generalized" form of the inverse Laplace transform, Eq. (15), useful for a slowly time-varying Hamiltonian. We expect ε to vary correspondingly slowly in time. Note that the exponent $\int d\eta\varepsilon(\eta)/\omega$ is proportional to the number of photons, ε/ω, and is indeed proportional to a large quantity that is slowly varying, in analogy with the situation in WKB or in geometrical optics.

Substituting the assumed form Eq. (21) into the governing integral Eq. (10), and neglecting the inhomogeneous term, as for the time-dependent case, we obtain

$$1 = \frac{i^{1/2}B}{\sqrt{2\pi}}\int_0^t dT\frac{1}{\sqrt{T}} \times \exp\left(i\left(S(t,t-T) - \int_{t-T}^t d\tau\varepsilon(\tau)\right)\right) , \tag{22}$$

from which we wish to develop an asymptotic expression for $\varepsilon(t)$. In order to make progress, we assume that, for sufficiently weak fields F, ε is nearly real. Then the argument of the exponential is large and nearly imaginary. This is the case since the two terms in the argument

$$\int_{t-T}^t \varepsilon(\tau)d\tau \cong B^2/\omega \qquad \text{and} \qquad S \cong F^2/\omega^3$$

are both assumed large. The ratio of these two terms is proportional to γ^2, where $\gamma \equiv B\omega/F$ is Keldysh's adiabatic parameter, defined above. Because of the $T^{-1/2}$ factor in the integrand, the largest contribution to the integral comes from $T = 0$. This forms the basis for approximation as follows:

We can write, for small T:

$$\int_{t-T}^t \varepsilon(\tau)d\tau \cong \varepsilon(t)T \tag{23}$$

We note, further, that both S and its derivative with respect to T approach zero as T approaches 0 (cf. Eq. (11)). These observations enable an evaluation of the dominant contribution to the integral in Eq. (22). The resulting evaluation of the "dispersion relation " yields, to lowest order,

$$\varepsilon(t) = B^2/2 . \tag{24}$$

In order to improve the result and calculate the ionization rate, the rest of the time integration needs to be included. Proceeding in the same spirit as in the time-dependent case, we write

$$\varepsilon(t) = \varepsilon_R(t) + i\delta(t) , \tag{25}$$

where the result, Eq. (24), is the first approximation to the real frequency $\varepsilon_R(t)$. The method for improving the result is analogous to the dc procedure. The contribution to the integral for small T is evaluated in terms of the unknown, exact $\varepsilon(t)$, whereas the remainder of the integration, to be evaluated via steepest descents, is computed using only the lowest-order result, Eq. (24), for the eigenvalue.

The difficulty of the problem is enhanced over the dc problem because the integral equation is no longer in convolution form, and there are many saddle-points because of the periodicity of the field. The detailed treatment of the problem is given in Ref. 11. However, we are interested in calculating the time average ionization

$$w = \lim_{T\to\infty} \int_0^T dt\, \delta(t) \tag{26}$$

It may be shown that contribution from w comes from the saddlepoint integrations and, further, one need only keep the lowest order result for ε, Eq. (24), in evaluating these integrals.

The general expression for w is complicated for arbitrary γ, but has the general feature that the ionization rate is a sum of processes involving an integral number σ of photons, where $\sigma \geq B^2/2\omega^2 + F^2/4\omega^3$. This is the so-called above threshold ionization. The ionization potential, as well as the ponderomotive potential $F^2/4\omega^2$, must be overcome by the ejected electron. (See Fig. 12.)

- Result for w is a complicated expression which has the following features:
  - Total rate is a sum over the (integer) number s of photons absorbed where
    $$s\omega \geq \frac{B^2}{2}(1 + \frac{1}{2\gamma^2})$$
    which is the sum of the ionization energy plus the ponderomotive energy acquired by an electron oscillating freely in the applied field.
  - In the adiabatic limit, w becomes equal to the period-averaged tunneling result,
    $$w_{\text{aver}} = \frac{\omega}{2\pi}\int_0^{2\pi/\omega} w(t)dt\,,$$
    $$= \frac{B^2\omega}{2\pi}\int_0^{2\pi/\omega} \exp(-\frac{2}{3}\frac{B^3}{F|\cos(\omega t)|})dt\,.$$
    as expected. **(Ionization occurs in 'bursts' at the instant of maximum field strength.)**
  - A comparison with a direct numerical solution of the integral equation shows good agreement for even a few-photon process.

FIGURE 12

In the limit $\gamma \to 0$ we obtain

$$w = (3F/\pi B)^{1/2} B^2 \exp\left(\frac{2}{3}\frac{B^3}{F}\right) \tag{27}$$

It is instructive to obtain this result in another way. If in the dc result Eq. (20) we let F have the time dependence $F \to F\,\cos(\omega\tau)$ and treat ω small, the system tracks the field adiabatically, and we can average the expression for w over one period of the applied field to obtain ω_{avg}. (See Fig. 12)

If $F \ll B^3$ the integral may be evaluated by the saddle-point method where we obtain two Gaussian contributions in the period $2\pi/\omega$. This corresponds physically to the fact that we have an ionization "burst" each time the barrier is low. Calculating w_{avg} in this way recovers the previous result.

For the case $\gamma \to 0$ we find

$$w \cong \frac{2}{\pi}\frac{\omega B}{k_s}\left(\frac{Fe^{1/2}}{2\omega B}\right) 2\sigma\ . \tag{28}$$

$\sigma \cong B^2/2\omega$ is the minimum number of photons to reach the continuum, and $k_e^2/2 = -\left(B^2/2\right)\left(1 + 1/2\gamma^2\right) + \left(1/2\gamma^2\right) + s\omega$ is the ejected electron energy.

In addition to the analytic work, a numerical solution of the integral equation was undertaken. Fig. 13 shows a plot of $|\chi(t)|^2$ for the parameters there indicated. The slope of $|\chi|^2$ versus t is directly proportional to the bound state probability.

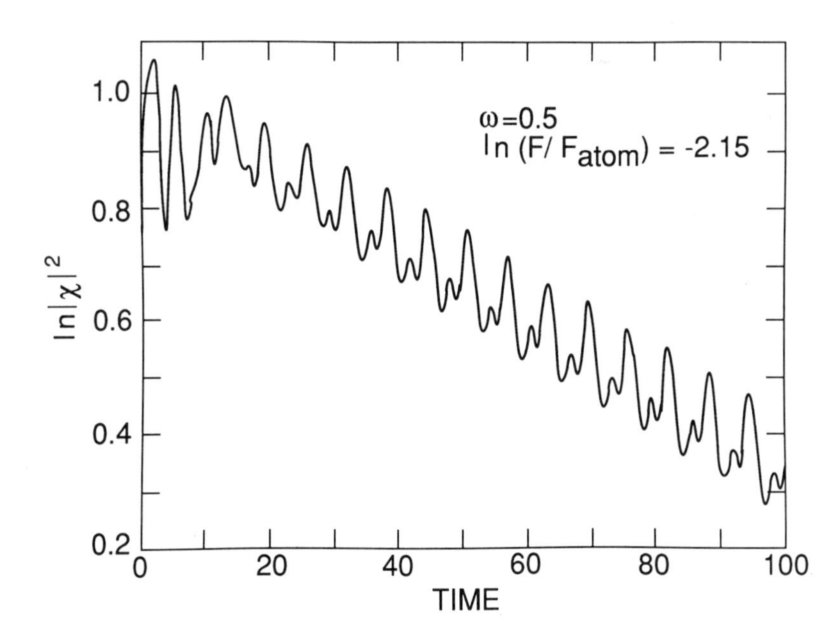

FIGURE 13

Figs. 14 and 15 show log-log plots of the ionization rate versus the value of F for the time dependent field normalized to the atomic field strength, $F_{\rm atm} = B^3$. The curve is an evaluation from the asymptotic formula, while the dots are results obtained from numerical solution of the integral equation. In Fig. 14, $\omega = .5$ and $B = 1.59$. The minimum of photons required for ionization as $F \to 0$ is 3, while for the larger fields' strengths an additional photon is needed to overcome the increasing ponderomotive potential, $F^2/4\omega^2$. The comparison is quite good, even though few photons are involved. In Fig. 15, ω has been reduced to 0.25. The minimum number of photons is now seven and increases to fifteen at the highest fields. The additional structure is due to the change of σ to $\sigma + 1$ photons with relatively little change in the field.

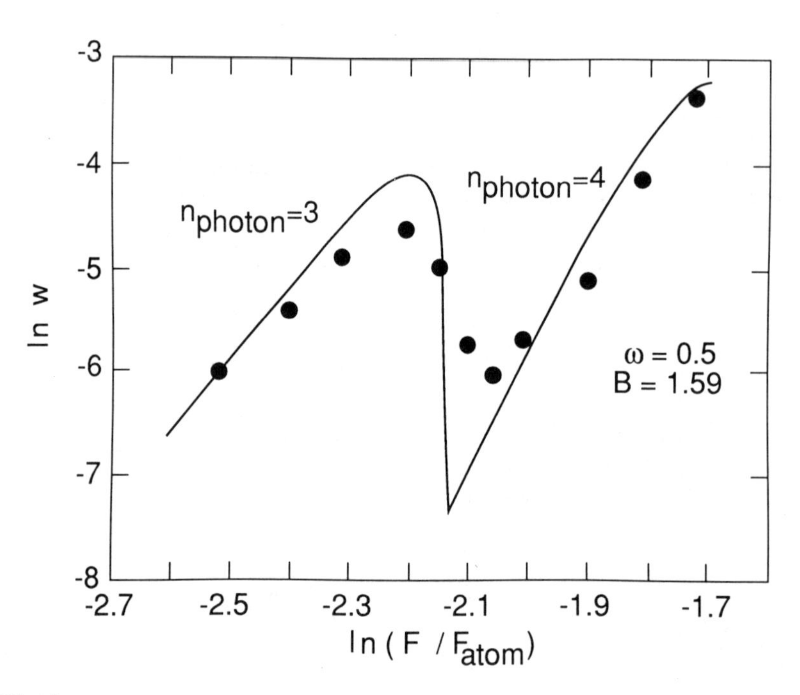

FIGURE 14

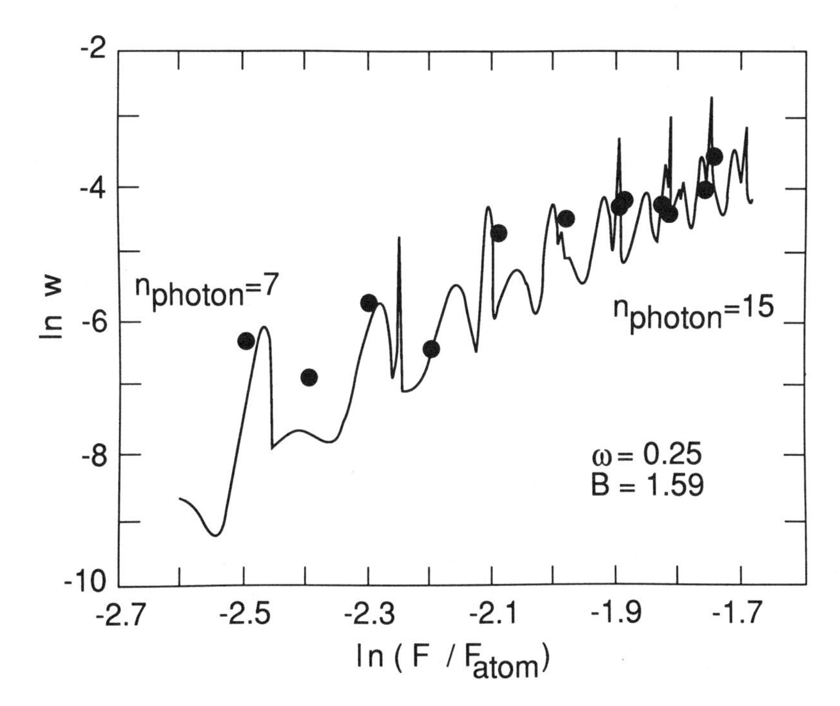

FIGURE 15

Charge States of Many Electron Atoms in Strong Fields

As we have discussed, modern laser systems can deliver electric field strengths near the focal region on the order of, or larger than, the nuclear electric field, as seen by the outer shell electrons. On the other hand, such intensities are available only at photon energies much smaller than typical binding energies, so that ionization requires absorption of a large number of photons. Recent experiments on noble gases[12–15] investigating multiphoton ionization (MPI) show markedly different behavior, depending on whether an intermediate state is excited (resonant ionization) or not (nonresonant), with the intensity required for observation of a given charge state (the threshold intensity) orders of magnitude lower in the former case. In contrast, a feature of the nonresonant experiments is the apparent absence of shell effects. Indeed, in Ref. 16, the authors note (see Fig. 16) that the observed threshold intensities depend very little on the details of the structure of the particular atom under study, but depend only on the binding energy of the extracted electron.

- Non Resonant MPI has generic features

- In the case of **nonresonant** MPI the threshold intensity for a given ionization stage is observed experimentally to depend most sensitively on the ionization potential of the preceeding stage.

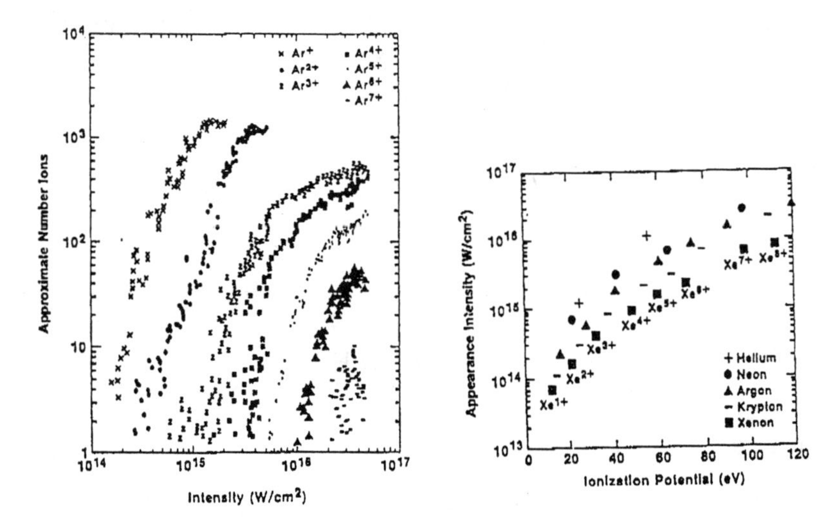

S. Augst, D. Strickland, D. D. Meyerhofer, S. L. Chin, and J. H. Eberly, Phys. Rev. Letters, **63**, 2212 (1989).

FIGURE 16

We limit ourselves here to a discussion of nonresonant ionization phenomena. Because of the high applied fields used in recent experiments, any theory purporting to explain the data has to be non-perturbative in character. In reference 12, the interpretation was based on the Keldysh single-electron model, which describes MPI from the ground state of hydrogen, ignoring effects of intermediate states. The multi-electron nature of the problem could, in principle, be addressed numerically at the level of a Hartree-Fock calculation. However, for a system such as a Xenon atom in the presence of a strong field, this appears to be prohibitive in practice. A simpler approach, motivated by the apparent absence of shell effects in the experimental results and by the large number of electrons, is to use a statistical model such as the Thomas-Fermi model (see Figs. 17 and 18).

- For many-electron ions this lack of dependence on shell structure suggests the possibility of a description by means of a statistical model, such as the Thomas-Fermi model:
  - Semiclassical model which accounts for screening properties of many-electron atoms.
  - Large Z expansion: appropriate for many electron systems.
  - However, it misses shell structure and tunneling.
  - Past successes include:
    * Calculation of atomic photoabsorption cross-section.
    * Computation of atomic properties such as volume vs Z and radial charge density.

FIGURE 17

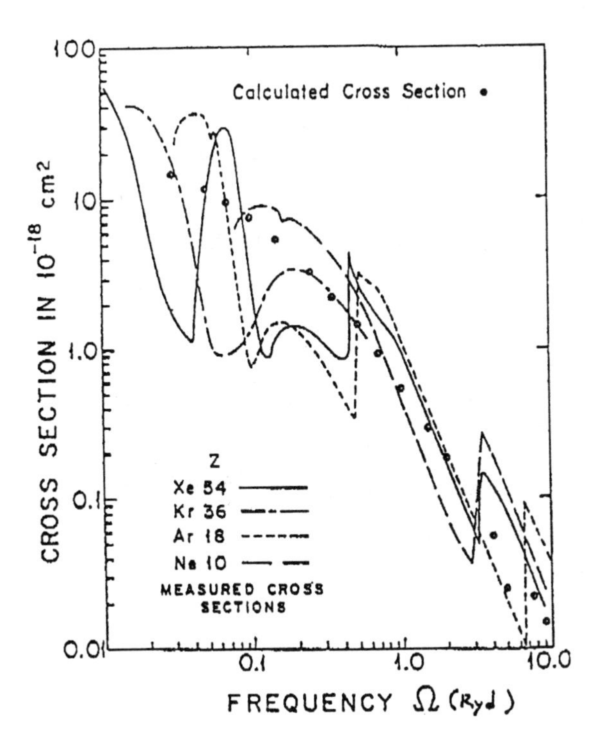

FIGURE 18

In the Thomas-Fermi model, the electronic cloud is considered to be a zero temperature Fermi gas whose density is calculated by filling the available phase space volume, taking account of the Pauli exclusion principle (Figs. 19, 20, and 21). The density is used in Poisson's equation to compute the self-consistent electrostatic potential of the atom. Alternatively, the Thomas-Fermi model can be shown to follow from the Hartree-Fock model upon the neglect of exchange effects, and by the semiclassical computation of the single-particle wave functions. For a large Z atom, exchange and quantum corrections are both of order $Z^{-2/3}$, compared to the terms retained in the Thomas-Fermi model (Fig. 22).

- **The TF model is relatively easy to solve**

- The model describes a zero-temperature, electrostatically confined (by the nuclear field) Fermi gas of electrons interacting through the self-consistent mean field.

- The dependence of electron density n_e on potential ϕ is computed by:

  - Dividing phase space into cells where

$$\# \text{ of cells} = \frac{\Delta p \Delta x}{(2\pi)^3} = \frac{1}{2} \# \text{ of electrons}.$$

  - Filling phase space completely for momenta

$$p \leq p_{\max} \equiv [2(\phi - \epsilon)]^{1/2},$$

    where ϵ is the Fermi energy.

- The electron density is then determined from the local potential ϕ and from ϵ

$$n_e = 4\pi \int_0^{p_{\max}} dp\, p^2 \frac{1}{4\pi^3} = \frac{p_{\max}^3}{3\pi^2} = \frac{2^{3/2}}{3\pi^2}(\phi - \epsilon)^{3/2}$$

  When $\phi < \epsilon$, $n_e \equiv 0$.

FIGURE 19

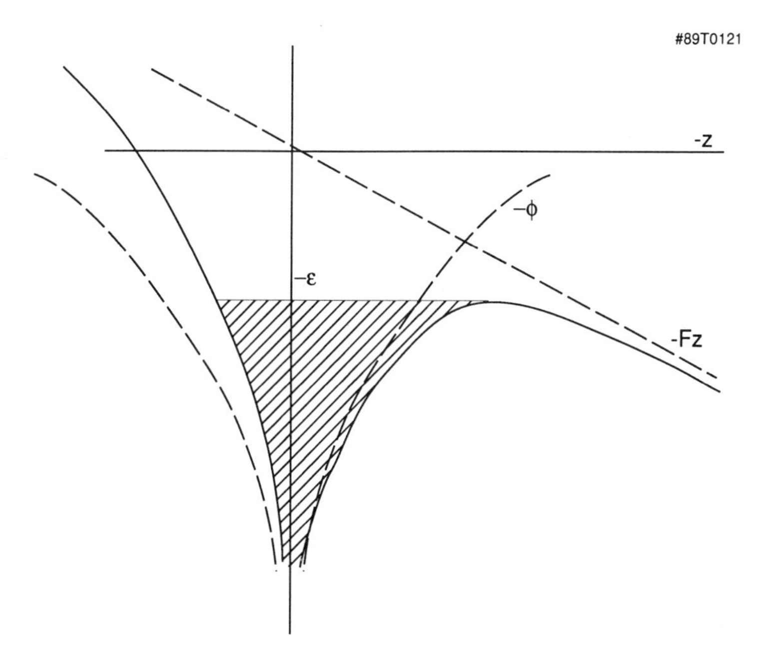

FIGURE 20

- Poisson's equation, together with the boundary conditions

$$r\phi(r,\mu) \to Z \quad \text{as} \quad r \to 0\,,$$
$$r\phi(r,\mu) \to Z - Q + Fr^2\mu \quad \text{as} \quad r \to \infty\,,$$

  where

$$Q \equiv \int d^3r\, n_e\,,$$

  completes the description.

- A numerical solution yields, besides Q, the dipole moment

$$\langle z \rangle \equiv Q^{-1} \int d^3r\, z n_e\,,$$

  vs F.

- To compare with experiment, we define the ionization potential as

$$I_p \equiv \epsilon|_{F=0}\,,$$

- In a separate sequence solve for I_p as a function of $Q|_{F=0}$.

- The model yields universal curves of Q/Z, $\langle z \rangle Z^{1/3}$ vs $FZ^{-5/3}$ and of Q/Z vs $I_p Z^{-4/3}$.

- Together, they yield $F(I_p)$.

- **There are no adjustable parameters.**

FIGURE 21

- **The TF model follows systematically from a large Z expansion**

Many-electron Schrodinger equation
neglect dynamical correlation $\sim Z^{-4/3}$

⇓

Hartree-Fock approach (independent particles)
neglect exchange $\sim Z^{-2/3}$

⇓

Hartree equations
neglect quantum effects $\sim Z^{-2/3}$

⇓

Thomas Fermi (keep terms up to $Z^{-1/3}$

FIGURE 22

One apparent inaccuracy of the model is its prediction of a weak, algebraic decay of electron density with radius as $r \to \infty$ for a neutral atom. However, for ions, the case of most interest here, the density is more realistically confined to a finite radius.

The disparity between the laser frequency and a characteristic atomic frequency suggests a straightforward modification of the model to study atoms in the presence of a low frequency electromagnetic field, viz., we add here an electric field $E(x,t) = \hat{z}F(x,t)\cos\omega t$ to the model. Here, ω is a wave frequency and F a slowly varying amplitude. The combined potential $\phi(x,t) = \phi_{\text{atomic}} + Fz$ has an instantaneous saddle point at $z = z_s$ and a value ϕ_s there (both z_s and ϕ_s are to be determined). Those electrons with energy $-\varepsilon > -\phi_s$ are ionized with a single field period near the peak of the field, so that it is sensible to freeze F in our calculations.

Within the model, it is only through increases in intensity, that is, F, that additional ionization occurs. The contribution to ionization from tunneling by electrons of energy, $-\varepsilon < -\phi_s$ through the barrier is ignored in the semiclassical limit. However, we will find that, for the short pulse experiments to which we compare, tunneling is negligible.

An estimate of the tunneling contribution to the ionization can be obtained from the well known Landau formula. This formula gives the tunneling rate for an electron through a barrier formed by a coulomb potential and a D.C. field, $w = 8\omega_a(F_a/F)\exp(-2F_a/3F)$ in atomic units. Here, F_a is the coulomb field exerted on the electron in the ground state, and ω_a is the orbital frequency. (This formula is valid when the barrier width is large compared to the de Broglie wavelength of the electron.) It is also known that, in the non-coulomb case, only the pre-exponential part of this formula will change, while the exponent remains the same. We can then rely on this formula for an order of magnitude estimate of the tunneling rate in our problem. ω_a and F_a for our ions can be estimated as I_p and $(2I_p)^{3/2}$, respectively. Clearly, the most favorable situation for tunneling is in the strongest possible field (largest F/F_a). In Augst et al.'s experiment, this occurs at the highest intensities ($\sim 10^{16}\,W/cm^2$), for which the ionization potential $I_p \sim 100$ eV. For a picosecond pulse, we have $\omega\tau \sim 10^{-4}$ for the tunneling contribution, which is negligible.

Since tunneling is negligible, the width of the above threshold ionization (ATI) spectrum is not resolved. This width arises, in the adiabatic (tunneling) limit, from the fact that tunneling ionization occurs not only at phases $\phi_n = n\pi$, corresponding to field maxima, but for small interval $\Delta\phi$ around ϕ_n. In the current situation, the width is likely set by other uncertainties. Furthermore, because all of the ejected electrons are freed from the atom at a local field maximum, they subsequently gain average oscillatory energy $\langle m v_{\rm osc}^2/2 \rangle = m(eF_{\rm ioniz}/m\omega)^2/4$, where $F_{\rm ioniz}$ is the field amplitude at the time of ionization, and $\langle\ \rangle$ indicates an average over the field period $2\pi/\omega$. Since these ponderomotive effects are simply additive, we ignore them below.

In detail, then, our system consists of Q_{el} electrons around a nucleus of charge Z, subject to an external potential $V_{\rm ext} = -Fz$, where the z-axis is chosen along the direction of the field. (Atomic units are used throughout.) If we are at position r and the maximum available energy of a single electron in the atom is $-\varepsilon$, then, classically, the momentum of the fastest electron at r is

$$p_f = \sqrt{2[\phi(r) + Fz - \varepsilon]}$$

where $-\varepsilon$ is the energy at the separatrix and $-\phi$ is the effective potential seen by the electron.

From the well known semiclassical phase-space quantization argument, the electron number density is

$$n(r) = \frac{p_f^3}{3\pi^2} = \frac{2^{3/2}}{3\pi^2}\,(\phi + Fz - \varepsilon)^{3/2}\,. \tag{29}$$

The electron density is taken to be zero outside the region delimited by the separatrix (outside the classical region). Equation (29) is two dimensional, due to the azimuthal symmetry around the field axis. This necessitates a numerical investigation.

It is convenient to change the dependent variable ϕ to χ, where $\phi + Fz = (Z/r)\chi$, and the independent variable r to the Thomas-Fermi scaled coordinates $r = bZ^{-1/3}x$ ($b \equiv 2^{-7/3}(3\pi)^{2/3}$). With these changes, Poisson's equation for ϕ, with the electron density Eq. (29) written in spherical coordinates ($\mu \equiv \cos\theta$), is

$$\frac{\partial^2\chi}{\partial x^2}+\frac{1}{x^2}\frac{\partial}{\partial\mu}\left[\left(1-\mu^2\right)\frac{\partial\chi}{\partial\mu}\right]=\begin{cases}\left[\chi-\left(\varepsilon b/Z^{4/3}\right)x\right]^{3/2}/\sqrt{x}\,, & \text{if } x<x_s(\mu)\\ 0 & \text{, otherwise}\end{cases} \tag{30}$$

where $x_s(\mu)$ is the separatrix curve (the curve that satisfies $\chi-\varepsilon Z^{-4/3}bx=0$). For $r\to 0$ $(x\to 0)$ the nucleus is unscreened, and for $r\to\infty$ $(x\to\infty)$ we determine the boundary condition by asking the total number of electrons to be Q_{el}. In terms of χ, these conditions result in

$$\begin{cases}\chi(0,\mu)=1\\ \chi\to Q/Z+\left(F/Z^{5/3}\right)b^2x^2\mu \quad \text{for } x\to\infty\end{cases}$$

where $Q=Z-Q_{el}$ is the total charge of the atom (nuclear + electronic). The electronic charge, Q_{el}, is in turn determined from χ by

$$\frac{Q_{el}}{Z}=\frac{1}{2}\int_{-1}^{1}d\mu\int_0^{x_s(\mu)}dx\,x^{1/2}\left(\chi-\frac{\varepsilon b}{Z^{4/3}}x\right)^{3/2}. \tag{31}$$

We can also obtain an expression for the dipole moment, D_z, induced by the external field F. In analogy with Eq. (31),

$$D_z=\langle z\rangle=\frac{\int dVn(r)a}{\int n(r)dV}=\frac{Z^{2/3}}{Q_{el}}\frac{b}{2}\int_{-1}^{1}d\mu\,\mu\int_0^{x_s(\mu)}dx\,x^{3/2}\left(\chi-\frac{\varepsilon b}{Z^{4/3}}x\right)^{3/2}, \tag{32}$$

In Eq. (29), with boundary conditions given by Eqs. (30, 31), ε appears only in the combination $\varepsilon Z^{-4/3}$ and the field strength F appears only in the combination $FZ^{-5/3}$. It is clear then that $\varepsilon Z^{-4/3}=g(FZ^{-5/3})$, where g is a function which is independent of Z; that is, universal for all (heavy) atoms. In the same way, from Eq. (31), Q_{el}/Z, and hence $Q/Z=1-Q_{el}/Z$, are universal functions of $\varepsilon Z^{-4/3}$, and hence of $FZ^{-5/3}$ i.e., $Q/Z=h(FZ^{-5/3})$. The same reasoning holds for the induced dipole moment $D_zZ^{1/3}$ given by Eq. (32). It is, therefore, sufficient to calculate the charge states, induced dipole moment, and ε curves versus applied field strength for $Z=1$ only, which produces the universal curves for all atoms.

We choose to solve Eq. (30) for $\chi(x,\mu)$ numerically by a relaxation method. We put an extra $\partial x/\partial t$ in Eq. (30) and obtain then a parabolic equation whose solution will converge to the solution of Eq. (30) as $t\to\infty$. To solve the two-dimensional partial differential equation, we use an alternate-direction-implicit scheme. We take as initial conditions for χ the well known Thomas-Fermi function χ_{TF} for the bare atom $(F=0)$. (This function satisfies $\chi_{TF''}=\chi_{TF}^{3/2}/x^{1/2}$). We take $\varepsilon=0$ initially. At each subsequent time step we find ε as follows: because of the azimuthal symmetry, we know that the saddle point of ϕ (hence χ/x) is at a point $z=z_s$ along the z-axis. We find this point numerically by solving $\partial(\chi/x)/\partial z=0$. Having z_s, we can calculate the energy at the separatrix as $\varepsilon=Z^{4/3}\chi(z_s,1)/(bz_s)$.

Once we know ε, we can generate the rest of the separatrix curve $x \equiv x_s(\mu)$ by solving $\chi(x,\mu) - \varepsilon Z^{-4/3}bx = 0$ numerically for $x = x_s(\mu)$. Knowing $x_s(\mu)$, we can construct the updated electron density in the right-hand side of Eq. (29) and solve gain for χ, repeating the whole process until we achieve self-consistency.

The resulting universal curves for the total charge of the ion, Q/Z, and dipole moment, $D_z Z^{1/3}$, versus field, $FZ^{-5/3}$, are presented in Figs. 19 and 20, respectively.

The physical meaning of ε is, as mentioned before, the maximum energy of a single bound electron in the presence of the external field. The amount of energy required to extract an electron from an ion X^{+q} for $F = 0$ is the ionization potential, I_p, of that charge state X^{+q}. It proves to be more convenient for comparison with experimental data to plot the field required to produce a certain charge state versus the ionization potential of that charge state, rather than versus ε. For that, we need to calculate the ionization potential of the Thomas-Fermi atom or ion of a certain charge and also the field required to produce that charge. The resulting universal curve of scaled intensity, $IZ^{-10/3}$, plotted versus scaled ionization potential $I_p Z^{-4/3}$, is shown in Fig. 23.

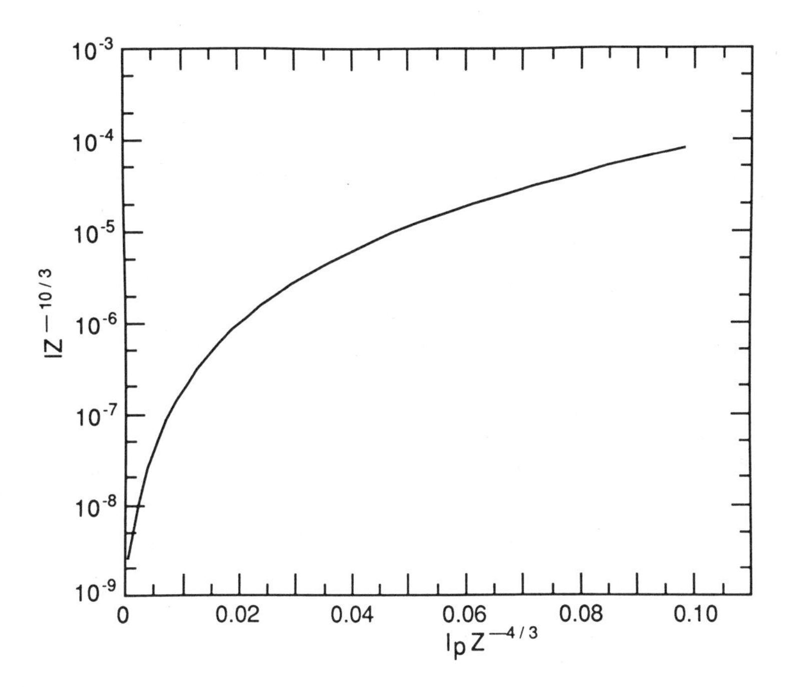

FIGURE 23

We compare our results with the data presented in Ref. 13. There the authors summarized the results of their experiments on multiphoton ionization of He, Ne, Ar, Kr, and Xe by high-intensity 1μm, 1psec laser pulse in a plot of the threshold intensity for the appearance of a charge state versus the ionization potential of the previous charge state. They defined the threshold or appearance intensity as the intensity at which a small number of ions is produced, which for the lowest charge states corresponded to a yield of $\sim 5 \times 10^{-3}$ (compared with $< 10^4$ for tunneling). In our semiclassical model, we cannot calculate the field which yields small ionization probability, but we can and do calculate the field for an ionization probability equal to 1. This provides an upper bound for the appearance intensity measured in experiments.

If we replot Fig. 23 for $Z = 2, 10, 18, 36$ and 54, and display the results in experimental units, a direct comparison can be made with data, as shown in Fig. 24. The sequence of theoretical curves progresses from left to right for increasing Z, as do the experimental points. It is clear that the model gives the right scaling with the atomic number Z and, moreover, that it is in quite good agreement with the observed appearance intensity of a given charge data.

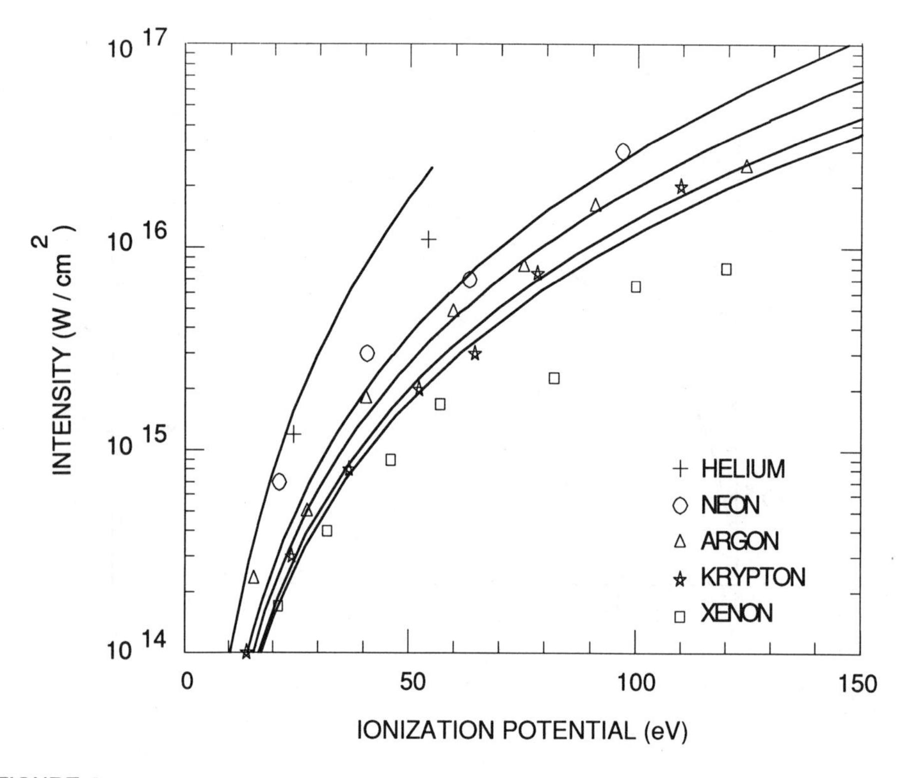

FIGURE 24

We remark that the interpretation of their data by the authors of Ref. 13 is similar in spirit to our approach. However, they ignore self-consistent deformation of the charge cloud in computing ε and use the unperturbed values of the energy levels in their computation of the field strength required for ionization. The fact that the charge cloud is indeed only slightly deformed at those field strengths which lead to ionization is a result of our self-consistent calculation, as seen from the maximum value $D_z Z^{1/3} \approx .094$ in Fig. 25.

We note that Perry et al.'s results for threshold using 1μm radiation are as much as an order of magnitude lower than those of Augst et al. One relevant difference between the two experiments may be the Keldysh adiabaticity parameter $\gamma \equiv \omega(2m\chi)^{1/3}/qE_0$, with ω, E_0 the laser frequency and amplitude, respectively, χ the ionization potential, and q, m the electron charge and mass, respectively.

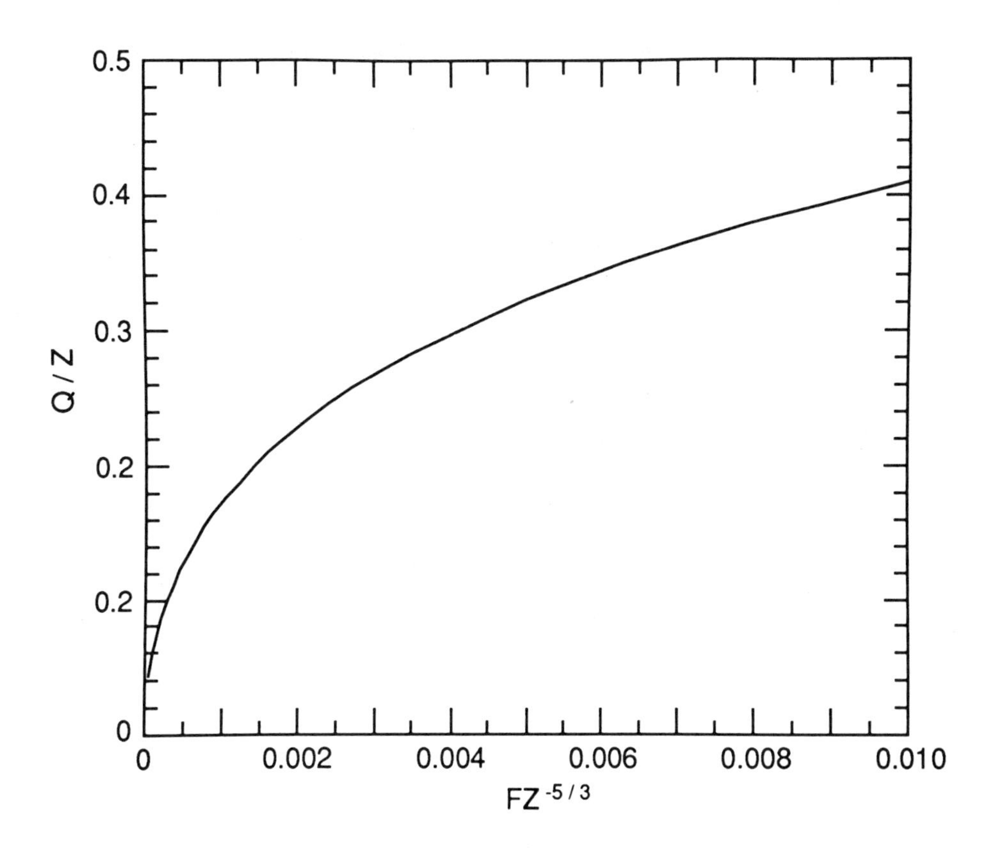

FIGURE 25

When $\gamma < 1$, the ionization rate is expected to be independent of field frequency and to be calculable using a static model. It ranges from 1-2 in the former case, and from .17 at the highest intensities to .95 at the lowest intensities for Augst et al.

In conclusion, the intensity required to reach a specified charge state by non-resonance multiphoton ionization of high Z atoms has been obtained by employing a statistical model of the atom subject to a constant electric field.

References

1. 1. C.K. Rhodes, *Science* **229**, 1345 (1985).
2. J.H. Shirley, *Phys. Rev.* **138**, B979, (1965).
3. V.L. Pokrovskii and I.M. Khalatnikov, *Sov. Phys.* JETP**13**, 1207 (1961).
4. I. Rabi et al., *Rev. Mod. Phys.* **26**, 167 (1954).
5. C.W. Clark et al., *J. Opt. S. Am.* B**3**, 371 (1986).
6. L.V. Keldysh, *Sov. Phys.* JETP**20**, 1307 (1965).
7. I.Sh. Averbukh and N.F. Perelman, *Sov. Phys.* JETP**61**, 665 (1985).
8. A.M. Perelomov et al., *Sov. Phys.* JETP**23**, 924 (1966).
9. I.J. Berson, *J. Phys.* B**8**, 3708 (1975).
10. L.D. Landau, *J. Phys.* (Moscow) **10**, 25 (1942).
11. S. Susskink, S. Cowley and E. Vales, *Phys. Rev.* A**42**, 3090 (1990).
12. M.D. Perry et al., *Phys. Rev. Lett.* **60**, 1270 (1988).
13. S. Augst et al., *Phys. Rev. Lett.* **63**, 2212 (1989).
14. O. Landau et al., *Phys. Rev. Lett.* **59**, 2558 (1982).
15. M.D. Perry et al., *Phys. Rev. Lett.* **63**, 1058 (1989).
16. S. Augst, D. Strickland, D.D. Meyerhofer, S.L. Chin, and J.H. Eberly, *Phys. Rev. Lett.* **63**, 2212 (1989).

W. Horton
University of Texas at Austin

Remarks of Professor W. Horton at the Dawson Symposium: Introduction of Professor Kyoji Nishikawa

The U.S. fusion community has looked to John Dawson for leadership in many areas, one of which is his knowledge of plasma physics in Japan.

If we think back to the fusion program in the 1970s, we will recall the horse race between the United States and the Soviet Union. The Soviets took the early lead with the invention of the tokamak, which produced an order of magnitude improvement in hot plasma confinement over that in the U.S. stellarator device. Princeton scientists quickly changed to the tokamak and surpassed the Soviet results with the operation of the Princeton Large Tokamak (PLT). All attention was focused on the race between the Soviets and Americans, and John Dawson was one of a very few who was following the growth of basic plasma physics research in Japan.

In 1978 to 1980 developments outside of the fusion program led President Carter and Premier Fukuda to sign an agreement calling for broad exchanges in the fusion programs between the United States and Japan. About the same time, the TV series of "Shogun" caught the imagination and interest of Americans in the mysteries of Japan. The U.S. fusion community, faced with planning a program of exchanges, needed a leading plasma physicist who understood both the scientific accomplishments and the culture of Japan. Luckily, John Dawson had already spent a sabbatical in 1968 at the Nagoya Plasma Physics Laboratory. He already understood when the Japanese "yes" means yes, means no, or means maybe.

In 1980 Professors John Dawson and Marshall Rosenbluth and Dr. David Nelson of the U.S. Department of Energy went to Japan and met with Professor Kyoji Nishikawa and other leaders of the Japanese fusion and plasma physics community, to develop what has turned out to be a very durable and popular exchange program between Japan and the United States. The program, called the Joint Institute for Fusion theory, or JIFT, has led to over forty workshops and about a hundred exchange visits over the past ten years. You will see the vitality of this activity with the US-Japan workshop at UCLA immediately following the Dawson symposium. This US-Japan workshop on "Advanced Simulation Techniques", as most, will contain the latest results presented by the leaders in the area of computer simulations with about fifty registered participants. We are grateful to John Dawson for his work in developing this durable and popular exchange program between the U.S. and Japanese plasma fusion scientists.

Our next speaker is Professor Kyoji Nishikawa of Hiroshima University, who is a leader and founding father of nonlinear plasma theory in Japan. At UCLA, even before the arrival of John Dawson, there was an appreciation of Japanese plasma theory leading to Professor Nishikawa's spending the summers of 1971 through 1975 at UCLA. There he was a leader in developing the theory of parametric instabilities with Burt Fried, Y.C. Lee, Predhiman Kaw and others. In Japan, Professor Nishikawa built a plasma group called the Hiroshima Institute for Fusion Theory, which produced a number of the leading second generation plasma scientists in Japan.

At Hiroshima University, Professor Nishikawa is a member of the Faculty Senate and a leader in the physics department. Today it is a pleasure to have him speak to us on his understanding of the "Dawson-Oberman Formula for High Frequency Conductivity and Anomalous Absorption."

Kyoji Nishikawa
Faculty of Science,
Hiroshima University

Dawson-Oberman Formula for High-Frequency Conductivity and Anomolous Absorption of Intense Electromagnetic Fields

Abstract

The Dawson-Oberman formula for the high-frequency conductivity of an isotropic, uniform and stationary plasma is derived by a standard linear response calculation. The result is expressed in terms of a given ion density fluctuation spectrum in the absence of the applied high-frequency field. Some features of the anomalous absorption power density expected from the formula are discussed. In particular, the anomalous absorption which is expected when the applied field frequency is close to the electron plasma frequency and which was predicted by Dawson and Oberman in 1963 is calculated and is shown to be consistent with the Manley-Rowe relation. Absorption power density due to stimulated scattering of a high-frequency electromagnetic wave by plasma waves is calculated in terms of both the scattered wave energy and the electron density fluctuation spectrum. Physical implications of the results are discussed.

Introduction

In 1962 and 1963, Dawson and Oberman published a series of papers on high-frequency conductivity of a plasma[1,2]. At that time, I had just started to get interested in the plasma kinetic theory and was very much impressed by these papers, since I had been slightly discouraged by being unable to follow the sophisticated formalism and highly advanced mathematics in the paper by Perel and Eliashberg on the same subject[3]. Dawson and Oberman's papers were different and were based on a physical intuition being full of interesting physics in them, as in the other papers by John Dawson. Then I encountered a new difficulty in judging the correctness of their reasoning of deriving the final results. I therefore started to work on the same problem by my own method which is a standard linear response calculation. Taking a long time and carrying out a tremendous amount of term by term calculations, I finally confirmed the results of Dawson and Oberman[4]. It was a surprising experience for me to learn that these authors have obtained the correct results just by physical arguments. The results are expressed in terms of the ion density fluctuation spectrum which is assumed to be given, but at the end of their papers they pointed out that the ion density fluctuation can be excited by the applied oscillating field itself to an enhanced level, so that an anomalous absorption of the electromagnetic field is possible. Motivated by this suggestion, I started a self-consistent calculation of the ion density fluctuation spectrum and this led to an instability which is now known by the name of Oscillating Two-Stream Instability[5]. Dawson himself has arrived at the same instability[6] and has demonstrated the anomalous absorption of intense electromagnetic waves by computer simulation[7].

Today, I am going to present my own derivation of the Dawson-Oberman formula and discuss some features of the anomalous absorption which can be expected from their formula. At the end of my talk, I will briefly discuss the anomalous absorption due to stimulated scattering of electromagnetic waves. Unless otherwise stated, we shall, throughout this paper, consider a spatially uniform and temporally stationary unmagnetized isotropic plasma.

The Dawson-Oberman Formula for High-Frequency Conductivity

Model and Formulation

We consider a situation in which a dipole oscillating electric field of frequency ω_0

$$\mathbf{E}_0 \exp[-i\omega_0 t] + \text{c.c.} \tag{1}$$

is applied to the plasma. We then calculate the linear response current density

$$\mathbf{J} \exp[-i\omega_0 t] + \text{c.c.}$$

in terms of the ion density fluctuation spectrum $\langle |n(\mathbf{k}, \omega)|^2 \rangle$ in the absence of $\mathbf{E}_0$. The high-frequency conductivity $\sigma(\omega_0)$ is then given by

$$\mathbf{J} = \sigma(\omega_0)\mathbf{E}_0 \ . \tag{2}$$

The current density can be written as the sum of the simple reactive part,

$$\mathbf{J}_0 = \frac{-e^2}{(im\omega_0)} n_{e0}\mathbf{E}_0 \ , \tag{3}$$

and the part induced by the fluctuation,

$$\mathbf{J}' = \frac{-e^2}{(im\omega_0)} \langle \tilde{n}_e \tilde{\mathbf{E}} \rangle \ , \tag{4}$$

as

$$\mathbf{J} = \mathbf{J}_0 + \mathbf{J}' \ , \tag{5}$$

where m and $-e$ are the electron mass and charge, n is the density, subscript e or i denotes the particle species, the wiggle stands for the fluctuation from the spatially uniform part denoted by the superscript 0 and the angled bracket depicts the spatial average.

We assume that the fluctuations are electrostatic, i.e., $\mathbf{E} = -\nabla\phi$. Then its Fourier components are written as

$$\mathbf{E}(\mathbf{k},\omega) = -i\left(\frac{\mathbf{k}}{k^2}\right)\left(\frac{e}{\varepsilon_0}\right)\left[\, Zn_i(\mathbf{k},\omega) - n_e(\mathbf{k},\omega)\,\right] . \tag{6}$$

The current density $\mathbf{J}'$ can then be calculated as follows:

$$\begin{aligned}\mathbf{J}' = {} & \frac{e^3}{(m\varepsilon_0\omega_0)}\sum_{\mathbf{k}}\left(\frac{\mathbf{k}}{k^2}\right) \\ & \times \int\frac{d\omega}{2\pi}\int\frac{d\omega'}{2\pi}\,\{Z\langle n_e(\mathbf{k},\omega+\omega_0)n_i(-\mathbf{k},-\omega')\rangle - \langle n_e(\mathbf{k},\omega+\omega_0)n_e(-\mathbf{k},-\omega')\rangle\} \\ = {} & \frac{Ze^3}{(m\varepsilon_0\omega_0)}\sum_{\mathbf{k}}\left(\frac{\mathbf{k}}{k^2}\right)\int\frac{d\omega}{(2\pi)^2}\,\langle n_e(\mathbf{k},\omega+\omega_0)n_i(-\mathbf{k},-\omega)\rangle ,\end{aligned} \tag{7}$$

where Z is the valency of the ion and ε_0 is the permittivity of vacuum. We note that in the present case of dipole oscillating field only the coupling of high-frequency electron density fluctuation and low-frequency electrostatic field fluctuation or ion density fluctuation can contribute to $\mathbf{J}'$. If the applied oscillating field has a finite wavenumber then the coupling of low-frequency electron density perturbation and high-frequency electrostatic field fluctuation dominates (see 4).

Oscillating Frame

As with Dawson and Oberman, we introduce the oscillating frame of reference. Namely, since in the presence of a dipole electric field all particles of the same species oscillate in phase with the same amplitude, we can eliminate the applied oscillating field from the basic equations by describing the system in the frame of reference moving with the oscillations.

For electrons, the oscillating frame is given by the following transformation from the rest frame:

$$\mathbf{r}\to\mathbf{r}-\mathbf{r}_0(t)\,, \qquad \mathbf{r}_0(t)\equiv\frac{e\mathbf{E}_0}{(m\omega_0^2)}\exp[-i\omega_0 t]+\text{c.c.}\,. \tag{8}$$

Let $A(\mathbf{r},t)$ be a quantity in the rest frame; then, following Arnush[8], we denote its expression in the electron oscillating frame by

$$\Delta A(\mathbf{r}-\mathbf{r}_0(t),t)\equiv A(\mathbf{r},t)\,. \tag{9}$$

In the approximation linear with respect to the applied oscillating field, we have the following relations for the Fourier components:

$$\Delta A(\mathbf{k},\omega) = A(\mathbf{k},\omega) + i\rho\ [\ A(\mathbf{k},\omega+\omega_0) + A(\mathbf{k},\omega-\omega_0)\ ] \tag{10}$$

$$A(\mathbf{k},\omega) = \Delta A(\mathbf{k},\omega) - i\rho\ [\ \Delta A(\mathbf{k},\omega+\omega_0) + \Delta A(\mathbf{k},\omega-\omega_0)\ ] \tag{11}$$

where $\rho \equiv \mathbf{k}\cdot\mathbf{r}_0(t) = \mathbf{k}\cdot\mathbf{E}_0\ \frac{e}{(m\omega_0^2)}$ which is the ratio of the electron excursion length due to the applied oscillating field to the wavelength of the fluctuation.

As for ions, we neglect the difference between the oscillating frame and the rest frame because of the large ion mass.

Linear Electrostatic Polarizabilities

We now introduce the linear electrostatic polarizability $\chi_s(\mathbf{k},\omega)$ in the absence of the applied oscillating field. For electrons, we use the oscillating frame representation and define it by the relation

$$e\Delta n_e(\mathbf{k},\omega) = \varepsilon_0\chi_e(\mathbf{k},\omega)k^2\Delta\phi(\mathbf{k},\omega)\ . \tag{12}$$

Similarly, for ions using the rest frame we have

$$Zen_i(\mathbf{k},\omega) = -\varepsilon_0\chi_i(\mathbf{k},\omega)k^2\phi(\mathbf{k},\omega)\ . \tag{13}$$

These polarizabilities can be calculated by using the linear Vlasov equation as

$$\chi_s(\mathbf{k},\omega) = \frac{\omega_{Ps}^2}{(k^2 n_{s0})}\int d^3\mathbf{v}\left[\frac{1}{(\omega-\mathbf{k}\cdot\mathbf{v}+i\delta)}\right]\mathbf{k}\cdot\left[\frac{\partial F_s(\mathbf{v})}{\partial\mathbf{v}}\right]\ , \tag{14}$$

$(\delta\to+0)$, where ω_{Ps} is the plasma frequency $\left(\omega_{Ps} = \frac{n_{s0}e^2}{[\varepsilon_0 m_s]}\right)$ and $F_s(\mathbf{v})$ is the velocity distribution function in the absence of applied oscillating field.

Combining (12) and (13) with the Poisson equation

$$\varepsilon_0 k^2\phi(\mathbf{k},\omega) = e\ [\ Zn_i(\mathbf{k},\omega) - n_e(\mathbf{k},\omega)\ ]\ , \tag{15}$$

we obtain the following relations:

$$\Delta n_e(\mathbf{k},\omega) = \left[\frac{\varepsilon_0\chi_e(\mathbf{k},\omega)}{\varepsilon_e(\mathbf{k},\omega)}\right] Z\Delta n_i(\mathbf{k},\omega)\ , \tag{16}$$

$$Zn_i(\mathbf{k},\omega) = \left[\frac{\varepsilon_0\chi_i(\mathbf{k},\omega)}{\varepsilon_i(\mathbf{k},\omega)}\right] n_e(\mathbf{k},\omega)\ , \tag{17}$$

where $\varepsilon_s(\mathbf{k},\omega) = \varepsilon_0[1+\chi_s(\mathbf{k},\omega)]$ is the longitudinal dielectric function of s^{-th} species of particle.

Linear Response Electron Density Fluctuation

We now calculate the driven linear response electron density fluctuation $n_e(\mathbf{k}, \omega + \omega_0)$ in terms of the ion density fluctuation in the absence of applied field, $n_i(\mathbf{k}, \omega)$. To this end we use the transformation relations (10) and (11) as follows:

$$n_e(\mathbf{k}, \omega + \omega_0) = \Delta n_e(\mathbf{k}, \omega + \omega_0) - i\rho \Delta n_e(\mathbf{k}, \omega)$$

$$= \left[\frac{\varepsilon_0 \chi_e(\mathbf{k}, \omega + \omega_0)}{\varepsilon_e(\mathbf{k}, \omega + \omega_0)} \right] Z \Delta n_i(\mathbf{k}, \omega + \omega_0) - i\rho \left[\frac{\varepsilon_0 \chi_e(\mathbf{k}, \omega)}{\varepsilon_e(\mathbf{k}, \omega)} \right] Z \Delta n_i(\mathbf{k}, \omega)$$

$$= i\rho \left\{ \left[\frac{\varepsilon_0 \chi_e(\mathbf{k}, \omega + \omega_0)}{\varepsilon_e(\mathbf{k}, \omega + \omega_0)} \right] - \left[\frac{\varepsilon_0 \chi_e(\mathbf{k}, \omega)}{\varepsilon_e(\mathbf{k}, \omega)} \right] \right\} Z n_i(\mathbf{k}, \omega)$$

$$= i\rho \left\{ \frac{\varepsilon_0}{\varepsilon_e(\mathbf{k}, \omega)} \frac{\varepsilon_0}{\varepsilon_e(\mathbf{k}, \omega + \omega_0)} \right\} Z n_i(\mathbf{k}, \omega)$$

$$= \frac{iZe\mathbf{k} \cdot \mathbf{E}_0}{(m\omega_0^2)} \left\{ \frac{\varepsilon_0}{\varepsilon_e(\mathbf{k}, \omega)} - \frac{\varepsilon_0}{\varepsilon_e(\mathbf{k}, \omega + \omega_0)} \right\} n_i(\mathbf{k}, \omega) \ . \tag{18}$$

Dawson-Oberman Formula

We now substitute the above result into eq. (7) to obtain

$$\mathbf{J}' = \frac{iZ^2 e^4}{(m^2 \omega_0^3)} \sum_{\mathbf{k}} \left(\frac{\mathbf{k}\mathbf{k} \cdot \mathbf{E}_0}{k^2} \right) \int \frac{d\omega}{(2\pi)^2} \left\{ \frac{1}{\varepsilon_e(\mathbf{k}, \omega)} - \frac{1}{\varepsilon_e(\mathbf{k}, \omega + \omega_0)} \right\} \langle |n_i(\mathbf{k}, \omega)|^2 \rangle \ . \tag{19}$$

If we ignore the ion density oscillation and set $\omega = 0$ in ε_e, then we obtain the most general form of Dawson-Oberman formula[2],

$$\mathbf{J}' = \frac{iZ^2 e^4}{(m^2 \omega_0^3)} \sum_{\mathbf{k}} \left(\frac{\mathbf{k}\mathbf{k} \cdot \mathbf{E}_0}{k^2} \right) \left\{ \frac{1}{\varepsilon_e(\mathbf{k}, 0)} - \frac{1}{\varepsilon_e(\mathbf{k}, \omega_0)} \right\} S_{ii}(\mathbf{k}) \ , \tag{20}$$

where we ignored ω in the expression for the electron dielectric functions and $S_{ii}(\mathbf{k})$ is the ion form factor defined by

$$S_{ii}(\mathbf{k}) \equiv \int \frac{d\omega}{(2\pi)^2} \langle |n_i(\mathbf{k}, \omega)|^2 \rangle \ . \tag{21}$$

Using this formula, Dawson and Oberman pointed out the possibility of anomalous absorption due to enhanced ion density fluctuations that can be produced by the applied oscillating field itself through its coupling with electron plasma oscillations.

Thermal Equilibrium

We next apply the above formula to the case of thermal fluctuations. For this purpose one needs to take into account the discreteness effect of the plasma. Here I will calculate the ion form factor due to thermal fluctuations following the method of Ichimaru[9].

We treat only those low-frequency ($|\omega| \ll kv_{Te}$ where v_{Te} is the electron thermal speed) fluctuations which are excited by the discreteness character of the ions. The source of the fluctuation can then be written as

$$R_i(\mathbf{k},\omega) = \sum_j \exp(i\mathbf{k}\cdot\mathbf{R}_j) 2\pi\delta[\omega - \mathbf{k}\cdot\mathbf{v}_j] \ , \tag{22}$$

where $\mathbf{R}_j$ is the position of the j^{-th} ion. The ion density fluctuation spectrum can then be calculated from the following set of equations:

$$\varepsilon_e(\mathbf{k},0) n_e(\mathbf{k},\omega) - \varepsilon_0 \chi_e(\mathbf{k},0) Z n_i(\mathbf{k},\omega) = 0 \ , \tag{23}$$

$$\varepsilon_i(\mathbf{k},\omega) Z n_i(\mathbf{k},\omega) - \varepsilon_0 \chi_i(\mathbf{k},\omega) n_e(\mathbf{k},\omega) = Z R_i(\mathbf{k},\omega) \ , \tag{24}$$

where we approximated $\chi_e(\mathbf{k},\omega)$ by its value at $\omega = 0$. Solving the above set of equations for $n_i(\mathbf{k},\omega)$, we obtain

$$n_i(\mathbf{k},\omega) = \left[\frac{\varepsilon_e(\mathbf{k},0)}{\bar{\varepsilon}(\mathbf{k},\omega)} \right] R_i(\mathbf{k},\omega) \ , \tag{25}$$

where $\bar{\varepsilon}(\mathbf{k},\omega) \equiv \varepsilon_e(\mathbf{k},0) + \varepsilon_0 \chi_i(\mathbf{k},\omega)$. Using the randam-phase approximation

$$\begin{aligned} &\langle \sum_i \sum_j \exp[i\mathbf{k}\cdot(\mathbf{R}_i - \mathbf{R}_j)] \ \delta[\omega - \mathbf{k}\cdot\mathbf{v}_i] \ \delta[\omega' - \mathbf{k}\cdot\mathbf{v}_j] \rangle \\ &= \langle \sum_j \delta[\omega - \mathbf{k}\cdot\mathbf{v}_j] \rangle \delta[\omega - \omega'] \end{aligned} \tag{26}$$

we obtain

$$\begin{aligned} \langle |n_i(\mathbf{k},\omega)|^2 \rangle &= \left| \frac{\varepsilon_e(\mathbf{k},0)}{\bar{\varepsilon}(\mathbf{k},\omega)} \right|^2 (2\pi)^2 \langle \sum_j \delta[\omega - \mathbf{k}\cdot\mathbf{v}_j] \rangle \\ &= \left| \frac{\varepsilon_e(\mathbf{k},0)}{\bar{\varepsilon}(\mathbf{k},\omega)} \right|^2 (2\pi)^2 \int d^3\mathbf{v} \delta[\omega - \mathbf{k}\cdot\mathbf{v}] F_i(\mathbf{v}) \ , \end{aligned} \tag{27}$$

which describes the ion density fluctuation spectrum due to Cerenkov emission by thermal motion of ions.

For a Maxwellian distribution of ions,

$$F_i(\mathbf{v}) = \frac{n_{i0}}{[\,(2\pi)^{\frac{3}{2}} v_{Ti}^3\,]} \exp\left[-\frac{v^2}{v_{Ti}^2} \right] , \tag{28}$$

we have

$$\begin{aligned} \int d^3\mathbf{v}\delta\,[\omega - \mathbf{k}\cdot\mathbf{v}]\; F_i(\mathbf{v}) &= -\left(\frac{v_{Ti}^2}{\omega}\right) \int d^3\mathbf{v}\delta\,[\omega - \mathbf{k}\cdot\mathbf{v}]\;\mathbf{k}\cdot \left[\frac{\partial F_i(\mathbf{v})}{\partial \mathbf{v}} \right] \\ &= 2n_{i0} \left[\frac{k^2\lambda_{Di}^2}{(2\pi\omega)} \right] \mathrm{Im}\{\chi_i(\mathbf{k},\omega)\} , \end{aligned} \tag{29}$$

where λ_{Di} is the ion Debye length. Using (29) and noting that $k^2\lambda_{Di}^2 = \dfrac{1}{\chi_i(\mathbf{k},0)}$, we can write (27) as

$$\langle\, |n_i(\mathbf{k},\omega)\,|^2 \rangle = -|\varepsilon_e(\mathbf{k},0)|^2 4\pi \left\{ \frac{n_{i0}}{[\varepsilon_0 \chi_i(\mathbf{k},0)\omega]} \right\} \mathrm{Im} \left\{ \frac{1}{\bar{\varepsilon}(\mathbf{k},\omega)} \right\} . \tag{30}$$

The ion form factor can then be calculated using the Krammers–Kronig relation as follows:

$$\begin{aligned} S_{ii}(\mathbf{k}) &= -\frac{n_{i0}|\varepsilon_e(\mathbf{k},0)|^2}{[\varepsilon_0\chi_i(\mathbf{k},0)]} \int \frac{d\omega}{(\pi\omega)} \mathrm{Im} \left\{ \frac{1}{\bar{\varepsilon}(\mathbf{k},\omega)} \right\} \\ &= \frac{n_{i0}|\varepsilon_e(\mathbf{k},0)|^2}{[\varepsilon_0\chi_i(\mathbf{k},0)]} \mathrm{Re} \left\{ \frac{1}{\bar{\varepsilon}(\mathbf{k},\infty)} - \frac{1}{\bar{\varepsilon}(\mathbf{k},0)} \right\} \\ &= n_{i0} \frac{\varepsilon_e(\mathbf{k},0)}{\bar{\varepsilon}(\mathbf{k},0)} , \end{aligned} \tag{31}$$

where in the last formula we used the relation

$$\bar{\varepsilon}(\mathbf{k},\infty) = \varepsilon_e(\mathbf{k},0) . \tag{32}$$

Substitution of (31) into (20) finally yields the original Dawson-Oberman formula[1].

Anomalous Absorption

Formulation

The absorption power density of the high-frequency wave due to the linear response current is given by

$$\mathbf{E}_0^* \cdot \mathbf{J} + \mathbf{E}_0 \cdot \mathbf{J}^* = 2\mathrm{Re}\{\mathbf{E}_0^* \cdot \mathbf{J}\} = 2\mathrm{Re}\{\mathbf{E}_0^* \cdot \mathbf{J}'\} \,. \tag{33}$$

Substituting (20) into this expression we obtain

$$2\mathrm{Re}\{\mathbf{E}_0^* \cdot \mathbf{J}'\} = \frac{2Z^2e^4}{(m^2\omega_0^3)} \sum_{\mathbf{k}} \left| \frac{\mathbf{k}\cdot\mathbf{E}_0}{k} \right|^2 S_{ii}(\mathbf{k}) \mathrm{Im} \left\{ \frac{1}{\varepsilon_e(\mathbf{k},\omega_0)} \right\} . \tag{34}$$

Effects of Enhanced Ion Density Fluctuations

Ion density fluctuations can be excited to an enhanced level by various instabilities. For instance, long wavelength ($k\lambda_{De} < 1$ where λ_{De} is the electron Debye length) ion acoustic waves can be excited by an electric current. More directly, the applied high-frequency field at frequency close to the electron plasma frequency, i.e., $\omega_0 \cong \omega_{Pe}$, or equivalently a very long wavelength ($k\lambda_{De} \ll 1$) electromagnetic wave, can excite a couple of electron plasma wave and a long wavelength ($k\lambda_{De} < 1$) static ion density fluctuation via oscillating two-stream instability[5,6] and ion acoustic wave via parametric decay instability[5].

Anomalous absorption of applied electromagnetic wave can be clearly seen from eq. (34), namely,the enhanced ion density fluctuation level is represented by $S_{ii}(\mathbf{k})$ and the frequency spectrum of the corresponding anomalous absorption is represented by the factor $Im\left\{\frac{1}{\varepsilon_e(\mathbf{k},\omega_0)}\right\}$. For instance, in the presence of long wavelength ion density fluctuations, i.e., $k\lambda_{De} < 1$, $\mathrm{Im}\left\{\frac{1}{\varepsilon_e(\mathbf{k},\omega_0)}\right\}$ has a sharp peak at $\omega_0 \cong \omega_{Pe}$, indicating excitation of an electron plasma wave.

Now, ion density fluctuations are weakly damped even at short wavelengths, i.e., $k\lambda_{De} \geq 1$, if the electron temperature T_e is much larger than the ion temperature T_i. These short wavelength ion density fluctuations have frequencies close to the ion plasma frequency and can be excited by an electromagnetic wave of frequency somewhat greater than the electron plasma frequency via kinetic parametric instability which was first pointed out by Silin[10]. If these fluctuations are enhanced, then the associated anomalous absorption has a broad frequency spectrum, since at $k\lambda_{De} \geq 1$ we have

$$\mathrm{Im}\left\{\frac{1}{\varepsilon_e(\mathbf{k},\omega_0)}\right\} = \frac{n_{e0}e^2}{[\,k^2 m|\varepsilon_e(\mathbf{k},\omega_0)|^2]}\int d^3\mathbf{v}\,\delta[\omega_0-\mathbf{k}\cdot\mathbf{v}]\,\mathbf{k}\cdot\left[\frac{\partial F_e(\mathbf{v})}{\partial\mathbf{v}}\right], \quad (35)$$

which is nonzero for $\omega_0 \sim kv_{Te} > \omega_{Pe}$. Physically, this absorption process corresponds to a stimulated mode conversion of the electromagnetic wave to an ion plasma oscillation via thermal motion of electrons.

Absorption Rate at $\omega_0 \cong \omega_{Pe}$

We now calculate the anomalous absorption rate for the case of $\omega_0 \cong \omega_{Pe}$. We start from eq. (7) and substitute it into (33) to obtain

$$2\mathrm{Re}\{\,\mathbf{E}_0^*\cdot\mathbf{J}'\,\} = \left[\frac{2Ze^3}{(m\omega_0\varepsilon_0)}\right]\sum_{\mathbf{k}}\left(\frac{\mathbf{k}\cdot\mathbf{E}_0^*}{k^2}\right)\int\frac{d\omega}{(2\pi)^2}\,\langle n_e(\mathbf{k},\omega+\omega_0)n_i(-\mathbf{k},-\omega)\rangle\,. \quad (36)$$

We express $n_i(-\mathbf{k},-\omega)$ in terms of $n_e(-\mathbf{k},-\omega-\omega_0)$ by using eq. (18). In doing so, we note the relations

$$\varepsilon_e(\mathbf{k},\omega) \cong \varepsilon_e(\mathbf{k},0) \gg |\varepsilon_e(\mathbf{k},\omega+\omega_0)|\,, \quad (37)$$

and neglect $\dfrac{\varepsilon_0}{\varepsilon_e(\mathbf{k},\omega)}$ to obtain

$$n_i(-\mathbf{k},-\omega) = \left[\frac{im\omega_0^2}{(Ze\varepsilon_0\mathbf{k}\cdot\mathbf{E}_0^*)}\right]\varepsilon_e(-\mathbf{k},-\omega-\omega_0)n_e(-\mathbf{k},-\omega-\omega_0)\,. \quad (38)$$

Substitution of (38) into (36) then yields

$$2\mathrm{Re}\{\mathbf{E}_0^*\cdot\mathbf{J}'\} = \left[\frac{2e^2\omega_0}{\varepsilon_0^2}\right]\sum_{\mathbf{k}}\left(\frac{1}{k^2}\right)S_{ee}(\mathbf{k},\omega_0)\mathrm{Im}\{\varepsilon_e(\mathbf{k},\omega_0)\}\,, \quad (39)$$

where $S_{ee}(\mathbf{k},\omega)$ is the electron dynamic form factor defined by

$$S_{ee}(\mathbf{k},\omega_0) \equiv \int\frac{d\omega}{(2\pi)^2}\,\langle|n_e(\mathbf{k},\omega+\omega_0)|^2\rangle\,. \quad (40)$$

For electron plasma waves at long wavelengths ($k\lambda_{De} < 1$), we have the relation

$$S_{ee}(\mathbf{k},\omega_0) = \left(\frac{\varepsilon_0^2}{e^2}\right)k^2|\mathbf{E}(\mathbf{k},t)|^2\ (\omega_0 \cong \omega_{Pe})\,, \quad (41)$$

where the time dependence has no contribution to the left-hand side because we are considering a stationary state.

Now the wave energy density $W(\mathbf{k})$ and the Landau damping rate $\gamma(\mathbf{k})$ of the electron plasma wave can be written as follows:

$$W(\mathbf{k}) = \omega_0 \frac{\partial \mathrm{Re}\{\varepsilon_e(\mathbf{k}, \omega_0)\}}{\partial \omega_0} |\mathbf{E}(\mathbf{k}, t)|^2 ,$$

$$\gamma(\mathbf{k}) = \frac{\mathrm{Im}\{\varepsilon_e(\mathbf{k}, \omega_0)\}}{\frac{\partial \mathrm{Re}\{\varepsilon_e(\mathbf{k}, \omega_0)\}}{\partial \omega_0}} , \tag{43}$$

where we approximated the real part of the electron plasma wave frequency by ω_0. Equation (39) can then be written as

$$2\mathrm{Re}\{\mathbf{E}_0^* \cdot \mathbf{J}'\} = \sum_{\mathbf{k}} 2\gamma(\mathbf{k}) W(\mathbf{k}) . \tag{44}$$

This relation implies that the power absorbed by the plasma is just balanced by the Landau damping of the excited electron plasma wave. This result is consistent with the Manley-Rowe relation[11] which states that only a small fraction (of order $\frac{|\omega|}{\omega_0}$) of energy absorbed goes to the low-frequency ion density fluctuations. We insist, however, that for an anomalous absorption of a dipole oscillating field to take place ion density fluctuations *must be* over-thermal.

Stimulated Scattering

Formulation

We now consider the case in which an electromagnetic pump wave of frequency well above the electron plasma frequency is coherently scattered by an electrostatic plasma wave. This stimulated scattering of an intense electromagnetic wave is receiving considerable recent interest in connection with the laser-plasma interaction in a large target plasma.

At very high-frequency, i.e., $\omega_0 \gg \omega_{Pe}$, the finite wavelength of the applied oscillating field or pump wave has to be taken into account and we write the pump wave as

$$\mathbf{E}_0 \exp[\, i(\mathbf{k}_0 \cdot \mathbf{r} - \omega_0 t)\,] + \text{c.c.} . \tag{45}$$

This pump wave resonantly excites a couple of high-frequency electromagnetic wave (Stokes light) and a low-frequency electrostatic wave. The high-frequency electromagnetic wave is represented by the electric field denoted by

$$\mathbf{E}_- \exp[\, i(\mathbf{k}_- \cdot \mathbf{r} - \omega_- t)\,] + \text{c.c.} \,, \tag{46}$$

while the low-frequency electrostatic wave is represented by the electron density perturbation denoted by

$$\delta n_e \exp[\, i(\mathbf{k} \cdot \mathbf{r} - \omega t)\,] + \text{c.c.} \,, \tag{47}$$

where we used the notation

$$\mathbf{k}_- = \mathbf{k}_0 - \mathbf{k} \,, \qquad \omega_- = \omega_0 - \omega \,, \tag{48}$$

with $k_- \sim 2k_0 \sim 2k$ and $\omega_- \sim \omega_0 \gg \omega$.

The linear response current density can then be written as follows:

$$\mathbf{J} \exp[\, i(\mathbf{k}_0 \cdot \mathbf{r} - \omega_0 t)\,] + \text{c.c.} = \mathbf{J}_0 + \mathbf{J}' \,, \tag{49}$$

$$\mathbf{J}_0 = \frac{-e^2}{(im\omega_0)}\, n_{e0} \mathbf{E}_0 \,, \tag{50}$$

$$\mathbf{J}' = \sum_{\mathbf{k}} \int \frac{d\omega}{(2\pi)^2} \left[\frac{ie^2}{(m\omega_-)} \right] \langle \delta n_e \mathbf{E}_- \rangle \,. \tag{51}$$

Current Density Due to Fluctuations

The Stokes light equation is the standard electromagnetic wave equation with current modulation due to the electron density perturbation:

$$\begin{aligned} \overleftrightarrow{D}(\mathbf{k}_-, \omega_-) \cdot \mathbf{E}_- &= \frac{-e\delta n_e^* \mathbf{u}_0}{(i\omega_-)} \\ &= \frac{e^2}{(m\omega_- \omega_0)}\, \delta n_e^* \mathbf{E}_0 \,, \end{aligned} \tag{52}$$

where

$$\overleftrightarrow{D}(\mathbf{k}, \omega) \equiv \varepsilon_0 \left(\frac{kc}{\omega} \right)^2 \left[\frac{\mathbf{k}\mathbf{k}}{k^2} - \overleftrightarrow{1} \right] + \overleftrightarrow{\varepsilon}_e (\mathbf{k}, \omega) \,, \tag{53}$$

$\overleftrightarrow{\varepsilon}_e\,(\mathbf{k},\omega)$ being the dielectric tensor of the plasma in the absence of the pump wave. Solving (52) we have

$$\mathbf{E}_- = \frac{e^2}{(m\omega_-\omega_0)}\,\delta n_e^*\,\overleftrightarrow{D}\,(\mathbf{k}_-,\omega_-)^{-1}\cdot\mathbf{E}_0\,. \tag{54}$$

Substitution of (54) into (51) yields

$$\begin{aligned}\mathbf{J}' &= \frac{ie^2}{m}\sum_{\mathbf{k}}\int\frac{d\omega}{(2\pi)^2}\left[\frac{\langle\delta n_e\mathbf{E}_-\rangle}{\omega_-}\right]\\ &= \frac{ie^4}{(m^2\omega_0)}\sum_{\mathbf{k}}\int\frac{d\omega}{(2\pi)^2}\left[\frac{\langle|\delta n_e|^2\rangle}{\omega_-^2}\right]\overleftrightarrow{D}\,(\mathbf{k}_-,\omega_-)^{-1}\cdot\mathbf{E}_0\,. \end{aligned}\tag{55}$$

In the last expression the coefficient of $\mathbf{E}_0$ gives the high-frequency conductivity tensor due to the fluctuations. We note that in the present case the conductivity tensor is determined by the electron density fluctuation spectrum alone in contrast to the case of the dipole high-frequency field where the presence of the ion density fluctuation is absolutely necessary.

Absorption Power

The absorption power density due to the stimulated scattering of an electromagnetic pump wave can be calculated in the same way as in the case of the dipole pump. That is, we first have

$$2\mathrm{Re}\{\mathbf{E}_0^*\cdot\mathbf{J}'\} = -\left(\frac{2e^2}{m}\right)\sum_{\mathbf{k}}\int\frac{d\omega}{(2\pi)^2}\left[\frac{\langle\delta n_e\mathbf{E}_0^*\cdot\mathbf{E}_-\rangle}{\omega_-}\right]. \tag{56}$$

From (52) we have the relation

$$\delta n_e\mathbf{E}_0^* = \left(\frac{m\omega_0\omega_-}{e^2}\right)\mathbf{E}_-^*\cdot\overleftrightarrow{D}\,(-\mathbf{k}_-,-\omega_-)\,. \tag{57}$$

Substitution of (57) into (56) gives

$$\begin{aligned}2\mathrm{Re}\,\{\mathbf{E}_0^*\cdot\mathbf{J}'\} &= -2\omega_0\sum_{\mathbf{k}}\int\frac{d\omega}{(2\pi)^2}\,\mathrm{Im}\{\,\langle\mathbf{E}_-^*\cdot\overleftrightarrow{D}\,(-\mathbf{k}_-,-\omega_-)\cdot\mathbf{E}_-\rangle\,\}\\ &= 2\omega_0\sum_{\mathbf{k}}\int\frac{d\omega}{(2\pi)^2}\,\mathrm{Im}\,\{\,\langle\mathbf{E}_-^*\cdot\overleftrightarrow{D}\,(\mathbf{k}_-,\omega_-)\cdot\mathbf{E}_-\rangle\,\}\\ &= \sum_{\mathbf{k}}\left[\frac{\omega_0}{\omega_H(\mathbf{k}_-)}\right]2\gamma(\mathbf{k}_-)W(\mathbf{k}_-)\,, \end{aligned}\tag{58}$$

where $\omega_H(\mathbf{k}_-)$, $\gamma(\mathbf{k}_-)$ and $W(\mathbf{k}_-)$ are, respectively, the frequency, damping rate and wave energy density of the Stokes light. Since $\frac{\omega_0}{\omega_H(\mathbf{k}_-)} \cong 1$, equation (58) indicates that the absorbed power is balanced by the damping of the Stokes light or the scattered electromagnetic wave, again consistent with the Manley-Rowe relation. In fact, the ratio $\frac{\omega_0}{\omega_H(\mathbf{k}_-)}$ precisely reflects the Manley-Rowe relation, as it should be. Since the damping rate of the electromagnetic wave is very small, the absorption efficiency due to the stimulated scattering is generally very low, provided that the linear response calculation of the induced current density be justified, and the plasma can be treated as uniform and stationary. In real situations of spatially nonuniform plasma, we have to take into account the convection loss term of the Stokes light,

$$\sum_{\mathbf{k}} \left[\frac{\omega_0}{\omega_H(\mathbf{k}_-)} \right] \left\{ \left[\frac{\partial \omega_H(\mathbf{k}_-)}{\partial \mathbf{k}_-} \right] \cdot \frac{\partial W(\mathbf{k}_-)}{\partial \mathbf{r}} - \left[\frac{\partial \omega_H(\mathbf{k}_-)}{\partial \mathbf{r}} \right] \cdot \frac{\partial W(\mathbf{k}_-)}{\partial \mathbf{k}_-} \right\}$$

on the right-hand side of the power balance equation (58). Then the absorption power density can become larger than that expected from the damping rate $\gamma(\mathbf{k}_-)$ of the electromagnetic wave.

Absorption Power in Terms of Density Fluctuation Spectrum

Let us finally express the absorption power density in terms of the electron density fluctuation spectrum. To this end, we introduce the equation for the low-frequency (in the sense that $\omega \ll \omega_0$) electron density perturbation. The latter is produced by the ponderomotive force due to the coupling of the pump wave and the Stokes light wave as follows[12]:

$$\delta n_e = -\left[\frac{\varepsilon_i \mathbf{k}, \omega) \varepsilon_0 \chi_e \mathbf{k}, \omega)}{\varepsilon \mathbf{k}, \omega)} \right] \frac{i\mathbf{k} \cdot \mathbf{F}_P(\mathbf{k}, \omega)}{e^2} , \tag{59}$$

where $\mathbf{F}_P(\mathbf{k}, \omega)$ is the ponderomotive force being given by

$$\mathbf{F}_P(\mathbf{k}, \omega) = \frac{-i\mathbf{k}e^2}{(m\omega_0\omega_-)} \mathbf{E}_0 \cdot \mathbf{E}_-^* . \tag{60}$$

Solving eq. (59) for $\mathbf{E}_0 \cdot \mathbf{E}_-^*$, we have for its complex conjugate the following expression:

$$\mathbf{E}_0^* \cdot \mathbf{E}_- = -\left(\frac{m\omega_0\omega_-}{k^2} \right) \left[\frac{\varepsilon^*(\mathbf{k}, \omega)}{\varepsilon_i^*(\mathbf{k}, \omega) \varepsilon_0 \chi_e^*(\mathbf{k}, \omega)} \right] \delta n_e^* . \tag{61}$$

Substituting (61) into (56) we finally get

$$2\mathrm{Re}\{\mathbf{E}_0^* \cdot \mathbf{J}'\} = 2\omega_0 \sum_{\mathbf{k}} \left(\frac{e^2}{k^2}\right) \int \frac{d\omega}{(2\pi)^2} \mathrm{Im} \left\{ \frac{\varepsilon^*(\mathbf{k},\omega)}{\varepsilon_i^*(\mathbf{k},\omega)\varepsilon_0 \chi_e^*(\mathbf{k},\omega)} \right\} \langle |\delta n_e|^2 \rangle \ . \quad (62)$$

Our final task is to confirm that this expression is consistent with the Manley-Rowe relation. For this purpose we consider two special examples, namely, we consider the cases that the low-frequency wave is an electron plasma wave (stimulated Raman scattering) and an ion acoustic wave (stimulated Brillouin scattering).

First, for the case of stimulated Raman scattering, we have the relations

$$\varepsilon^*(\mathbf{k},\omega) \cong \varepsilon_e^*(\mathbf{k},\omega) \, , \ \varepsilon_i^*(\mathbf{k},\omega) \cong \varepsilon_0 \, , \ \chi_e^*(\mathbf{k},\omega) \cong -1 \ , \quad (63)$$

and

$$\langle \, |\delta n_e|^2 \, \rangle = \left(\frac{k\varepsilon_0}{e}\right)^2 |\mathbf{E}(\mathbf{k},\omega)|^2 \ . \quad (64)$$

Use of these relations in (62) gives

$$\begin{aligned} 2\mathrm{Re}\{\mathbf{E}_0^* \cdot \mathbf{J}'\} &= -2\omega_0 \sum_{\mathbf{k}} \int \frac{d\omega}{(2\pi)^2} \mathrm{Im}\{\varepsilon_e(\mathbf{k},\omega)\} |\mathbf{E}(\mathbf{k},\omega)|^2 \\ &\cong \left(\frac{\omega_0}{\omega_{Pe}}\right) \sum_{\mathbf{k}} 2\gamma_L(\mathbf{k}) W_L(\mathbf{k}) \ , \end{aligned} \quad (65)$$

where $\gamma_L(\mathbf{k})$, and $W_L(\mathbf{k})$, are, respectively, the damping rate and wave energy density of the electron plasma wave. Obviously, the result is consistent with the Manley-Rowe relation.

We next consider the stimulated Brillouin scattering. In this case we have

$$\varepsilon_i^*(\mathbf{k},\omega) \cong -\varepsilon_0 \chi_e^*(\mathbf{k},\omega) \ \cong \ \frac{-\varepsilon_0}{(k\lambda_{De})^2} \ , \quad (66)$$

$$\langle \, |\delta n_e|^2 \, \rangle \cong \left(\frac{n_0 e}{T_e}\right)^2 \langle |\phi|^2 \rangle \ \cong \ \left[\frac{\varepsilon_0}{(ek\lambda_{De}^2)}\right]^2 |\mathbf{E}(\mathbf{k},\omega)|^2 \ , \quad (67)$$

form which we obtain

$$\begin{aligned} 2\mathrm{Re}\{\mathbf{E}_0^* \cdot \mathbf{J}'\} &= -2\omega_0 \sum_{\mathbf{k}} \int \frac{d\omega}{(2\pi)^2} \mathrm{Im}\{\varepsilon(\mathbf{k},\omega)\} \, |\mathbf{E}(\mathbf{k},\omega)|^2 \\ &= \sum_{\mathbf{k}} \left[\frac{\omega_0}{\Omega(\mathbf{k})}\right] 2\gamma_L(\mathbf{k}) W_L(\mathbf{k}) \ , \end{aligned} \quad (68)$$

where $\Omega(\mathbf{k})$, $\gamma_L(\mathbf{k})$ and $W_L(\mathbf{K})$ are, respectively, the frequency, damping rate and wave energy density of the ion acoustic wave. Again, it is clearly consistent with the Manley-Rowe relation.

We remark that the absorption power can be enhanced if the electron density fluctuation level is by some means enhanced above the thermal level. This, however, does not mean that the energy i absorption by the plasma can be enhanced. Because of the Manley-Rowe relation, only a small fraction of the energy goes to the scattered electromagnetic wave (Stokes light) and is lost out of the plasma by convection. In real situations one has to take into account the nonlinear modification of the polarizabilities, particularly the growth rates [13].

Concluding Remarks

It is clear that John Dawson's early work with Carl Oberman on high-frequency conductivity of a plasma has initiated many theoretical and experimental works in both basic plasma physics and nuclear fusion research. In this connection, I wish to recall that the annual symposium on "Anomalous Absorption of Intense Electromagnetic Waves" have started in 1971 when the anomalous absorption of high-frequency wave at $\omega_0 = \omega_{Pe}$ was observed by a computer simulation, as had been predicted by Dawson and Oberman in 1963. This series of symposia have stimulated many interesting works in the field of nonlinear plasma physics as well as in the physics of laser-plasma interaction in connection with the laser fusion. Without hezitation I would claim that John Dawson's pioneering contributions to these fields are among the most important in the recent progress in plasma physics.

References

1. J. Dawson and C. Oberman, *Phys. Fluids*, **5**, 517 (1962).
2. J. Dawson and C. Oberman, *Phys. Fluids*, **6**, 394 (1963).
3. V.I. Perel and G.M. Eliashberg, *Zh. Eksperm. Teor. Fiz.*, **41**, 886 (1961); *Sov. Phys. - JETP* 14, 633 (1962).
4. K. Nishikawa and Y.H. Ichikawa, *Phys. Fluids* **12**, 2563 (1969).
5. K. Nishikawa, *J. Phys. Soc. Japan*, **24**, 916, 1152 (1968).
6. P.K. Kaw and J.M. Dawson, *Phys. Fluids* **13**, 472 (1970).
7. W.L. Kruer, P.K. Kaw, J.M. Dawson and C. Oberman, *Phys. Rev. Lett.*, **24**, 987 (1970).

8. D. Arnush, B.D. Fried, C.F. Kennel and P.K. Kaw, *Bull. Am. Phys. Soc.*, **16**, 1257 (1971).

9. S. Ichimaru, *Ann. Phys.*, **20**, 78 (1962).

10. V.P. Silin, *Zh. Eksperm. Teor. Fiz.*, **51**, 1842 (1966).

11. J.M. Manley and H.E. Rowe, *Proc. IRE*, **44**, 904 (1956).

12. J. Drake, P.K. Kaw, Y.C. Lee, G. Schmidt, C.S. Liu and M.N. Rosenbluth, *Phys. Fluids*, **17**, 778 (1974).

13. See, for instance, H.A. Rose, D.F. DuBois and B. Bezzerides, *Phys. Rev. Lett.*, **58**, 2547 (1987).

W. Horton
University of Texas at Austin

Remarks of Professor W. Horton at the Dawson Symposium: Introduction of Professor Kodi Husimi

When the U.S. delegations of scientists began arriving in Japan in the early 1980s under the exchanges called for by the Carter-Fukuda agreement, they discovered a vigorous and well planned research program covering all the main approaches to fusion power, along with supporting basic university plasma experiments. We are honored to have here today the physicist who is largely responsible for the creation of this farsighted research program in Japan, that is, Professor Kodi Husimi. Professor Husimi was the founder and the first director of the Institute of Plasma Physics, or IPP Nagoya, from 1962 to 1973. The creation of this institute gave Japan a valuable national resource for a large scale plasma research center in the Nagoya area. For over twenty-five years the Ministry of Education has funded research in many areas of plasma confinement and heating experiments, theory, and simulations at IPP Nagoya. The importance of this center has recently been elevated to a national laboratory by the major expansion and construction of a new campus at Toki, outside of Nagoya, as the National Institute for Fusion Studies,or NIFS.

Professor Husimi recently celebrated his eightieth birthday and comes to science policy from solid state and nuclear physics. He also worked briefly in statistical physics deriving some important results. He is a Professor Emeritus of physics at

Osaka University and Nagoya University. He is President of the Japanese Science Council. In the 1980s he was a member of the Japanese upper house of the Diet, which is Japan's Parliament. We are all lucky to know Dr. Kodi Husimi, who, in the best sense, is a real "Shogun" in our modern world of plasma physics. We ask Dr. Husimi to share his thoughts with us on the development of the plasma and fusion research programs in Japan.

Kodi Husimi
University of Nagoya and University of Osaka

The Early Days of the I.P.P., Nagoya, and Professor Dawson

The IPP (Institute of Plasma Physics), Nagoya, was established in the spring of 1961 after long and arduous discussions about what we shall do in the field of nuclear fusion research. Even before the famous declaration by H. Bhabha on the possibilities of nuclear fusion along with nuclear fission on the occasion of the 1955 Geneva Conference on Peaceful Uses of Atomic Energy, Japanese physicists had become aware of the classified activities in this field in the US and USSR. It cannot be denied that restricted knowledge was most effective in exciting Japan's interest in fusion research[1].

Just before the Second Geneva Conference, while in the Physics Department of Osaka University, I listened to a report made by a colleague who happened to have had the chance of attending a special lecture by L. Spitzer on stellarator principles. This was my first experience of considering seriously the possibility of confining a completely ionized plasma. But active accelerator physicists, T. Ohkawa, S. Mori, and other students at University of Tokyo were in more advanced stages of research. There were also people who tried pinch experiments using high current.

[1] *Yome, toome, kasa- no- uchi.* (She looks prettier) at night, at a distance and under the umbrella.

The discussions centered around the plan of immediately building a plasma machine comparable in size with the C-stellarator. But the majority preferred to set up a research laboratory where basic plasma physics experiments can be performed with high grade apparatus and measuring instruments. The decision was made by Prof. H. Yukawa as Chairman of the Special Committee of the Atomic Energy Commission.

I was elected as the first director of the newly born IPP, affiliated to Nagoya University. Nagoya was selected as the center of the mass of the Japanese plasma scientists, having sympathetic professors such as Prof. K. Yamamoto, Prof. S. Sakata, and Prof. S. Hayakawa.

One of my first jobs was to attend the first International Conference on Plasma Physics and Controlled Nuclear Fusion Research, held in Salzburg. I listened to discussions by, among others, Artzimovich and Spitzer on the minimum B principle. After the conference I made a round-world trip, observing many European and American plasma physics laboratories. I had some trouble in getting a visa from the US, which enabled me to enjoy life in London for a month. When permitted to enter the US, I, of course, visited Princeton, where two Japanese scientists, S. Yoshikawa and J. Fujita, were already working with stellarators. I must have met Dr. Dawson at this first visit, but I am not sure. I *am* sure, however, that on my second visit Dawson guided me through laboratories and showed me, among others, a magnetic coil of complicated shape, called a circularizer. He seemed to be skeptical about the function of this magnet.

About this time, IPPN decided to invite Dr. Dawson to instruct and to cooperate with the theoretical group of the Institute. This plan materialized in 1964. From September of that year to July of the next year, Dawson and his family stayed in Nagoya.

Originally Dr. Dawson seemed to have the idea of executing computer experiments based on particle models, but the computer available at that time at IPPN, the OKITA 5090, turned out to be too small for such purposes. Dawson was compelled to work in more analytical ways. In this work he was assisted by Dr. T. Nakayama, leading to the papers, "Kinetic Structure of a Plasma"[2], and "Derivation of Hierarchies of N-Particle Systems and Vlasov Systems by Means of Functional Calculus"[3].

Prof. Y. Mizuno, who had worked in IPPN for a long time, recalls pleasant discussions with Dr. Dawson, particularly on the nature of Landau Damping. Mizuno attended the Summer School at Trieste in the fall of 1964 and learned from A. Simon about the various aspects of Landau Damping. Simon concluded with the remark that the true nature of Landau Damping is not so clear. Dawson immediately refuted that the physics of Landau Damping is already quite clear, citing

[2] J.M. Dawson and T. Nakayama, "Kinetic Structure of a Plasma," *Physics of Fluids*, **9**, 252-264 (1966).

[3] T. Nakayama and J.M. Dawson, "Derivation of Hierarchies of N-Particle Systems and Vlasov Systems by Means of the Functional Calculus," *J. Math. Phys.*, **8**, 553-560 (1967).

his paper "Landau Damping from a Multistream Model from the Plasma" [*Phys. Rev.*, **118**, 381 (1960)]. But Mizuno preferrs Dawson's second paper, "On Landau Damping" [*Phys. Fluids*, **4**, 869 (1961)].

On the last days of Dr. Dawson's stay in Japan, they visited Sapporo, where the Physical Society of Japan held their annual general meeting. There must be a photograph in which Dawson appears in front of the statue of Dr. Clark[4], together with his Japanese colleagues in Plasma Physics.

Thank You!

Meeting on Collective Motions and Instability (November 12-13, 1964) Given by Dr. Dawson

Plasma Turbulence

Definition: Any Random Collective Motion in a Plasma

Plasma Turbulence contains many problems of varying degrees of complexity. We may subdivide the subject of Plasma Turbulence into the following:

1. *Weak Turbulence* , such as the nonlinear interaction of waves of moderate amplitude. Fluctuations density, temperature, etc., are small compared to their mean values.
2. *Strong Turbulence* , such as might arise when a large current is passed through a plasma or as might be produced by a collisionless shock. Here the density, temperature, etc., fluctuate by amounts comparable to their mean values.
   (a) *Steady* or *Quasi-Steady* , where the process goes on for a long time compared to the time required for growth and decay of the collective motions
   (b) *Non-Steady* , where the process lasts for times of the order of that required for growth and decay of the motion.
3. *High Frequency Turbulence* : Here the frequency might be comparable to the electron of ion plasma frequencies on the electron or ion cyclotron frequencies.
4. *Low Frequency Turbulence* : Hydromagnetic motions, acoustic motions, resistive driven motions are included here.

[4] William Smith Clark (1826-1886), American educator. Came to Japan and taught at Sapporo Agricultural School for about a year. Had influence on Kanzo Uchimura, Inazo Nitobe and so many other young people. "Boys, be ambitious!" was his famous well-wishing words for the students.

5. *Microturbulence* : Here the motions are characterized by wave lengths of the order of a few Debye lengths or a few ion Larmor radii.
6. *Macro Turbulence* : Here the characteristic lengths are much larger than the Debye length or the ion Larmor radius.

Example of Weak Turbulence

Longitudinal Instability: Due to a small bump in the tail of the electron distribution waves grow until nonlinear effects limit them. Either $f(v)$ distorts so as to become stable, or wave-wave interaction prevents growth.

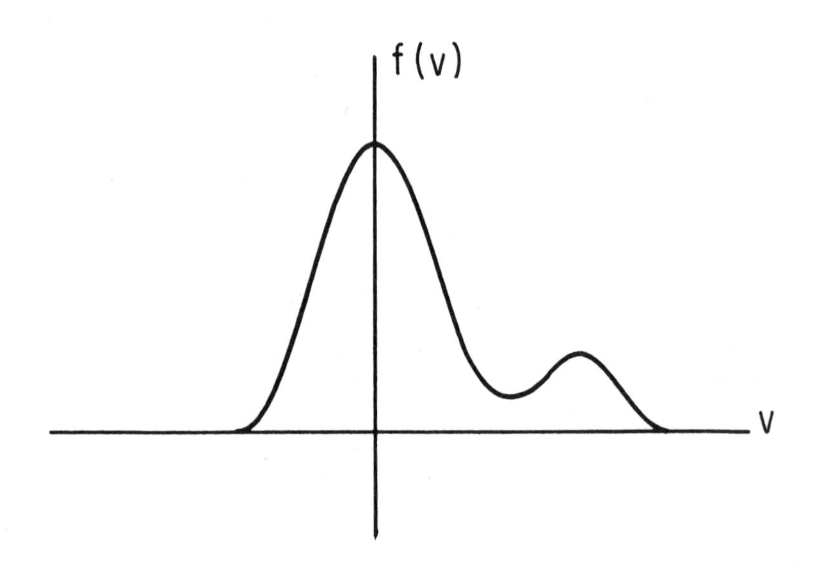

FIGURE 1

1. The so-called Quasi-Linear Theory of Drummond, Pines, Vedenov, Velikov and Sagdeev applies to this problem. This theory is based on the adiabatic approximation $\gamma \ll \omega$. This is not a Galilean invarient relation. We should require $\gamma \ll \omega - k, v$. This cannot be satisfied for all particles. It fails for the resonant particles. The theory may not treat these particles correctly. Since they supply the energy to the wave this may be important.

   Problem: The foundation and regions of validity of this theory should be further investigated.

2. We wish to know how much energy can be converted to wave energy. An argument by C. Gardner gives a bound on this energy. The Vlasov equation conserves f following the motion. The motion can only redistribute f. Divide f into slices. The motion interchanges slices. If an interchange of slices decreases $\int v^2 f dv$ then energy is available for waves. Thus, a double humped distribution can be unstable but a monotonic decreasing function of v^2 is stable.

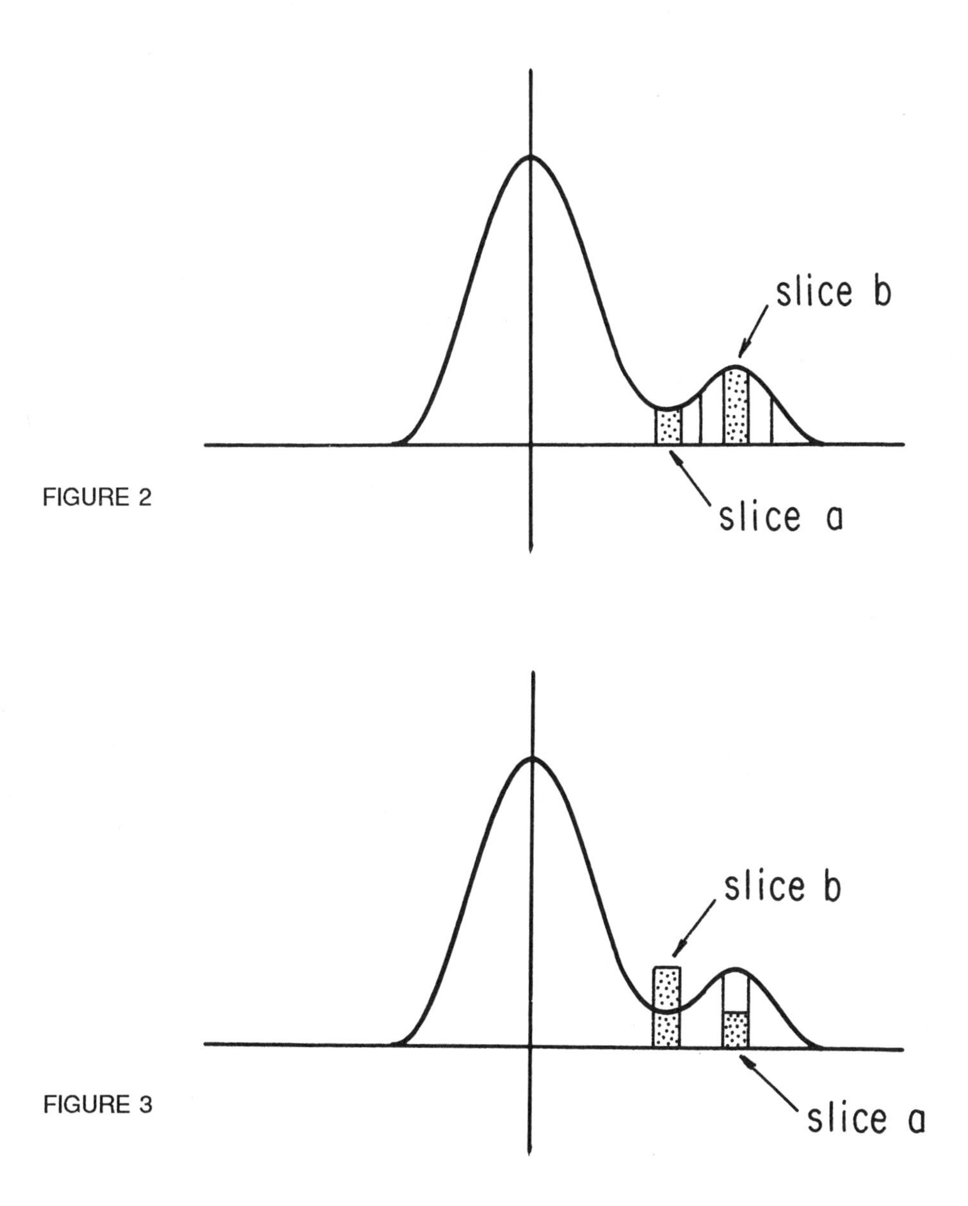

FIGURE 2

FIGURE 3

All such interchanges of slices may not be allowed. There may be some constraints. In this case there is one, conservation of electron momentum. The theory as given above would predict a displaced Maxwellian to be unstable because we could gain energy by moving it back to the origin. Requiring conservation of momentum prevents us from doing this and gives results consistent with the theory of longitudinal plasma oscillations. The bound on the energy which can be converted to wave energy agrees within a factor of 2 with quasilinear theory.

Problems:

Is the limit obtained above the best we can obtain?

Can similar arguments be applied to other unstable situations and other types of waves?

Is there a type of Free Energy for a given distribution function and for a given type of wave generation?

3. The Generation of Ion Waves and Anomalous Resistance resulting from the passage of a large current through a plasma. When an E field is applied to a plasma so as to produce a current, the electrons drift with respect to the ions. With large enough drift, ion waves are generated.

   Problems:

   (a) Find the nonlinear conductivity for voltages lower than are required for the ion wave instability.

   (b) Find the quasi-steady state of the plasma for voltages only slightly larger than are required for ion wave instablities. Find the conductivity and the fluctuating density and E fields.

   (c) Repeat (b) for a plasma containing a magnetic field. In this case compute the random walk of a particle across the magnetic field.

4. Theory and experiment need to work together on the difficult problems associated with plasma turbulence. Here a new technique is the use of numerical experiments. One makes a mathematical model of the plasma and carries out numerical experiments. The results of such experiments can be compared with theory. One can check the theory and also gain insight into the plasmas behavior by this means.

References

Quasi-Linear Theory:

1. W.E. Drummond and D. Pines, also A.A. Vedenov, E.P. Velikov and R.Z. Sagdeev, *Proceedings of the Salzburg Conference on Plasma Physics and Controlled Nuclear Fusion Research*, Sept. 4-9, 1961; also appear in *Nuclear Fusion* about 1963.
2. W.E. Drummond and D. Pines, *Annals of Physics*, **28**, 478 (1964).
3. E. Frieman and P. Rutherform, "Stability Criteria and Energy Available for Wave Generation", *Annals of Physics*,**28**, 134 (1964).
4. C. Gardner, *Phys. Fluids*, **6**, 839, (1963).
5. T.K. Fowler, "Lyapuhoff's Stability Criteria for Plasmas" Oak Ridge National Laboratory Report, about 1962-1963.
6. M. Kruskal and C. Oberman, *Phys. Fluids*, **1**, 275 (1958).
7. C. Oberman and J. Dawson, "Ion Wave Instabilities and Anomalous Resistance", *Phys. Fluids*, **7**, 773 (1964)
8. 0. Buneman, *Phys. Rev.*, **115**, 504 (1959).
9. I.B. Barnstein and R.M. Kulsrud, *Phys. Fluids*, **3**, 937 (1960); also *Phys. Fluids* **4**, 1037 (1961).
10. C. Smith and J. Dawson, Princeton Plasma Physics Laboratory Report, Matt-151, (1963).

Numerical Experiments:

11. J. Dawson, *Phys. Fluids*, **5**, 445 (1962) and **7**, 419 (1964); Conference on Plasma Physics and Controlled Nuclear Fusion Research, Sept. 4-9, Salzburg, Austria 1961, Appears in Nuclear Fusion (1963).
12. 0.C. Eldridge and M. Feix, *Phys. Fluids*, **6**, 398 (1963).
13. A. Hasegawa and C.K. Birdsall, Electronics Research Laboratory, University of California, Berkeley, California, Report No. 64-5 (1964).
14. P.L. Auer, H. Hurwitz, and R.W. Kilb, *Phys. Fluids*, **5**, 298 (1962), [also References 8 and 10, Scattering and Coupling of Waves.]
15. Y. Terashima and J. Yajima, *Progress of Theoretical Physics*, **30**, 443 (1963).
16. D.A. Tidman and G.H. Weiss, *Phys. Fluids*, **4**, 703 (1961).
17. R. Aamodt and W. Drummond, *Plasma Physics* (Journal Nuclear Energy Part C) **6**, 147 (1964).
18. T. Birmingham, J. Dawson, C. Oberman, Princeton Plasma Physics Lab., Princeton University, Princeton, N. J., Matt-266 (1964).

19. H.L. Berk, *Phys. Fluids*, **7**, (1964).
20. N. Kroll, A. Ron and N. Rostoker, *Phys. Rev. Lett.*, **13**, 83, (1964)
21. P. Platzman and N. Tzor, *Phys. Rev. Lett.*, **12**, 573 (1964).

J.M. Kindel, J.M. Wallace, D.W. Forslund* and G. Olson
Applied Theoretical Division, Los Alamos National Laboratory
*Advanced Computing Laboratory, Los Alamos National Laboratory

Very Short Pulse Laser-Plasma Interactions†

Introduction

We would like to say how Joe Kindel came to know John Dawson and how they began to work together. In 1970 Kindel received his Ph.D. from UCLA under Charles Kennel. He then went to Princeton as a Research Associate, primarily working with Francis Perkins at Princeton on ion cyclotron heating of tokamak plasmas. At the time, John was head of the Theoretical Division and already in 1970 was quite famous. Kindel wanted to collaborate with John and in a real sense, as a recent graduate student, he regarded Rip Perkins and John Dawson as a second thesis advisor. John, as you probably know, is very unassuming, very personable, and nice to people. One of the things that he has done, in addition to graduating 30 thesis students, is to help the careers of hundreds of people; he has done this by his ideas, his willingness to give of himself and he did that for Kindel. They had two collaborative efforts at Princeton. The first was work on trapped particle instabilities of ion waves with Kaw and John. Then Okuda and Kindel worked with John on nonlinear instabilities of lower hybrid waves. This work lead to publication in *Physical Review Letters*[1]. In the years following, Forslund and Kindel worked briefly with John on laser heating of solenoid plasmas. The latter effort was something that John was pursuing with Abe Hertzberg, George Vlases, and Hans Bethe. Dave Forslund and Kindel looked at potential laser plasma stimulated scattering instabilities associated with this scheme. In the late '70s, Kindel again collaborated with John on the Dawson Isotope Separation Process. John, Bob Martinez, Tony Lin, and Kindel carried out the first two dimensional plasma simulations of the process while Kindel was at TRW[2].

Warren Mori, Chan Joshi, Dave Forslund, John Dawson, Frank Chen, and Joe Kindel collaborated on the beat wave accelerator over nearly a five year period in the mid '80s. This resulted in many publications[3-7]. In the beat wave accelerator one of the key items is a laser; in fact, two lasers beating, but separated by an appropriate frequency. The very short pulse length in this process served as one of the pieces of foundation work for research that we are going to talk about today. Today, we will discuss our ongoing theoretical plasma physics research at Los Alamos National Laboratory on very short laser plasma interactions.

As many of you know, particularly those who work in laser fusion, the one thing you do not want to do is to model all the relevant physics during the period when the plasma is forming. Because of energy considerations, it is a good approximation to ignore this initial period in laser fusion. However, for high brightness laser plasma problems, i.e., for very high intensity lasers with very short pulse length, incident on a partially ionized or un-ionized plasma, we are in this difficult modeling regime. We are going to motivate the problem of high brightness source laser-matter interaction and describe a little bit of what, in our view, the important physics is. We will emphasize the modeling tools that we have developed at Los Alamos. Below you will see one example of a process that is quite ubiquitous in laser-plasma interaction, i.e., resonant absorption[8]. A key point to be emphasized is that in many experiments which people believe to be completly classical or collisional, collective affects can be very important.

High brightness source lasers exist at many facilities throughout the world. In particular, there are two short-pulse lasers at Los Alamos: the krypton fluoride laser and the xenon fluoride laser[9,10]. We will not describe detailed modeling of experiments involving these lasers, but simply indicate relevant laser parameters in Table 1 which motivate what we are modelling. We note that when we talk about high brightness laser systems we are talking about pulse lengths less that a picosecond. We include intensities, such that the electric field is large enough that the quiver energy of an electron in the laser electric field can cause the electron to become free. Electrons at the highest intensities can have relativistic energies.

To model this process we need to simultaneously include atomic physics, hydrodynamics, collisional effects including ionization effects, and plasma physics. It turns out that time scales for collisions, ionization, and collective processes are within a factor of 10 of being all the same time scale. First of all, what Gordon Olson, Jack Comly, and co-workers have developed at Los Alamos is a radiation hydrodynamics code, ZAP, that can have a large number of atomic levels[11]. What is contained in ZAP is listed in Table 2. We are going to give one example, which includes a tabulation of electron density as a function of density and temperature which Olson has adapted from this code. One needs to do all of the plasma physics that one is familiar with in laser plasma interactions, and we simply list them for completness in Table 3. The important part is at the bottom; we need to do all of the plasma processes plus have collisional models for the electron-ion scattering, electron-electron scattering, inverse Bremsstrahlung, and ionization.

TABLE 1 Los Alamos High Brightness Laser Systems

	LABS I - KrF	LABS II - XeC1
Wavelength (nm)	248	308
Energy (mj)	30	250
Pulsewidth (ps)	0.7	0.33
Intensity (w/cm^2)	8×10^{17} @ f/1.7	4.6×10^{18} @ f/3.7 6.4×10^{19} @ f/1.0
Repetition Rate	5	1

TABLE 2 ZAP 1D Radiation Hydrodynamics Code

- Fluid Treatment of Target
- Multigroup Radiation Transfer
- Detailed Atomic Physics Treatment
- Inverse Bremsstrahlung Laser Light Absorption
- Ray-Trace Resonant Absorption Model (Under Development)
- Treats Long Space and Time Scales
- Calculates Density, Temperature, and Ionization Profiles
- Calculates Kev Line Raidation Emission
- Requires Long Machine Run Time

TABLE 3 Wave 2D Plasma Simulation Code

- Solves Maxwell-Vlasov System
- Relativistic PIC Treatment of Plasma
- Properly Models Laser Propagation in Plasma
- Treats Short Space and Time Scales
- Calculates High-Energy Particle Production
- Calculates Collective Plasma Phenomena Especially Laser Emission
- Separate Models for Electron-Ion Scattering, Inverse Bremsstrahlung, and Ionization

For the collisional models developed a number of years ago for the WAVE code by Cranfill, Brackbill, and Goldman[12], one does deterministic Rutherford scattering in the polar angle and random scattering in Φ, the azimuthal angle. The scattering equation is as follows:

$$\langle\theta_S^2\rangle = \frac{e^4 8\pi N_i Z(Z+1) ln\Lambda\Delta t}{P_e^2 V_e}$$

where θ_s is in the scattering angle, P_e is the electron momentum, V_e is the electron velocity and all other definitions are standard.

Three points should be made:

- The PIC electrons are scattered particle by particle off the background ion density at each time step.
- Electron kinetic energy is unchanged.
- The PIC ions are not affected.

In electron-ion collisions, because of the phase difference between the electrons and the electric field, one can have absorption. For the examples we show here, Bremsstrahlung is chosen to be unimportant because we want to emphasize the collisional ionization interplay with collisionless effects. For ZAP we choose parameters for an aluminum atom with its 168 atomic levels. To mold this complicated non-LTE atomic physics code with the plasma simulation code WAVE, we generate a table which calculates the time dependent Z in the plasma as a function of electron temperature, density, and Z. Also implemented in WAVE is a very simple above threshold ionization model from Landau and Lifshitz[13] There are better models by Keldysh[14] and Reiss[15] which will be implemented later. Our ionization model as implemented in WAVE is as follows.

A phenomenological ionization-rate model determines the injection of PIC electrons and ions into the computational mesh. The time dependent Z is given by the follwing equation.

$$\frac{dZ}{dt} = F(Z, \theta_e, n_e)$$

where F represents the tabular function as gleaned from ZAP calculations for aluminum. The Landau-Lifshitz model is

$$\frac{dZ}{dt} = \frac{4M_e e^4}{\hbar^3}\left(\frac{\varepsilon_i}{\varepsilon_h}\right)^{\frac{5}{2}}\frac{\varepsilon_a}{E(t)}\exp\left[-\frac{2}{3}\left(\frac{\varepsilon_i}{\varepsilon_h}\right)^{\frac{3}{2}}\frac{\varepsilon_a}{E(t)}\right],$$

where

$$\varepsilon_a = \frac{M^2 e^5}{\hbar^4}, \varepsilon_h, \varepsilon_i \ = \ \text{Ionization Potential.}$$

Our ionization parameter region is predominantly collisional in that multi- photon ionization will pull off the first couple of electrons and the subsequent electron interactions will dominate the ionization process. In the example below this will become clearer. In the particular case we will show, the KrF laser has an intensity of a few times 10^{17} watts per cm^2. The laser pulse is shown schematically in Fig. 1.

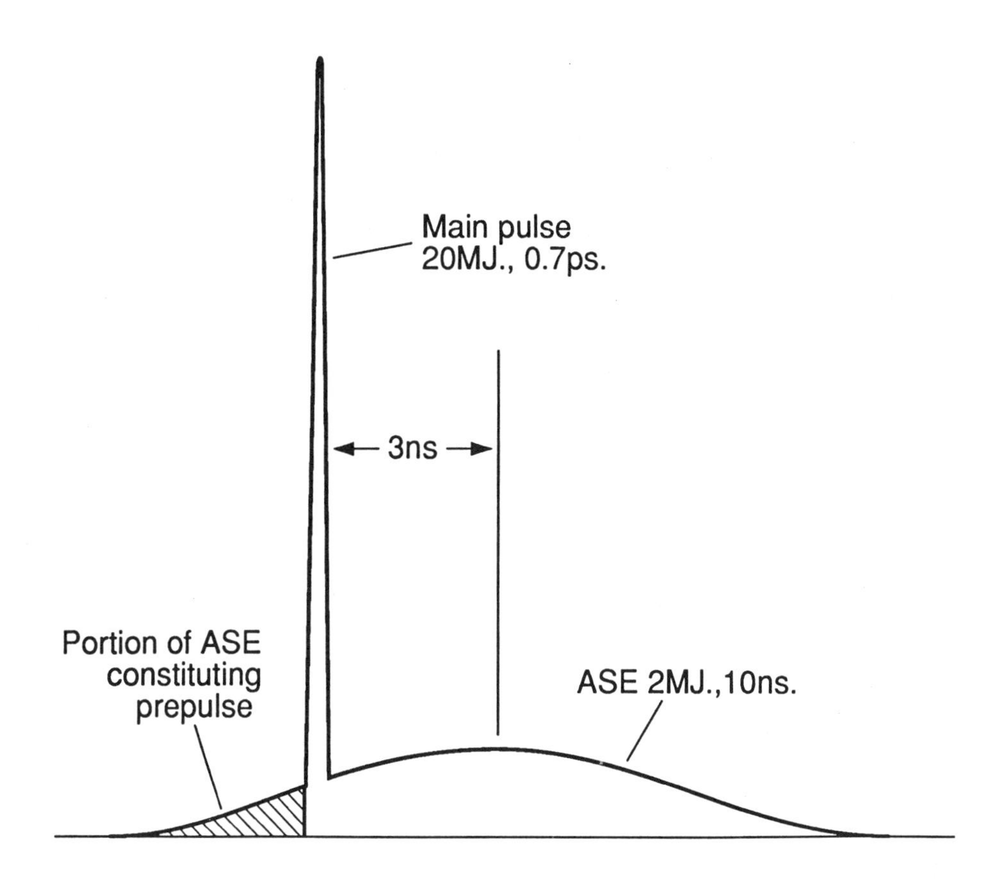

FIGURE 1 Typical Krypton Fluoride (Krf) laser pulse for the LABS I laser at Los Alamos

We use ZAP hydrodynamics radiation transport code described in Table 2 to calculate the early part of the pulse, and in this example we have actually picked up the high intensity laser interaction in the middle of the main pulse. Our objective was to first understand the dynamics of the ionization. The plasma formation is shown in Fig. 2 by the curves of density vs. x at various times. The lowest curve is the density profile at the end of the prepulse as calculated by ZAP for the conditions as shown in Fig. 2. As the main pulse is turned on, there is very rapid ionization as shown by the subsequent density versus position curves. An interesting thing to note is that the critical density has a velocity which is a significant fraction of the velocity of light during the early part of the pulse. One would expect that this velocity would have an effect on any critical surface collective processes which might occur.

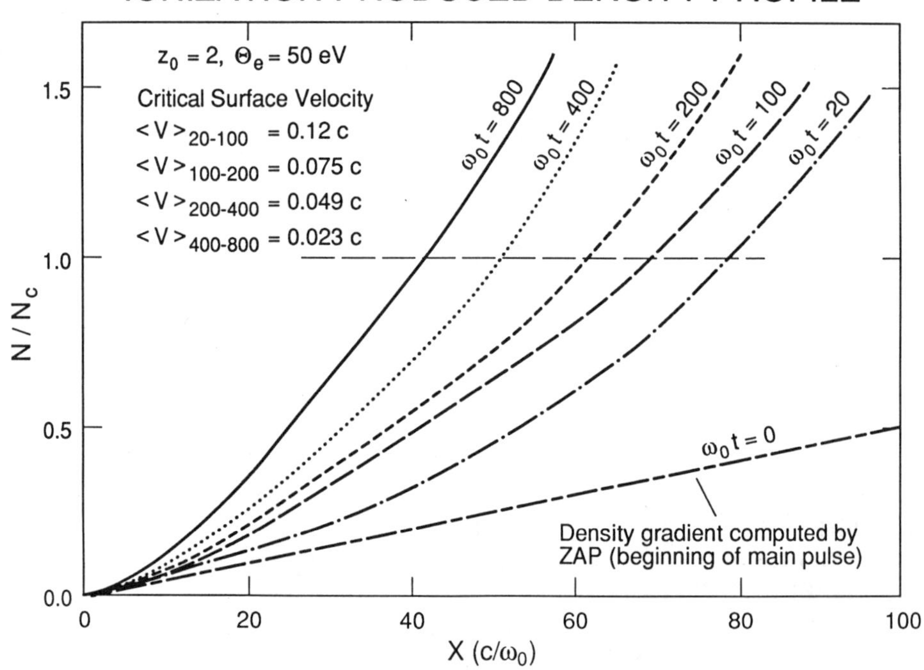

FIGURE 2 Evolution of the electron density (in units of the KrF critical density) versus position x (in units of the velocity light c divided by the laser frequency ω_0. The prepulse density profile is the bottom curve in this figure and corresponds to the laser parameters of Fig. 1. Peak laser intensity is as shown in Fig. 1. Ionized density profiles at various times are calculated by the ZAP code.

We have set up a WAVE simulation in which we have the collisional ionization model activated and in which the incident laser electric field is polarized in the plane of incidence. The laser field is incident on the linear density profile as shown in Fig. 3. This is density as a function of position corresponding to the $t = 0$ profile in Fig. 3. The laser propagation is along the x direction and there is a small E_x

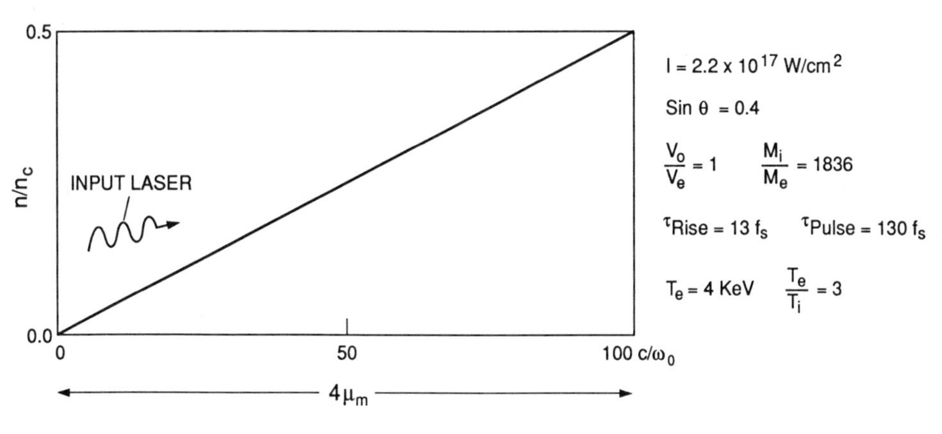

FIGURE 3 Schematic of the incident laser and density profile for our Particle-in-Cell WAVE Simulations. Plasma and laser parameters are as indicated. The laser pulse is allowed to transit and reflected light to return to the left-hand launching boundary before ionization is turned on and allowed to progress according to Fig. 2.

electric field although the electric field is primarily in the y direction. The laser magnetic field is in the z direction. The sine of the incident angle is 0.4. Note that the computational box is $100c/\omega_0$ long. This problem assumes an initial ionization and proceeds as follows. The laser is turned on in a faster time than in an actual experiment, i.e., in 100 ω_0^{-1}. The laser is allowed to propagate across the computational box and any reflection returns before ionization is turned on at time $200\omega_0^{-1}$. The density profile develops as the ionization proceeds. In this run we observed that the critical density initially moves at a 10th of the velocity of light or less. We expect for this simulation that resonant absorption should occur[8,18,19]. For this simulation (BR-025) and the next simulation (BR-026), we show in Fig. 4 the Poynting flux

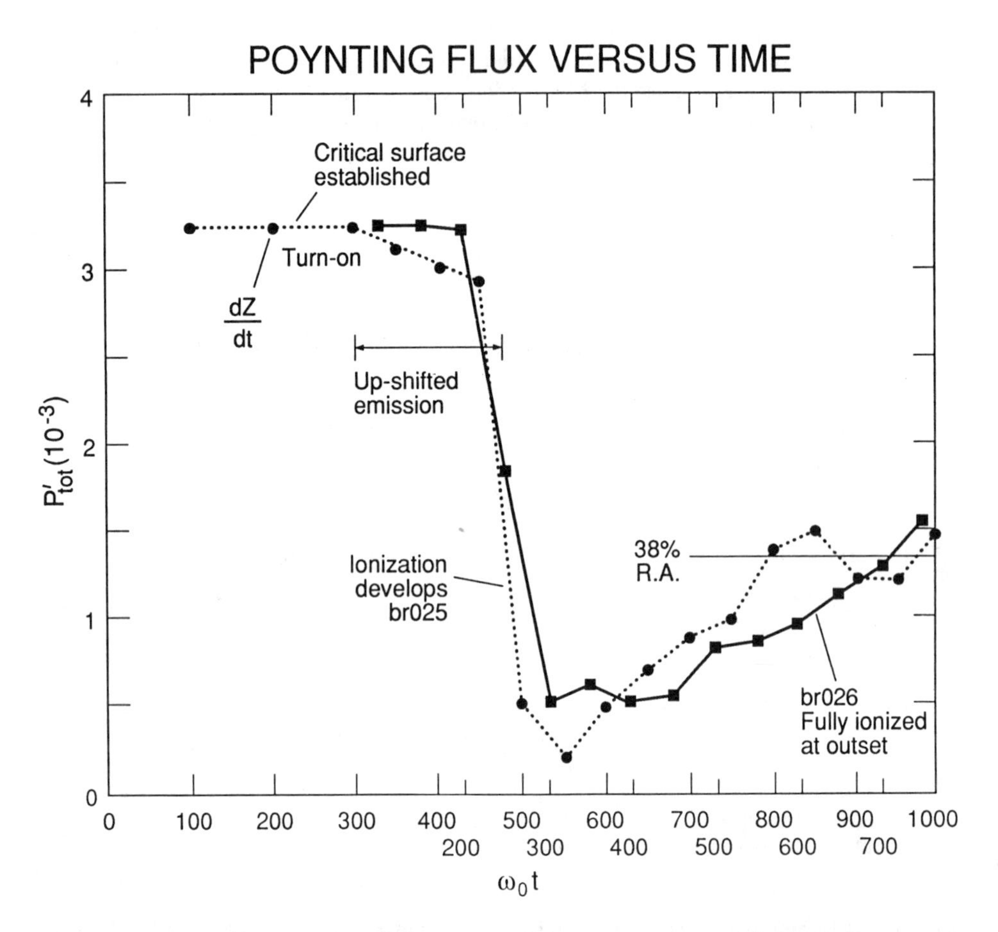

FIGURE 4 For both the simulation with small initial ionization (BR-025) and with full ionization (BR-026) chosen close to the final ionization of BR-026, we plot the Poynting flux P^t_{tot} at the left boundary as a function of time. The top $\omega_0 t$ scale corresponds to the BR-025 run and the bottom $\omega_0 t$ scale to the BR-026 run. Both simulations show a minimum in absorption ($P^t_{tot} = 0$ means no absorption) and a rise to nearly 40 percent absorption.

as a function of time. At time $200\omega_0^{-1}$ ionization is turned on. If there were no absorption in the plasma, then initially the Poynting flux would be shown as in Fig. 4 and then it would drop to zero and stay at zero. The minimum absorption is not zero, but is a few percent, then it rises from its minimum and settles down to something approaching 40 percent. The evolution of the density profile including the development and subsequent motion of the critical surface position can be seen by noting the nine frames in Fig. 5 including the $N/N_c = 1$ position.

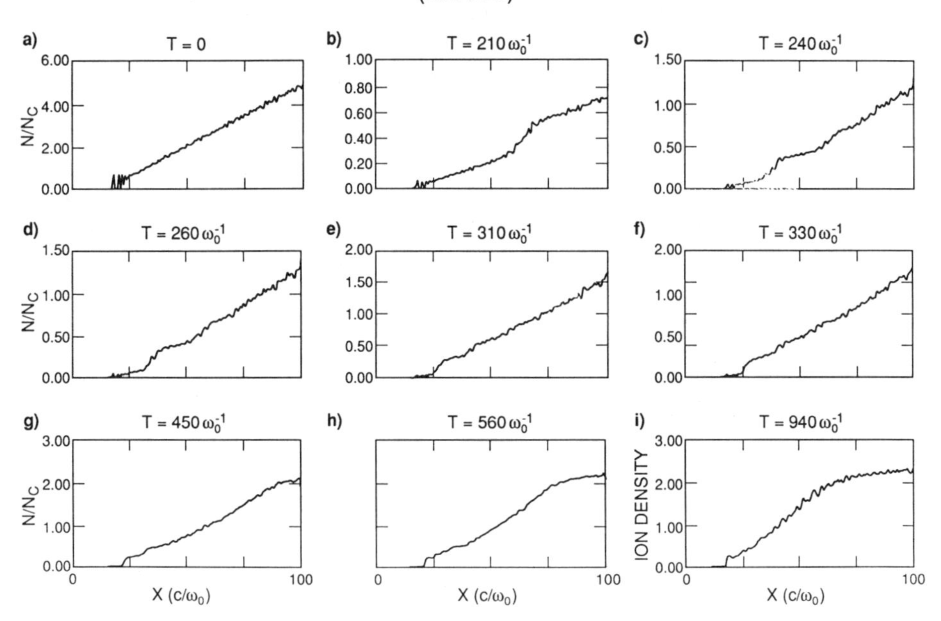

FIGURE 5 The initial pre-ionized density profile for BR-025 is shown in Fig. 5(a). The subsequent ionization following the prescription shown in Fig. 2 is shown in Figures 5(a)-5(i). Note the motion of the critical density.

For comparison there is a simulation BR-026 in which the density profile is run initialized at the outset close to the final profile in the BR-025 simulation. In this simulation, also, the absorption rises to about 40 percent. The density gradient for this particular profile, would allow from simple resonant absorption theory approximately 10 percent absorption[8].

Having summarized absorption and density profile information for these two simulations, we see that a closer look is required to see what is happening. First of all, for the case where rapid ionization occurs (BR-025), it is not obvious that resonant absorption can even occur; but it does. Fig. 6 shows E_x, the electric field along the density gradient. Fig. 6 also shows the incident and reflected electromagnetic waves. Most importantly, you see the characteristic plasma wave peak in E_x near critical density. Notice in Fig. 6 the other Fourier plots of the E_x electric field. $k_y = 0$ is the normal incidence component, and $k_y = 1$ is the incident laser component. Note that the other k_y are components barely out of the noise. The point is that there actually is a very weak instability, the parametric decay instability of the resonant absorption wave[18].

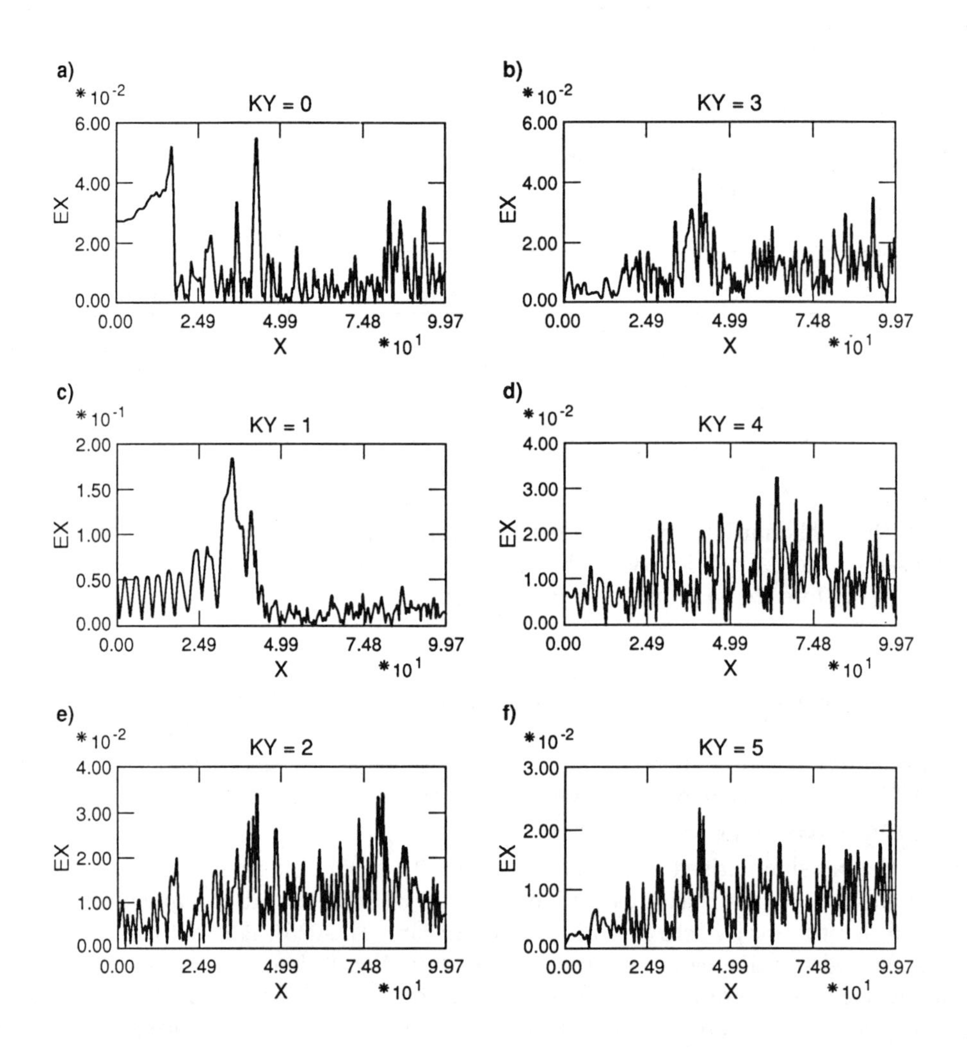

FIGURE 6 The Fourier Components of the electric field component E_x along the density gradient at the time $\omega_0 t = 950$. The $ky = 0$ component is at normal incidence. The $k_y = 1$ component shows the incident and reflected E_x field and the plasma wave near critical density.

Based on the next simulation with very similar parameters but not dynamic ionization, we believe that the weak instability exhibited in Fig. 6 is being reduced by the ionization. The other point, which is a point about high brightness source lasers, is that if resonant absorption occurs, then there should be hot electrons produced[16,17]. Indeed, we see hot electrons in the electron phase space and in the electron distribution function in Fig. 7. However, for this ionizing simulation with weak decay instability, the hot electron spectra is softened compared to previous results. We can say, however, a few words to describe what is happening. If you are creating new electrons while the laser light is being absorbed, the resonant absorption plasma wave is being changed because of these new electrons. If the ionization front were moving out too quickly you might imagine you would never drive up a resonant absorption plasma wave. For our parameters, there is enough time for a plasma wave to be driven up. The E_x electric field appears wider than the corresponding case without ionization. The ionization might be viewed as analogous to adding thermal effects to resonant absorption[8] in that ionization acts like an effective convection process. It can contribute to reducing the hot electron spectra of the resonant absorption wave. The group at Ecole Polytechnique has noted that parametric instabilities associated with the resonant absorption wave can lower the hot electron spectrum[19].

To understand this better we did a run (BR-026) shown in Figs. 8 and 9 in which we have full ionization at the outset. We have shown in Fig. 4 that the absorption eventually reaches the point where it does in (BR-025). Noting the Fourier plots of the E_x electric field in Fig. 7, we observe that compared to Fig. 6 we see very strong electric field spikes at various k_y. For this fully ionized plasma, we notice the enhanced E_x electric field near the critical density in Fig. 8 indicative of resonant absorption. In particular in Fig. 8 notice the $k_y = 3$ and the $k_y = 5$ spikes. The density Fourier plots in Fig. 9 show that the ion density surface is being rippled by a very strong instability. Eventually the self-consistent state is a high absorption state. Professor Nishikawa talked about parametric instabilites and this is a parametric decay instability of the plasma wave at the critical density. In fact, you can do a linear theory of this, and it needs to be done numerically[18]. It turns out that one predicts a very wide spectrum of unstable waves in the critical surface and indeed, we observe that. The instability reduces the maximum size of the electric field and it obviously has some affect on the absorption[19]. Lastly, as shown in Fig. 10, the electron spectrum for this case without ionization is harder.

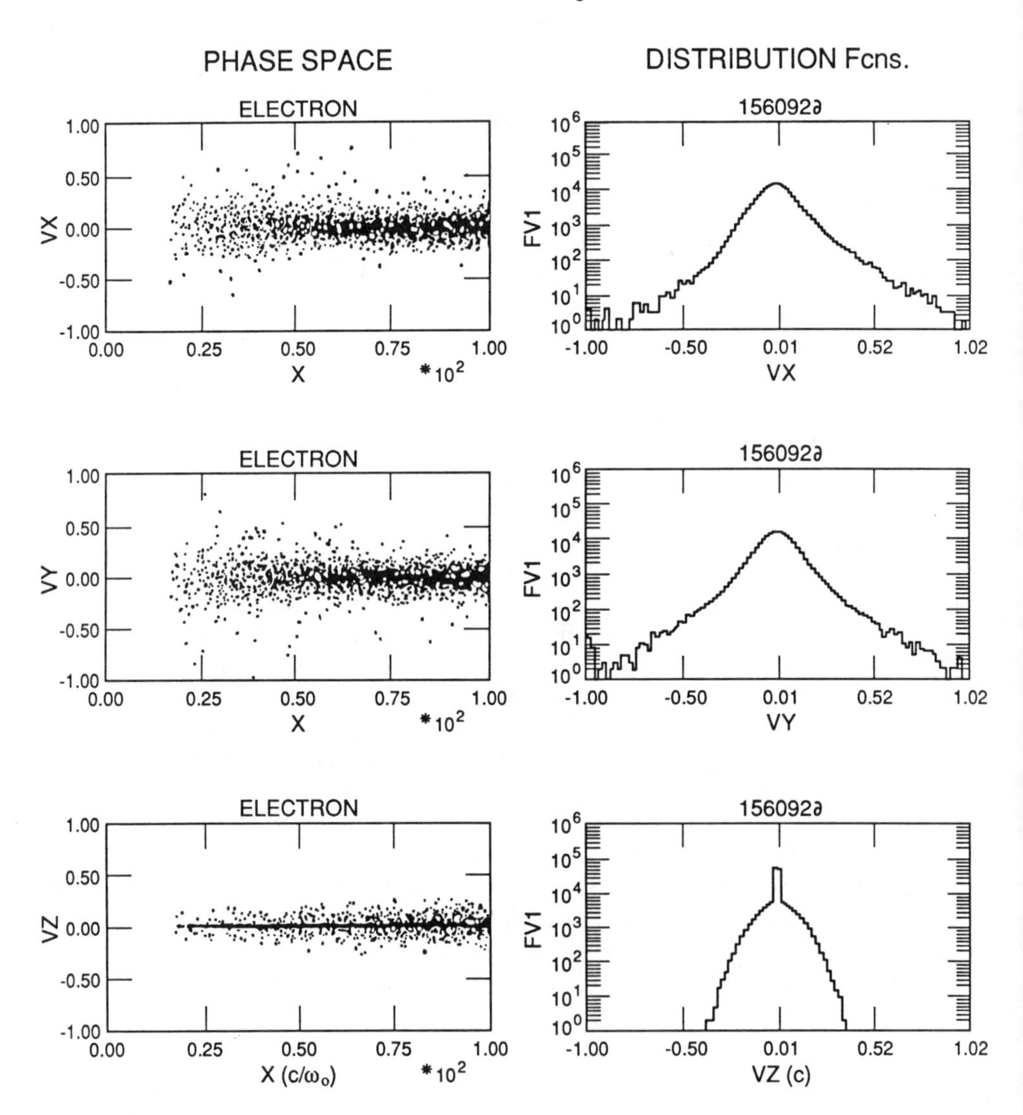

FIGURE 7 For the previous ionizing simulation at the $t = 950\omega_0^{-1}$ the electron phase space V_x,V_y, and V_z integrated over Y. The v's are relativistic momenta. The electron distribution is d) x, e) y, and f) z.

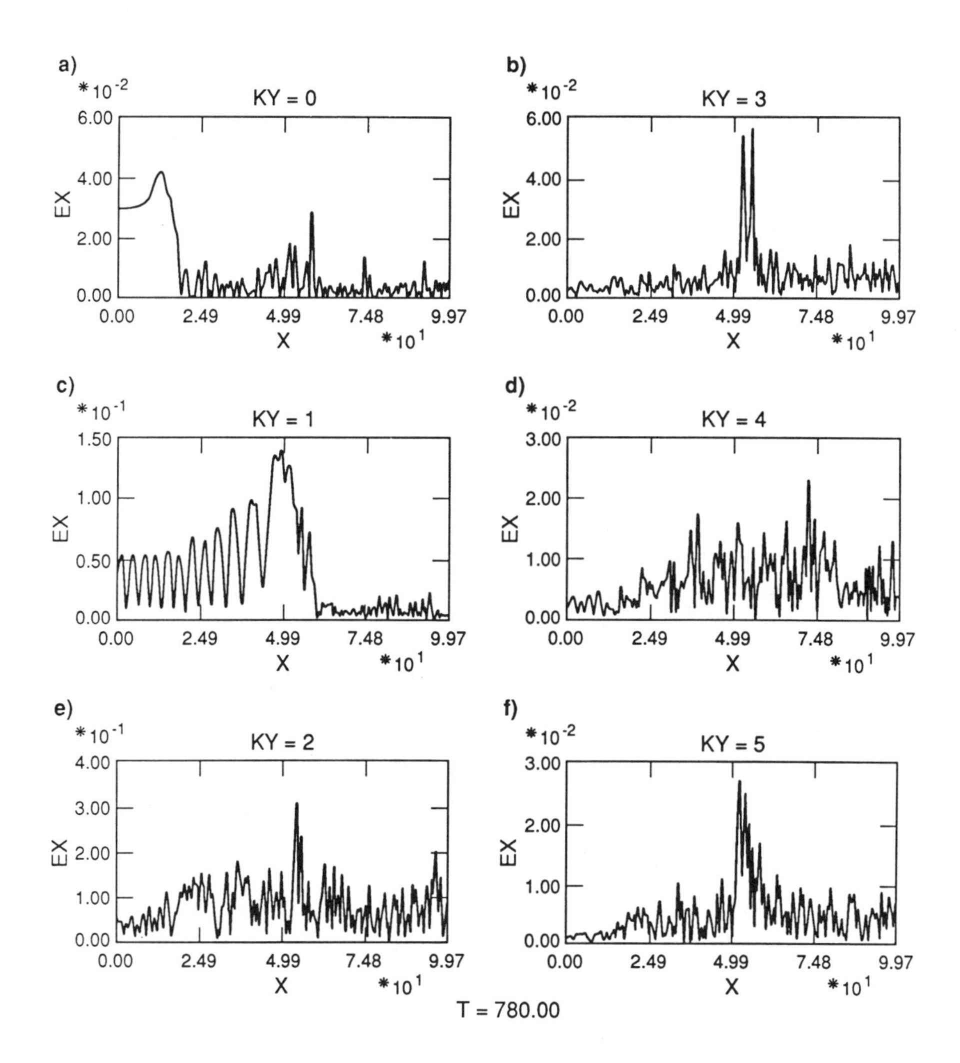

FIGURE 8 The Fourier components of the electric field component E_x along the density gradient at $t = 780\,\omega_0^{-1}$ for BR-026 which is initialized fully ionized. Labels and definitions are the same as in Fig. 6.

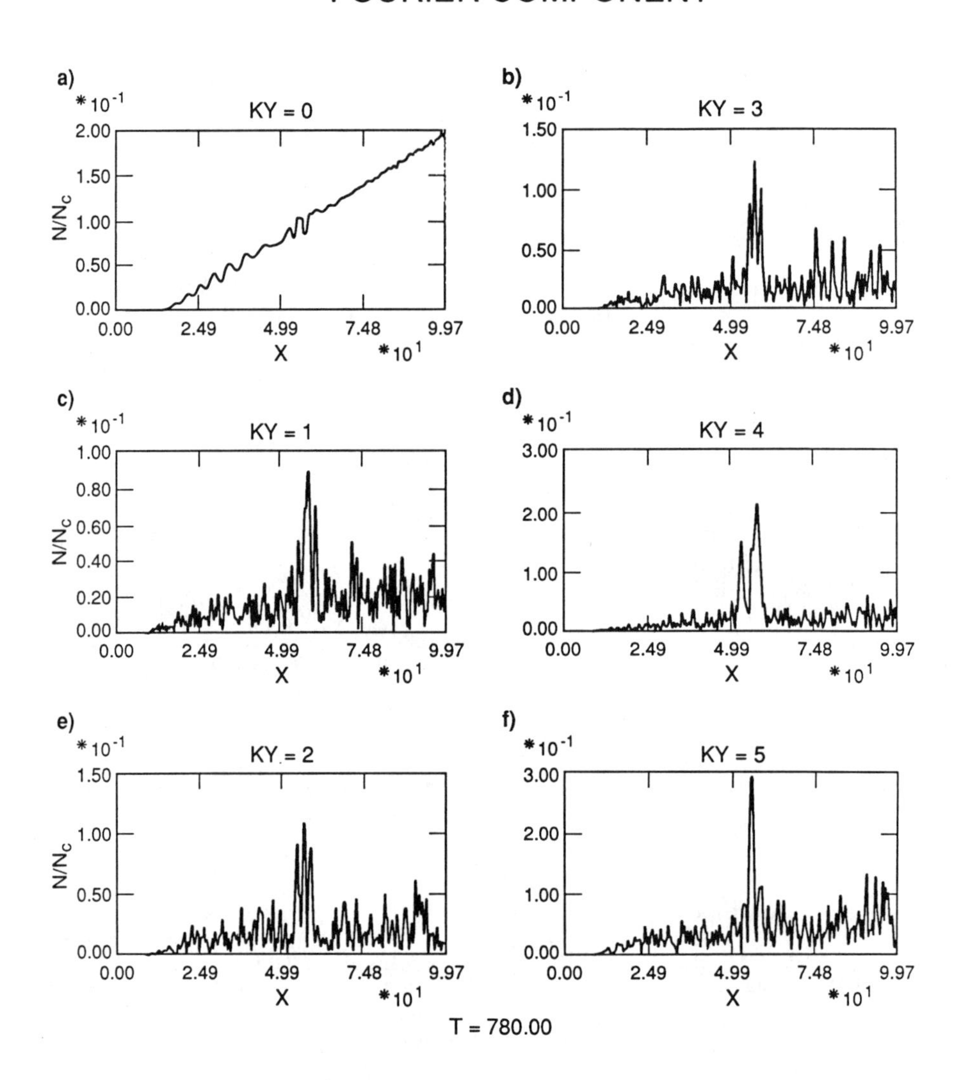

FIGURE 9 For the previous simulation at time $t = 780\omega_0^{-1}$, the Fourier components of the ion density as a function of position.

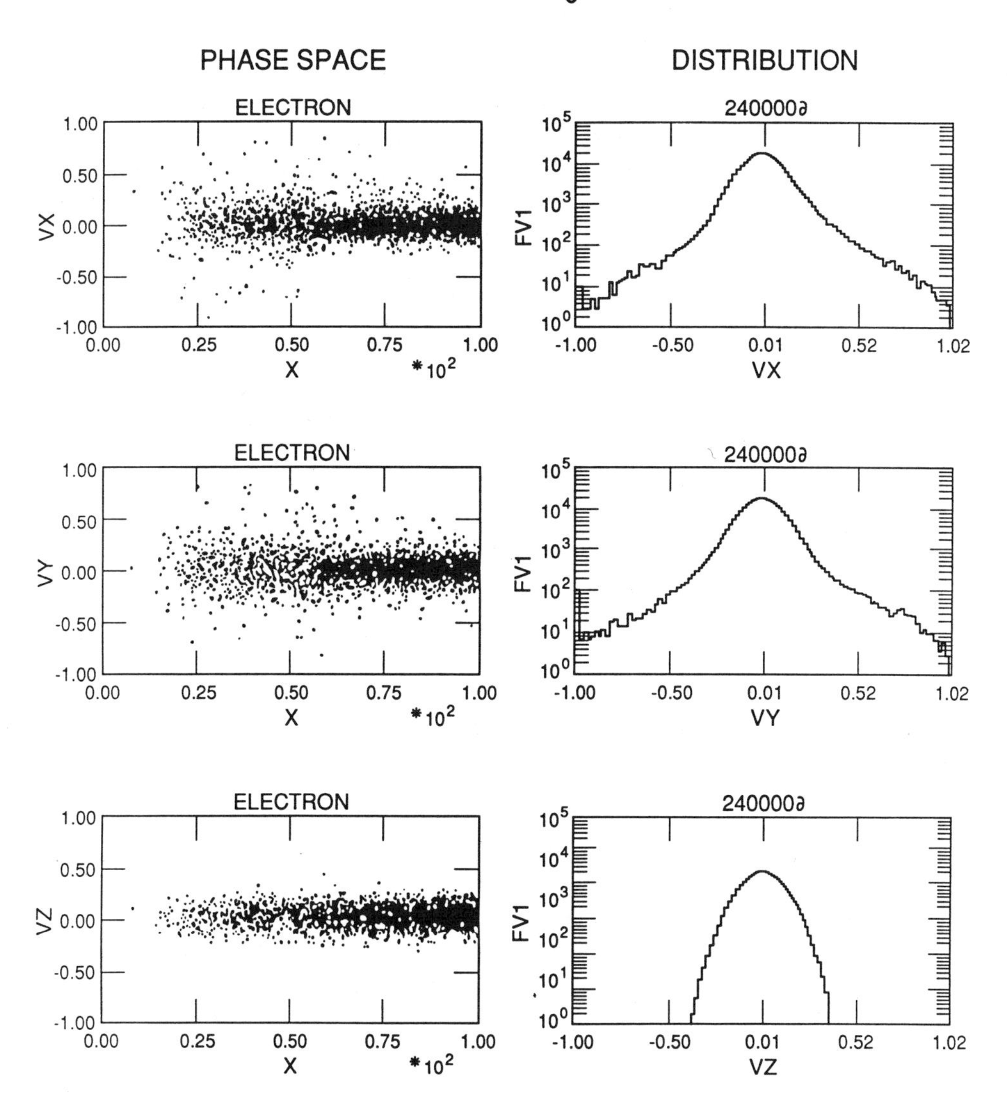

FIGURE 10 The electron phase space and electron distribution functions at $t = 780\omega_0^{-1}$ for the previous simulations. Labels and definitions are the same as in Fig. 7.

Summary

What we wanted to convey here was the current status of our developing modeling tools for high brightness laser problems and to give you some early results associated with collective processes when one has to include collisions, inverse Bremstrahling and ionization simultaneously. We have heard a number of talks on high brightness lasers and we want to emphasize that for many parameters of interest, resonant absorption is important; it produces hot electrons; it produces greatly enhanced absorption. One of the dilemmas in this community[20] has been to explain the observed high absorption in experiment. We have shown that in the presence of ionization that resonant absorption occurs; that the absorption wave is subject to weak parametric decay instability, that the ionization reduces the hot electron temperature and also reduces the parametric decay instabilities. Typical resonant absorption values reach 40%.

Acknowledgments

Special thanks are due for helpful discussions with J.C. Comly, L. Jones, G. Kyrala, G. Schappert, D. Casperson, and J. Cobble.

† Work supported by the U. S. Department of Energy.

References

1. J.M. Kindel, H. Okuda, and J.M. Dawson, "Parametric Instabilities and Anomalous Heating of Plasmas near the Lower Hybrid Frequency," *Phys. Rev. Lett.* **29**, 995 (1972).
2. J.M. Kindel, A.T. Lin, J.M. Dawson, and R.M. Martinez, "Nonlinear Effects of Ion Cyclotron Heating of Bounded Plasmas," *Phys. Fluids* **24** (13), 498 (1981).
3. D. Forslund, J. Kindel, W. Mori, C. Joshi, and J. Dawson, "Two Dimensional Simulation Studies of Single Frequency and Best Wave Laser Plasma Heating," *Phys. Rev. Lett.* **54**, 558 (1984).
4. C. Joshi, et al., "Ultra-high Gradient Particle Acceleration by Intense Laser Driver Plasma Density Waves," *Nature* **311**, 525 (1984).
5. J.M. Kindel, D.W. Forslund, W.B. Mori, C. Joshi, and J.M. Dawson, "Two Dimensional Beat Wave Acceleration Simulation," *Proc. of International Conference on Lasers, San Francisco, 1984*, a publication of the Soc. for Optical Quantum Electronics; also Los Alamos Report (LA-UR-85-233).

6. Hauer et al., “Current New Applications of Laser Plasmas,” in *Plasmas and Applications* ed. by L.J. Radziemski and D.A. Cremers, 1989.
7. W.B. Mori, C. Joshi, J.M. Dawson, D.W. Forslund, and J.M. Kindel, “Evolution of Self-Focusing of Intense Electromagnetic Waves in a Plasma,” *Phys. Rev. Lett.* **60**, 1298 (1988).
8. D.W. Forslund, K. Lee, J.M. Kindel, E.L. Lindman, and R.L. Morse, “Theory and Simulation of Resonant Absorption in a Hot Plasma,” *Phys. Rev.* A 11, 2, 679-683 (1975).
9. J.A. Cobble, G.A. Kyrala, A.A. Hauer, A.J. Taylor, C.C. Gomez, N.D. Delemater, and G.T. Schappert, “Kilovolt X-Ray Spectroscopy of a Subpicosecond-Laser-Excited Source,” *Phys. Rev.* A **39**, 454 (1989).
10. J.A. Cobble, G.T. Schappert, L.A. Jones, A.J. Taylor, G.A. Kyrala, and R.D. Fulton, “The Interaction of a High Inadiance, Subpicosecond Laser Pulse with Aluminum: The Effects of the Prepulse on X-Ray Production,” submitted to *Applied Phys. Lett.*
11. G.L. Olson, J.K. LaGattuta, and J.C. Comly, *Proceedings of the 2nd International Colloquium on X-Ray Lasers, York U.K. 17-21 September 1990* (to be published).
12. C.W. Cranfill, J.U. Brackbill, and S.R. Goldman, “A Time-Implicit Monte Carlo Collision Algorithm for Particle-in-Cell Electron Transport Models,” *J. Comput. Phys.* **66**, 239 (1986).
13. L.D. Landau and E.M. Lifshitz, *Quantum Mechanics*, 2nd ed. Pergamon, New York, p. 276, (1965).
14. L.Y. Keldysh, “Ionization in the Field of a Strong Electromagnetic Wave,” *Soviet Phys. JETP* **20**, 1307 (1965).
15. H.R. Reiss, “Effect of an Intense Electromagnetic Field on a Weakly Bound System,” *Phys. Rev.* **A22**, 1789 (1980).
16. K. Estabrook and W.L. Kruer, “Properties of Resonantly Heated Electron Distrubution,” *Phys. Rev. Lett.* **40**, 42 (1978).
17. K.G. Estabrook, E.J. Valeo, and W.L. Kruer, “Two-Dimensional Relativistic Simulations of Resonance Absorption,” *Phys. Fluids* 18, 1151 (1975).
18. D.W. Forslund, J.M. Kindel, K. Lee, E.L. Lindman, “Two-Dimensional Stability of an Electromagnetic Wave Obliquely Incident on a Nonuniform Plasma,” *Phys. Rev. Lett.* **34**, 193-197 (1975).
19. J.C. Adam and A. Heron, “Parametric Instabilities in Resonant Absorption,” *Phys. Fluids* **31**, 2602 (1988).
20. J. Delettrez, H. Chen, E. Epperlein, D.D. Meyerhofer, and S. Uchida, *Proceedings of the 20th Annual Absorption Conference, Traverse City, MI 9-13 July, 1990.*

B. Lembège,* J.N. Leboeuf, P. Liewer†, and M. Ashour-Abdalla††**
*CRPE/CNET, 92131 Issy-les-Moulineaux Cedex (France)
**Oak Ridge National Laboratory, Oak Ridge (USA)
†JPL, Pasadena (USA) ††UCLA, IGPP, Los Angeles (USA)

Professor Dawson's Pioneering Work in Computer Simulations of Space and Astrophysical Plasmas

Abstract

This paper is a non-exhaustive review of John Dawson's research work in the domain of space plasma simulation. Its includes some of his most significant contributions to the physics of auroral phenomena, magnetospheric physics, solar physics, and astrophysics. Professor Dawson has pioneered the development of new global MHD and particle codes and their application to the physics of space plasma phenomena. Finally, Professor Dawson's interest in the new architectures of highly parallel super-computers will be presented and illustrated by simulation results obtained recently.

Introduction

This paper presents a review of some of the illustrative simulation results that Professor Dawson brought to space plasma physics. Participating in this symposium to honor John Dawson is a significant honor in itself, but such a review is not an easy task because of the abundant and continuous flow of innovative ideas that Professor Dawson has provided, and of studies achieved through many years in various domains of space plasma simulations. Thus, this review looks more like a typical "cocktail" offered for Professor Dawson's birthday, than like a systematic review of all his work developed in this domain. We hope that the reader will forgive us for such a choice and for the "non-exhaustive" aspect of this review. We have chosen to concentrate only on a few important results that Professor Dawson has obtained and/or which he has largely contributed to, in order to emphasize his pioneering work in simulations of space plasmas; moreover, several of these works illustrate quite well his continuous effort spent on establishing a link between the fusion and space communities.

The content of this review has been structured as follows :

1. Global simulations of the time-dependent magnetosphere
2. Particle simulations of kinetic instabilities in space plasmas
3. "Forays" in particle simulations of astrophysical plasmas
4. Particle simulations of nonlinear magnetosonic waves and magnetosonic shocks in space plasmas.
5. Simulations of relativistic particles in a strongly magnetized plasma
6. Particle simulations on new architectures of highly parallel supercomputers: an application to magnetospheric shocks

Conclusions, illustrating the particular importance of numerical simulation in the domain of space plasma physics, will close this review.

Global Simulations of the Time-Dependent Magnetosphere

As far as the earth's magnetosphere is concerned, the lack of global information characteristic of laboratory experiments stems from the experimental inability to photograph the magnetosphere, and the theoretical inability to solve the nonlinear time dependent three-dimensional (3D) magnetohydrodynamic (MHD) and plasma equations describing the magnetosphere. Laboratory and (more appropriately here) numerical simulations can potentially bridge this gap and allow theorists to go beyond what was once referred to, by Roald Sagdeev, as the "Cartoon Approximation" in constructing global models of the earth's interaction with the solar wind.

In point of fact, circa 1977, Prof. Charles Kennel came back to UCLA from the USSR, where he had occasion to discuss results from laboratory experiments performed by Prof. Podgorny's group which attempted to reproduce the interaction of the solar wind with the earth[49]. Unexpectedly, these laboratory experiments were showing a different magnetospheric topology depending on the solar wind Alfven Mach number M_A. For $M_A < 2.5$, the classic Dungey (1961) picture of the magnetosphere could be reproduced, but for $M_A \geq 10$, the dayside reconnection region was moving to the polar cusps. Prof. Kennel then asked the seminal question: why not carry out similar experiments on the computer?

It so happened that we had developed a one-of-a-kind particle magneto-hydro-dynamic simulation code[30] which we thought could be readily modified to handle such situations. The first model was two-and-one-half dimensional (2-1/2D or two space and three velocities dimensions) with the solar wind modeled as a plasma stream in the $x-y$ plane, carrying its own northward or southward magnetic field if desired. The earth was modeled by a dipole magnetic field produced by a pair of oppositely flowing currents out of the plane in the z-direction, which were kept constant throughout the time-dependent simulations. In the particle MHD model, elements of the fluids are treated as finite size particles. The particle quantities such as position, mass and momentum are advanced in a Lagrangian way while the magnetic fields are calculated in an Eulerian manner. The position of each j particle making up the fluid is calculated by:

$$d\vec{r}_j/dt = \vec{v}_j \tag{1}$$

$$d\vec{v}_j/dt - -1/\rho\left[\vec{\nabla}P + 1/8\pi\vec{\nabla}B^2 - 1/4\vec{\nabla}\cdot\vec{B}\vec{B}\right] \tag{2}$$

The plasma is modeled as an ideal two-dimensional (2-D) MHD fluid with an adiabatic equation of state so that

$$T/n^{\gamma-1} = \text{constant} \tag{3}$$

The density is in turn expressed as

$$n(r_g) = \sum_j f(r - r_j) \tag{4}$$

where $f(r - r_j)$ is the form factor of the particle, typically a Gaussian one. The magnetic fields are updated at the mesh points as follows:

$$\partial\vec{B}/\partial t = \vec{\nabla}\times\left(\vec{v}_g\times\vec{B}\right) + \eta_{\text{num}}\nabla^2\vec{B} \tag{5}$$

where the fluid velocity $\vec{v}_g$ is defined as

$$\vec{v}_g(r) = \sum_j \vec{v}_j(r)f(r - r_j)/\sum_j f(r - r_j) \tag{6}$$

The resistive term η_{num} in the magnetic field equation (Eq. 5) is of numerical origin and is due to the fact that we use the diffusive Lax scheme to integrate the magnetic field in time. It gives rise to an effective Lundquist number of 10-20 for the magnetospheric simulations discussed here. This numerical resistivity was all but eliminated in subsequent improvements to the particle MHD model[7]. It turns out to be a blessing here, because it allows for easy reconnection of magnetic field lines.

Deceptively simple and highly resistive as it was, the computer model yielded a wealth of physics results[29,30], at that time unexpected. In the absence of solar wind magnetic field, an essentially closed "teardrop" magnetosphere is produced. Appearance of the density compression associated with the bow shock, the density maximum in the magnetosheath over the polar caps, the formation of a low density cavity in the downstream region, and the inhomogeneous high density region near and within the magnetopause, are all within theoretical expectations and in agreement with in-situ observations. More importantly, the simulations reveal that the shock front, magnetosheath, and magnetopause were fluttering with a time period and scale length that could be explained by a Kelvin-Helmholtz instability generated at the shock, the magnetopause, or both. This prediction is now generally accepted as an experimental fact.

With a steady southward solar wind magnetic field, the first question the computer model answered, by letting the time-dependent simulations evolve to a quasi-steady state, was the classical Dungey picture of the reconnected, open magnetosphere with neutral lines in the nose and tail persisted for $M_A \geq 10$, contrary to the laboratory experiments. With a steady-state northward solar wind magnetic field, a teardrop like magnetospheric configuration was produced with neutral lines over the polar cusps.

Since the topology of the magnetosphere depended on the solar wind field orientation, we postulated that a "substorm" was the time-dependent adjustment of magnetospheric topology resulting from the sudden rotation of the upstream solar wind magnetic field. A "substorm" would correspond to a southward rotation. We then proceeded to exploit fully the time-dependent nature of the simulations and modeled a substorm as the passage of a rotational continuity over the magnetosphere, which switches the solar wind field from east-west to southward. Results from such a simulation are shown in Figure 1, where magnetic field lines are plotted in the $x-y$ plane of the calculations. The initial solar wind magnetic field is everywhere uniform and parallel to the geomagnetic equatorial plane. At time $t=0$, the solar wind field suddenly rotates southward at $x=0$. The rotational discontinuity thereafter propagates inward along x at the speed of the fluid, which has a fast Mach number $M_F = 2.5$. In the top frame of Figure 1, the discontinuity has approached the nose of the magnetosphere and apparently initiated reconnection there. In the middle frame of Figure 1, the discontinuity is passing over the polar caps. Note that tail reconnection has been initiated and that a bubble of closed flux has formed downstream. In the bottom frame of Figure 1, the discontinuity has passed beyond the far edge of the diagram, and the magnetosphere has arrived at a Dungey configuration.

The most interesting features of our time-dependent simulations were the prediction of the formation of a bubble of closed flux which would propagate downstream during a magnetospheric substorm. This prediction was dramatically confirmed in 1984 by observations with the ISEE 3 spacecraft in the magnetotail 220

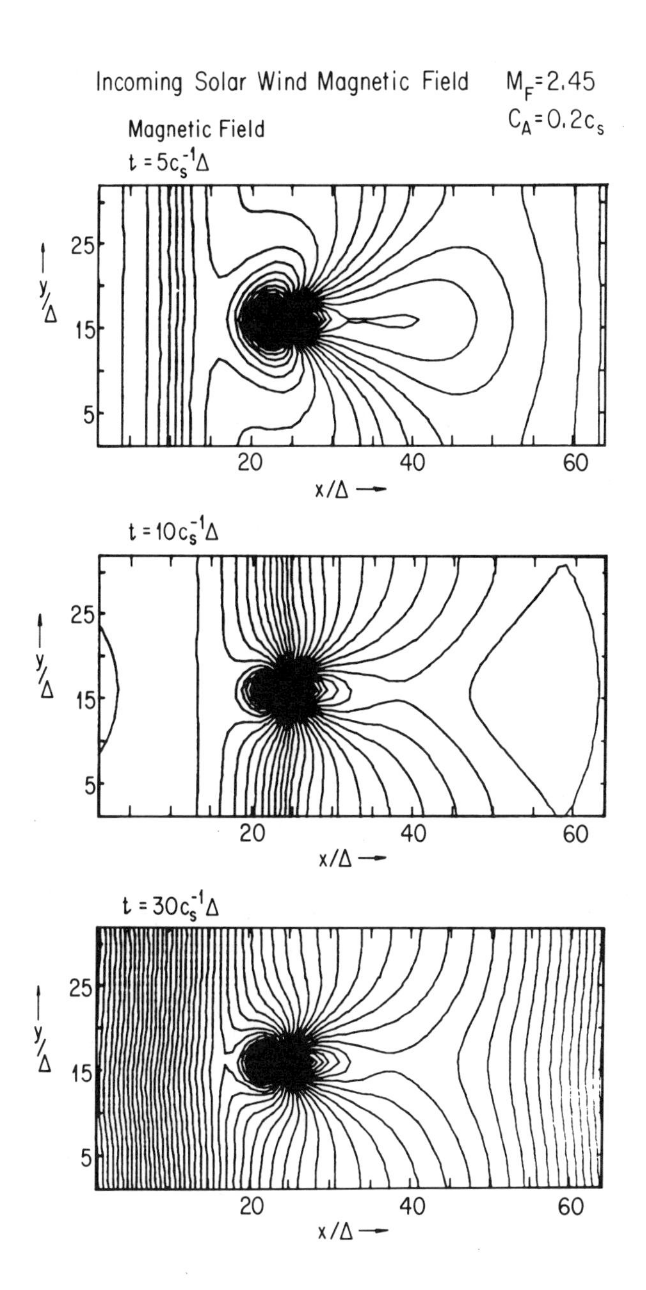

FIGURE 1 Global magnetospheric "substorm" simulation. These three insets show the evolution of the magnetospheric topology resulting from the passage over the magnetosphere of a rotational discontinuity which switches the solar wind field from east-west to southward.

RE from earth, of plasmoids (bodies of plasma and closed magnetic loops) passing that location in clear delayed response to substorms[21].

Our ability to model the time dependent magnetosphere in three dimensions was severely limited by the available computational resources. Our necessarily low resolution first attempt[32], using a three-dimensional extension of our particle MHD model, did, however, demonstrate the feasibility of such calculations. It was successful in reproducing the steady-state Dungey topology in three dimensions. It revealed some interesting features which were absent in our 2D simulations. In particular, it showed the formation of a compression zone downstream of the tail neutral line, probably bounded by wake shocks, whose cross-section was also changing with distance downstream.

These first global simulations of the magnetosphere have been followed by many others at UCLA and elsewhere. The activity has in fact grown into a field of research in its own right, supplementing today's more sophisticated means of experimental observations and more elaborate nonlinear theoretical methodology.

Particle Simulations of Kinetic Instabilities in Space Plasmas

One of the inherent difficulties of interpreting spacecraft observations related to waves and particle distribution functions in space plasmas is that the time resolution is too coarse to resolve the much shorter time scales over which kinetic instabilities might occur in the particular region of space under scrutiny. The ingenuity of the theorist is put to the test since one has to be able to infer from postulated initial conditions what the nonlinear saturated states of the plasma might be, given that the spacecrafts probe the plasma in its steady state. These problems can, however, readily be attacked by means of particle simulations.

Heating of cold electrons by odd half-harmonic cyclotron waves

Such a challenge was presented to us by what have come to be known as "odd half-harmonic" electron cyclotron emissions, since their frequency $\omega \approx (n + 1/2)\omega_{ce}$. Satellite observations of electric field fluctuations in the magnetosphere were showing intense oscillations at these frequencies during the diffuse aurorae[20].

Linear theory had shown that a combination of a weak loss cone distribution for hot electrons (of plasma sheet origin) and a Maxwellian distribution for the less dense, cold electron population (of ionospheric origin) was unstable for the growth of electrostatic electron cyclotron waves at frequencies $\omega \approx (n + 1/2)\omega_{ce}$.

Two saturation scenarios had also been advanced. The first relied on resonance broadening to neutralize the effect of the hot electron free energy source. The second advocated that the cold electrons would be heated nonlinearly and would change the nature of the instability from non-convective to convective, and that the waves would saturate by propagating out of the unstable region.

Our particle simulation study of these cyclotron harmonic waves was undertaken with two objectives in mind: first, to determine the saturation mechanism, and second, to see if the cold electrons could be heated and, if so, by what mechanism[1].

A standard 2-1/2D electrostatic finite size particle simulation code was used, with the electrons described by particles moving in the $x-y$ plane under the action of the Lorentz force and the ions constituting a neutralizing background of charges. The particle equations of motion for the electrons are:

$$dx/dt = v_x$$

$$dy/dt = v_y$$

$$d\vec{v}/dt = -|e|/m\left(\vec{E} + \vec{v}/c \times \vec{B}_0\right)$$

with $\vec{v} = (v_x, v_y, v_z)$ supplemented by a solution of Poisson's equation on a grid in (x, y)-space:

$$\vec{\nabla} \cdot \vec{E} = -4\pi\,|e|(n_e - n_0)\ ,$$

with the electronic number density accumulated in the same way as in Eq. 4 for the particle MHD code described earlier.

The simulation was started with a distribution known to be unstable to the growth of electron cyclotron harmonic waves. The hot electrons' free energy source was taken to be a hot ring distribution in velocity space, perpendicular to the magnetic field, with ring velocity $v_b = 5v_{te}$, i.e., five times the thermal velocity of the hot electrons (with Maxwellian distribution parallel to the magnetic field). The cold electrons component was chosen to have a density 25% that of the hot electrons and a Maxwellian velocity distribution function in all three velocity dimensions, with temperature 100 times colder than for the hot electrons. The magnetic field strength was chosen such that the ratio of electron cyclotron frequency to electron plasma frequency was $\omega_{ce}/\omega_{pe} = 0.3$.

The many results from these simulations are perhaps best illustrated by Figure 2, where the location of hot and cold particles are plotted in the velocity space $vz - vy$, perpendicular to the magnetic field in the x-direction. In the top frame of Figure 2, the dense ring of hot electrons and the dilute cold electrons (the central white dot) are clearly visible at the initial time. Between the top and middle frame of Figure 2, the wave energy increases linearly. Time series analysis shows that two waves are growing with frequencies $\omega_0 \approx 3/2\omega_{ce}$. Heating of the cold electrons, giving rise to the expanding white cloud, is apparent in the middle frame of Figure 2. It is difficult to detect any change at that point in the hot electrons' population. The waves saturate at that time, and thereafter decay slightly. The bottom frame of Figure 2 shows the distribution of magnetospheric electrons in the nonlinear saturated state, with the hot electrons having lost some of their free energy and the cold electrons having been heated further. There is still enough of a positive slope left in the hot electron distribution function for it to be unstable according to linear theory, indicating that growth is being balanced by some nonlinear damping mechanism.

The identification of the heating mechanism for the cold electrons and of the wave saturation mechanism led to some interesting detective work. Heating of the particles can only take place if the particles resonate with the waves. Initially, the cold electrons have a thermal velocity $v_{Tc} \approx 0.1 v_{Te}$, and the smallest resonant velocity with the waves (the cyclotron one) is $v_{\parallel} = (\omega_0 - \omega_{ce})/k_{\parallel} = 1.5 v_{Te}$, with $k_{\parallel}$ the component of the wavevector parallel to the magnetic field, so that the cold particles do not immediately resonate with the waves. The following scenario does, however, emerge. Initially the cold electrons are set into oscillations (no temperature increase) in both the perpendicular and parallel directions by the electric field of the growing wave. When $v_{\parallel}$ reaches a value such that cyclotron resonance can occur ($v_{\parallel} = 1.5 v_{Te}$), then the cold temperature should increase in the direction perpendicular to the magnetic field. The hot electrons do not participate in this interaction since the effect of the waves is averaged out over their larger Larmor radius. Careful temperature and particle orbit diagnostics actually confirmed this picture, and the calculated wave amplitude needed to bring the cold electrons into resonance did indeed match the measured value of the electric field in the simulation. We therefore concluded that the nonlinear onset of cyclotron resonance was the cause of saturation of the unstable electron cyclotron harmonic waves, and the cause of the heating of the cold particles.

Velocity Space Shell Instabilities

We applied essentially the same techniques to a study of electrostatic velocity space shell instabilities in magnetized plasmas[54]. This work was motivated by satellite observations which showed that dayside magnetospheric electron distributions detected at low latitudes, simultaneously with the occurrence of intense upper hybrid (UHR) noise, could take the form of hot shells in velocity space[28]. While these are more stable than ring distributions, addition of a colder background can render otherwise stable shell distributions unstable to cyclotron waves.

Our particle simulation model was identical to the one used for studying ring distributions with a colder background, and the same for the hot electron distribution which was modeled as a shell in velocity space. We found that both resonant and nonresonant (i.e., for which $k_{\parallel} = 0$) instabilities could occur. Saturation of the resonant instabilities was caused by nonlinear cyclotron resonance with the colder background (as for the ring distributions simulated previously). We ultimately found that the non-resonant instabilities were saturating due to non-stochastic cyclotron harmonic damping by the cold background.

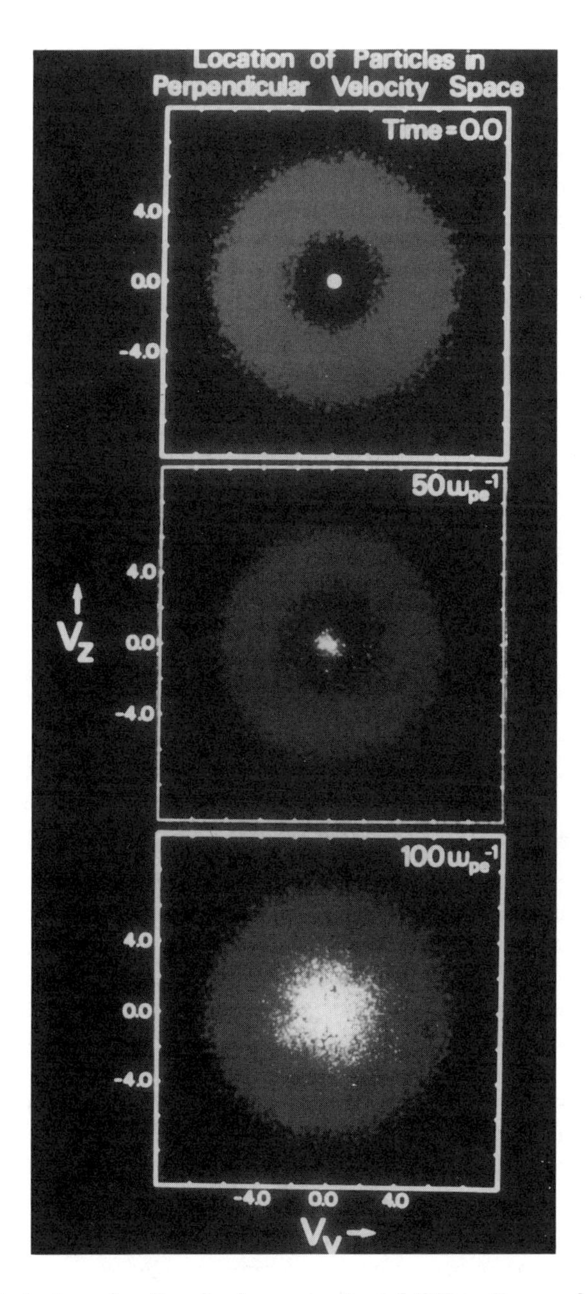

FIGURE 2 Cold electron heating by loss cone instabilities. Perpendicular velocity distribution of the cold (center) and hot particles: their $v_z - v_y$ phase space (a) at initial time, (b) at time of maximum wave amplitude, and (c) in the nonlinear saturated state.

While these simulations have intrinsic scientific merit, the reason they are mentioned here is somewhat different. It is mainly because in the search and in the identification of the saturation mechanism for the nonresonant shell instabilities, we literally stumbled across the $\vec{v} \times \vec{B}$ detrapping mechanism[11], which has had enormous impact on other works at UCLA. Even though $\vec{v} \times \vec{B}$ detrapping turned out to be irrelevant in the end for the shell instabilities we were studying, a detailed investigation of the mechanism in its relativistic rendition sparked the invention of the Surfatron concept of laser plasma acceleration by Katsouleas and Dawson[23]. Identification of the $\vec{v} \times \vec{B}$ detrapping mechanism as the dominant limiting factor in wave amplitude[34] has also lead to significant advances in the understanding of magnetosonic shocks in space and astrophysical plasmas as detailed in the following sections.

Application of particle simulation techniques to space plasmas has been pursued vigorously ever since at UCLA. The work is having an ongoing impact in the interpretation of spacecraft observations and in predicting the outcome of particular events or perturbative spacecraft experiments.

"Forays" in Particle Simulations of Astrophysical Plasmas

There is probably no other medium in plasma physics where experimental verification of theoretical ideas is more difficult, if feasible at all, than in astrophysical plasmas. It is also a medium where theoretical inventiveness can be exerted freely, and where one's intuition could be best guided or confirmed by numerical experiments.

One area of interest in the early 1980s concerned ultrarelativistic waves. They are ultrarelativistic in the sense that the quiver velocity, imparted by their electromagnetic field E oscillating at frequency $\omega \approx k_{\|}c$ to a particle of rest mass m_j, $eE/m_j\omega$, far exceeds the speed of light c. These waves are thought to occur in the outer magnetosphere of pulsars and it has also been argued that their magnetospheres may be largely positronic plasmas. The plasma can in any case be treated as an electron-positron one, since for ultrarelativistic waves the ion quiver momentum is large, and the difference between electron and ion rest masses becomes insignificant.

For reasons of analytical tractability, theoretical work had been limited to using nonlinear fluid equations, which only have a closed form when the plasma is assumed cold (zero temperature) in the presence of relativistic waves[24]. It is, however, unrealistic to assume this, since such strong waves are expected to cause highly nonlinear wave-particle interactions. It thus appeared to us that computer simulations using particles would provide a powerful tool to investigate such problems, particularly since no time-dependent kinetic analysis of nonlinear relativistic waves had yet been carried out.

Our fully self-consistent electromagnetic particle codes with relativistic dynamics were readily put to the task. These types of particle simulation models consist of the following particle equations :

$$d\vec{p}_j/dt = q_j\left[\vec{E} + \vec{v}_j/c \times \vec{B}\right] \tag{7}$$

with the momentum for each j particle related to the velocity by

$$\vec{p}_j = \gamma_j m_j \vec{v}_j \tag{8}$$

$$\gamma_j = \left(1 + p_j^2/m_j^2 c^2\right)^{1/2} \tag{9}$$

$$d\vec{r}_j/dt = \vec{v}_j \tag{10}$$

and field equations

$$\vec{\nabla} \times \vec{E} = -1/c\partial\vec{B}/\partial t \tag{11}$$

$$\vec{\nabla} \times \vec{B} = 1/c\partial\vec{E}/\partial t + 4\pi/c\vec{J} \tag{12}$$

$$\vec{\nabla} \cdot \vec{E} = 4\pi\rho \tag{13}$$

$$\vec{\nabla} \cdot \vec{B} = 0 \tag{14}$$

To be truly relativistic, the finite size particles should undergo Lorentz contraction. This would make the form factor $f(r - r_j)$ a function of velocity, as used in Eq. 4, for example. This complicates the calculations, but can be done if the particles are not strongly relativistic. If they are strongly relativistic, the Lorentz contraction can become so large that the particle size in the direction of motion becomes much smaller than the grid spacing. However, the interpolation scheme of the particles to the grid, on which the field evaluations are carried out, automatically expands the particle to a grid size, so true contraction beyond this is possible. To our knowledge, such corrections have not yet been included in any calculations; however, for strongly relativistic situations, the codes appear to give reasonably accurate results agreeing with theory where checks can be made.

Our first particle simulation work on ultrarelativistic waves in electron-positron plasmas was stimulated by concurrent research in the beat wave laser plasma accelerator concept. In this acceleration scheme, an intense, spatially localized pulse of laser light impinging on the plasma leaves behind a wake of plasma waves which, in turn, accelerate particles to energies estimated to be in the GeV range[55]. In an electron-positron plasma, the absence of charge separation precludes the existence of the wake of plasma waves, and the acceleration mechanism, if present, would have to be different. Our simulations of the propagation of a linearly polarized ultrarelativistic pulse in positronic plasmas[2] were performed in one space dimension but three velocities, and electromagnetic field directions, or 1-2/2 D. They show that particles can still be accelerated to very high energies. The acceleration stops when the bulk of the wave energy is converted to particle energy, with the maximum momenta achieved scaling as the square of the wave amplitude. The pulse leaves behind as a wake a vacuum region whose length scales as the amplitude of the wave. The results can be quantitatively explained by a simple snow plow or piston-like action of the radiation on the plasma, thus confirming the applicability of particle simulations to ultrarelativistic situations.

The second application concerned the existence of a nonlinear sawtooth wave solution, predicted by nonlinear fluid theory to be the likely outcome in steady-state of the interaction of ultrarelativistic waves with overdense electron-positron plasmas[24]. The following simple argument leads one to expect a sawtooth wave in steady state: for relativistic amplitudes, the current density is a square wave, which by integration of Ampere's law gives rise to a triangular, or sawtooth-shaped, magnetic field. The same code as was used for pulses can be applied to such initial, spatially extended waveforms. An electron-positron plasma with equal initial temperatures, cold or warm, is considered. The initial transverse electric and magnetic fields and particle momenta are chosen self-consistently. The input waveform is a sawtooth. The wave thereafter evolves self-consistently in space and time. The simulations show that the initial sawtooth (the nonlinear solution of the fluid equations) is destroyed in roughly one wave period, and evolves into an unsteady wave modulated by spiky density and current perturbations. This is also true for a sinusoidal input waveform. Intense particle acceleration in the direction of propagation takes place, intense plasma heating is observed, and the system reaches a quasi-stationary state consisting of a heated forward-propagating plasma containing a complex wave spectrum. A snapshot in this evolution is displayed in Figure 3, where the top frame clearly shows forward acceleration in the x component of the particle momenta. The initial sawtooth wave shape is barely recognizable in the y component of the momenta and in the z component of the transverse magnetic field.

Preliminary though they may have been, the ultrarelativistic pulse and sawtooth simulations nonetheless showed that the original wave energy was rapidly converted to ultrarelativistic particles and waves. While pulses might be observable from the radiation of their accelerated particles in the pulsar environment, or, during star collapse to a neutral star or black hole, it seems likely that a single wave of immense amplitude will not exist long enough to be observable.

Application of particle simulation techniques to astrophysical situations clearly deserves to be pursued further. As limited as our attempts have been, the work is, however, proving useful in the interpretation of recent results showing acceleration of electrons and ions to relativistic energies by large amplitude, steepening magnetosonic waves.

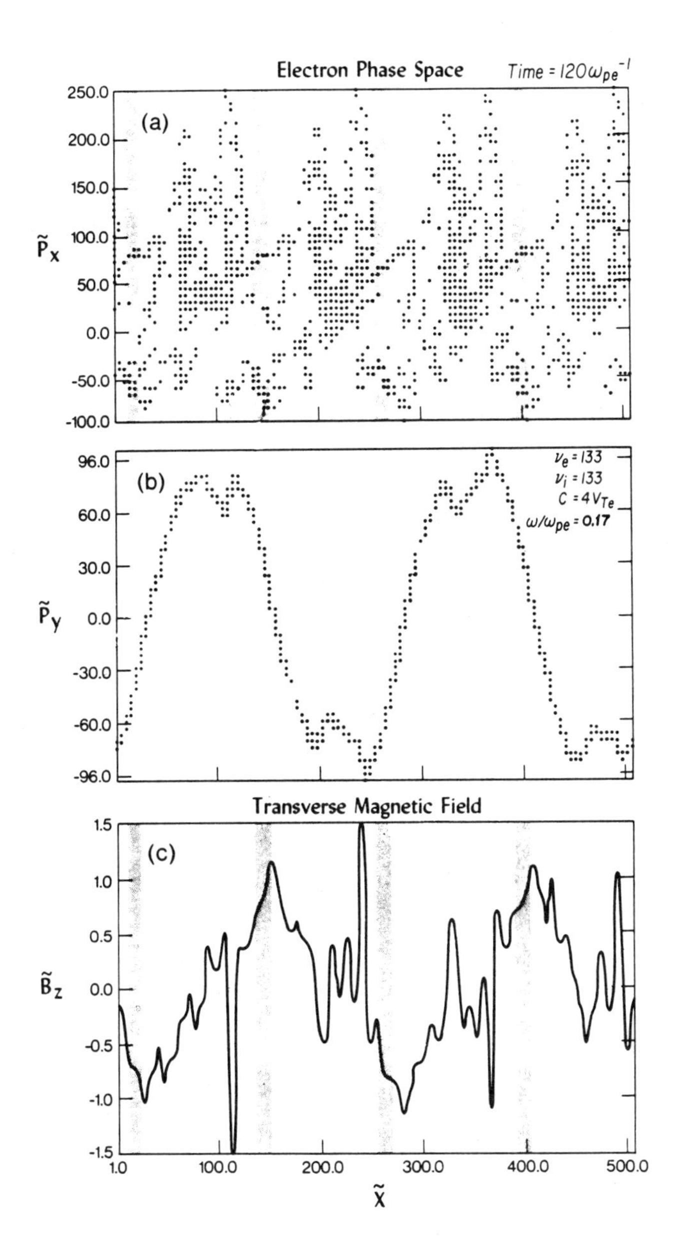

FIGURE 3 Ultrarelativistic sawtooth wave. Snapshots of the system at $\omega_{pe}t = 120$. (a) electrons $p_x - x$ phase space; (b) electrons $p_y - x$ phase space; (c) wave magnetic field Bz as a function of distance.

Particle Simulations of Nonlinear Magnetosonic Waves and Magnetosonic Shocks in Space Plasmas

Particle acceleration and heating by low frequency waves ($\omega_{ci} < \omega < \omega_{lh}$)

One of the main applications of this study (initially developed for fusion purposes) to magnetospheric physics consists in establishing a possible correlation between the presence of ULF waves and the high energy ion distribution function simultaneously observed on board the GEOS satellite[17]. Let us remember that GEOS was the first spacecraft on which a sensitive three-component ULF antenna operated in a waveform mode in the whole frequency range from $\sim$ 0.1 Hz to 10 Hz, i.e., above and below the proton frequency f_{ci} ($f_{ci} \simeq 1$ Hz at the geostationary orbit). In particular, two correlations have been found, as reported in Gendrin's review (1985) :

a. During some magnetospheric wave events, the energetic proton distribution shows a structure in which there is a flux maximum both at a certain energy (of the order of 10 kev) and at 90° pitch angle[48]. Such a distribution can be roughly modeled by a ring of energetic protons to which an isotropic cold proton distribution is added. The frequency increase which occurs during the emission is probably due to the inward diffusion of the ring-like distribution, the wave propagating outward to the spacecraft. This could be consistent with the fact the emission starts at the local proton gyrofrequency, provided that it could be demonstrated that such waves can propagate only towards regions of lower gyrofrequencies.

b. Korth et al.[26,27] have also observed magnetosonic waves (associated with wide band impulsive noise) in conjunction with displacements of energetic proton boundaries. The radial gradient in the energetic proton distribution is determined by the large east-west asymmetry observed during one spin period; its characteristic length is of the order of a few gyradii ρ_i for $\sim$ 20 keV protons.

In short, magnetosonic waves are often observed in the equatorial magnetosphere, though their intensity is usually not large ($\leq 0.3\,nT$). Since they are generated near the proton gyrofrequency, they could resonate with the thermal plasma provided they have a small but finite parallel wave number. Therefore, they could be the cause of the heating of the thermal plasma of ionospheric origin to a few eV, as is often observed[12].

In order to establish the possible correlation between the ion ring distribution and magnetospheric waves, a 1-2/2D fully electromagnetic, particle code (for both electrons and ions) has been used, where a magnetosonic wave is generated by applying an external (non-plasma) current $j_{oy} = j_o \sin(\omega t - kx)$ during the whole time length of the run and within the whole simulation box[34,37,38].

The frequency w and the wave-number k of the wave are chosen within the frequency range $\omega_{ci} < \omega < \omega_{lh}$, where ω_{lh} is the lower hybrid frequency. The nonlinear level of the pump wave is determined by the chosen value of the amplitude $|j_0|$. Two cases may be differentiated, due to the quite different dynamics of ions and electrons, according to the propagation angle $\theta = \left(\vec{k}, \vec{B}_0\right)$ between the wave vector and the magnetostatic field:

Perpendicular propagation ($\theta = 90°$)

Numerical results have shown that a longitudinal electrostatic field $\tilde{E}_{lx}$ grows during the build up phase of the wave (due to the self-consistent space charge effects), and attains a nonlinear level which strongly distorts its shape so that many higher harmonics are produced. Ions are accelerated by this field and become trapped as soon as this acceleration is large enough, so that the velocity of some accelerated ions reach the phase velocity $\tilde{v}_{\phi x}$ of the wave ($\tilde{v}_{xi} \simeq \tilde{v}_{\phi x}$), as evidenced in Figure 4; trapping loops clearly appear in Figure 4(i). This ion trapping has various consequences:

i. It enhances the wave overtaking both electrostatic and electromagnetic components (the wave crests overtaking the wave troughs).

ii. It produces a large wave damping.

iii. While trapped, ions suffer an important $\vec{E} \times \vec{B}$ drift parallel to the wave front (y direction), which becomes large enough so that the detrapping force $q\left(\vec{v} \times \vec{B}\right)$ balances the trapping force $q\vec{E}$; detrapped ions are ejected on a large Larmor orbit within the plane perpendicular to $\vec{B}_0$ and form a ring distribution.

iv. This global 3-steps mechanism , "acceleration-trapping-detrapping", is the source of an important ion heating perpendicular to $\vec{B}_0$; the energy transfer from the wave to the particles is well illustrated by the fact the ion heating becomes particularly large at the time the magnetosonic wave (both electrostatic and electromagnetic components, respectively named $\tilde{E}_{lx}$ and $\tilde{E}_{ty}$, $\tilde{B}_{tz}$) saturates because of the nonlinear damping (Figure 5). In contrast, electrons only suffer a poor heating by adiabatic compression, as seen in Figure 4.

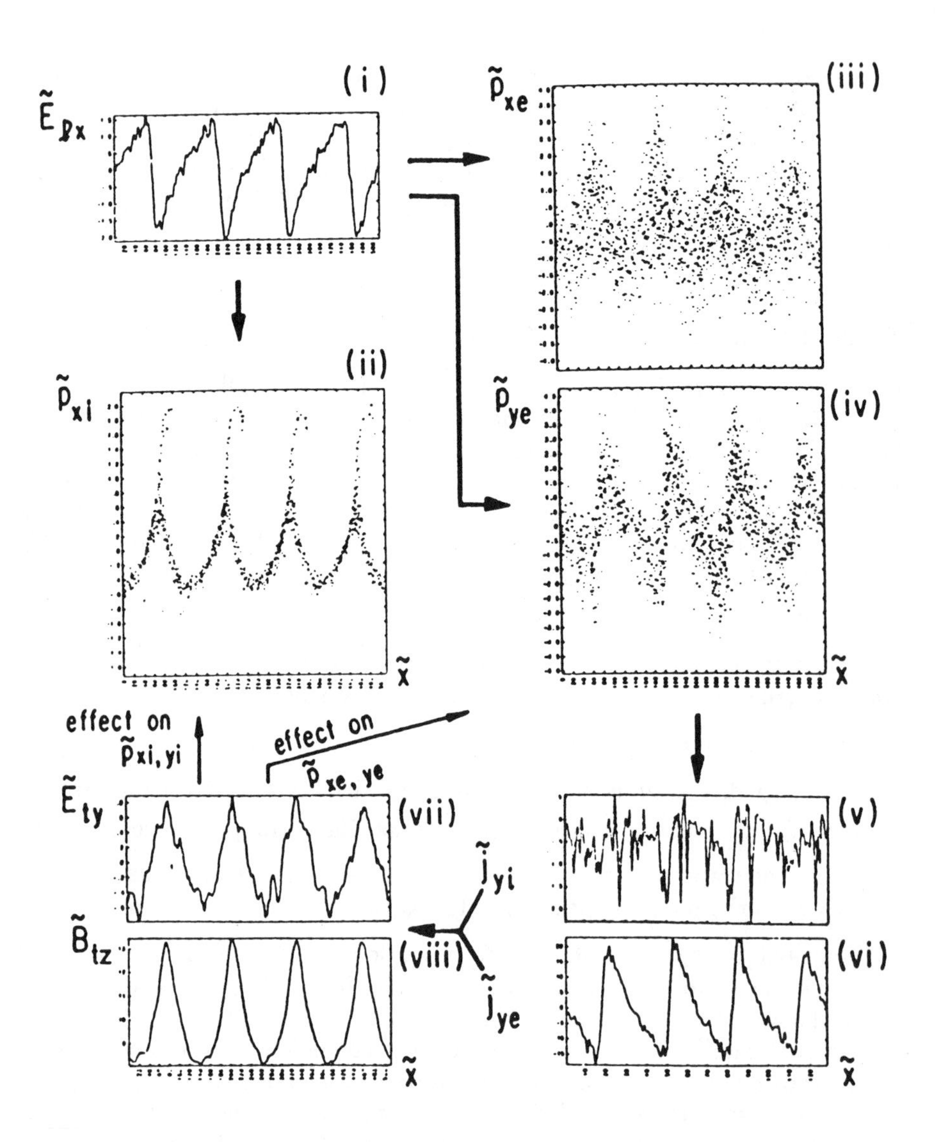

FIGURE 4 Numerical results illustrating the steepening of a large amplitude magnetosonic wave at time $\tilde{\tau}_{ci}$; $\tilde{E}_{lx}$ is the longitudinal electrostatic component; $\tilde{E}_{ty}$ and $\tilde{B}_{tz}$ are the transverse electromagnetic components. Ion trapping loop clearly appears in plot (ii); electrons ($\tilde{p}_{xe}$, $\tilde{p}_{ye}$) only suffer an adiabatic compression heating.

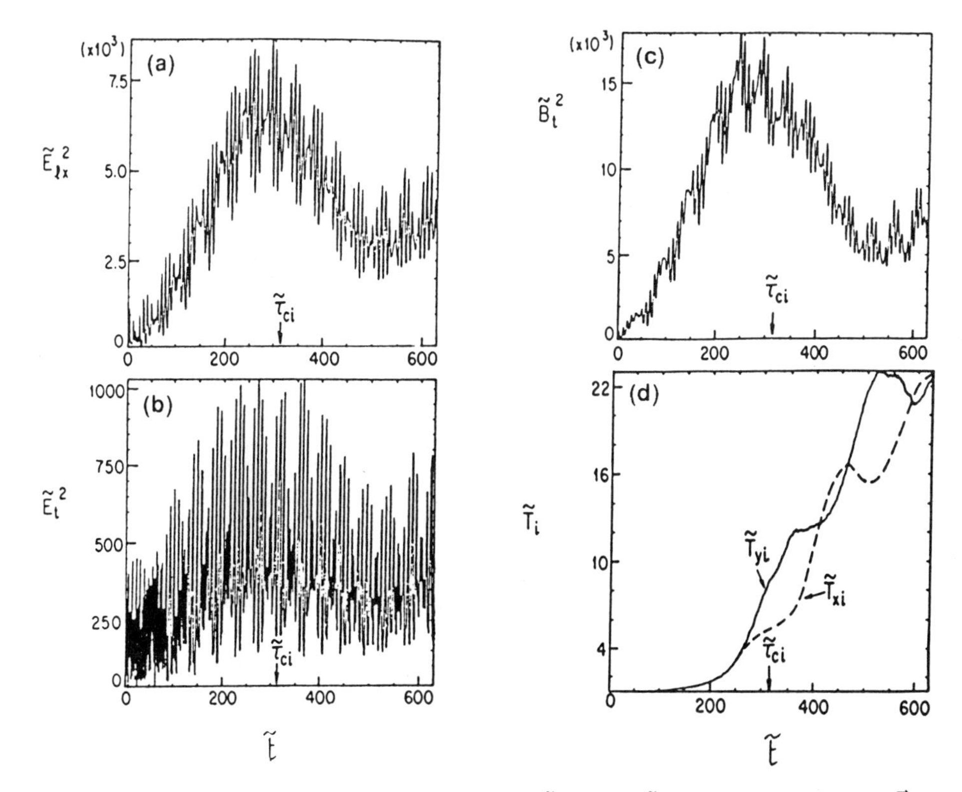

FIGURE 5 Time history of ion temperatures $\tilde{T}_{xi}$ and $\tilde{T}_{yi}$ perpendicular to $\vec{B}_0$ (a - c), and of field energies; subscripts "l" and "t" hold for longitudinal and transverse components, respectively. $\vec{B}_0$ is along z axis.

The formation of the ion ring is clearly evidenced in the 2D projection $(\tilde{p}_{xi}, \tilde{p}_{yi})$ of the 3-D plot of Figure 6 representing the ion phase space $(\tilde{p}_{xi}, \tilde{p}_{yi}, \tilde{x})$; the static field $\vec{B}_0$ is along z. The resulting ion heating is related to the coherent motion of ions (non-stochastic heating) while they gyrate along the large Larmor orbit. Numerical results show that this overall mechanism is still efficient for any wave frequency ω within the range $\omega_{ci} < \omega < \omega_{lh}$ (without any resonance constraint as $\omega = n\omega_{ci}$ or $\omega = \omega_{lh}$), provided a sufficient intensity of the wave is reached. Such results emphasize that an ion ring can easily result from the interaction of magnetosonic waves with an ambient plasma. However, experimental results obtained from the GEOS have stressed that possible sources of magnetosonic waves could be the ion ring distribution itself (or steep radial gradients in the hot particle distribution) as noted by Gendrin (1985). Then, one wonders if the resulting ion ring observed in the numerical simulations can lead to a "feed back" effect by relaxation, and maintain a certain level of the magnetosonic wave. This further point is still an open question, since this mechanism could not be observed in the present results obtained with a 1-2/2 D code (i.e., 1D real space); further simulation using 2-1/2 D code (i.e., 2D real space) would be necessary to study the consequences of the ion ring relaxation and the new mechanisms of energy dissipation related to the additional dimension in the real space.

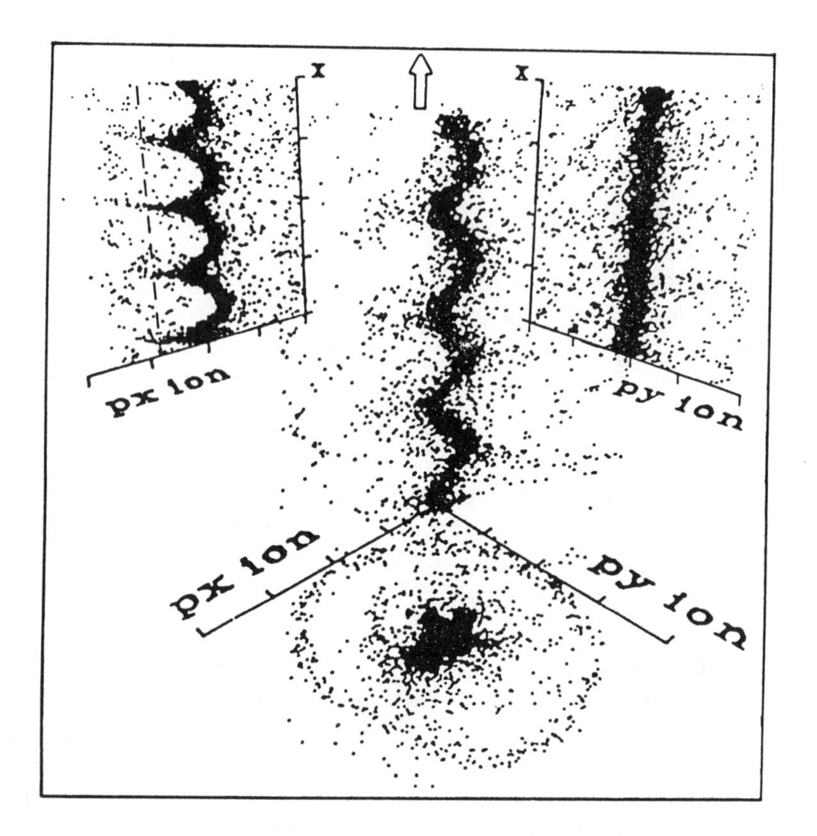

FIGURE 6 Three dimensional plot of the ion phase space at time $\tilde{t} = 1.8\,\tilde{\tau}_{ci}$ (i.e., after the ion trapping ad detrapping have started), where $\tilde{\tau}_{ci}$ is the ion gyroperiod. The 3-D plot itself is in the central part of the picture where each point represents the position $(\tilde{p}_{xi}, \ \tilde{p}_{yi}, \ \tilde{x})$ of an ion. The three other plots represent the projections of the central plot on 2-D planes respectively $(\tilde{p}_{xi}, \tilde{x})$, $(\tilde{p}_{xi}, \tilde{x})$, and $(\tilde{p}_{xi}, \tilde{p}_{yi})$. $\tilde{p}_{yi}$ and $\tilde{p}_{yi}$ are the momenta ($\simeq$ velocities) components. The arrow on the top indicates the direction of the wave propagation along x (from the bottom to the top of the figure); the dashed line is the phase velocity of the wave.

Oblique propagation ($90° > \theta > 45°$)

The next step, of course, was to determine the efficiency of the mechanism quoted above when q departs from 90°. Similar plasma conditions have been used with the same simulation code (1-2/2 D) in order to study the interaction of a "pseudo-oblique" magnetosonic wave with the ambient plasma[35]. The overall mechanisms may be summarized in terms of two new results:

i. A strong ion heating and acceleration perpendicular to $\vec{B}_0$ will persist within an angular range $90° > \theta > \theta_{ti}$, where θ_{ti} is a critical angle below which the ion dynamics drops to very low level; θ_{ti} was found to vary as $\pi/2 - \sqrt{m_e/m_i}$.

ii. Intense electron heating and acceleration take place for θ below a second critical angle θ_{te} (Figure 7). Such particle dynamics may be explained both by the dispersion properties and by geometrical effects. Indeed, both the phase velocity $v_{\phi x}$ and the group velocity defined for a given wave number increase when θ decreases; this makes the ion trapping conditions $(v_{xi} \simeq v_{\phi x})$ more stringent than for $\theta = 90°$, and implies stronger ion heating if ions are trapped. For $\theta \leq \theta_{ti}$, ion trapping cannot take place any more and resulting ion heating drops to very low values. However, the situation largely differs for the electrons; as θ decreases, the phase velocity parallel to $\vec{B}_0$, $v_{\phi\|}$, (which is infinite for $\theta = 90°$) decreases, while a small but finite component $E_{l\|}$ of the electrostatic field builds up. This results in appropriate conditions for electron acceleration parallel to $\vec{B}_0$ and further trapping. For $\theta < \theta_{te}$, more and more electrons begin to be trapped, and some electron heating increases as θ decreases ($v_{\phi\|}$ also decreases, which makes easier electron trapping), at the expense of the wave amplitude, which suffers an important additional damping. Then the electrons to a large extent short out the E_{lx} field which, combined with the increasing $v_{\phi x}$ (more difficult ion trapping) and an increasing $E_{l\|}$ (easier electron trapping), means that ion trapping is no longer possible below the critical angle θ_{ti}. As a consequence, three regimes may be defined according the angle θ (Figure 7): the main part of the wave energy is transferred, (a) to ions for θ around 90°, (b) to both electrons and ions for $\theta_{te} > \theta > \theta_{ti}$ (with a particularly large energy transfer to electrons parallel to $\vec{B}_0$ since $\tilde{T}_{ze} \gg \tilde{T}_{xe}, \tilde{T}_{Ye}$) and (c) only to electrons (isotropic heating, since $\tilde{T}_{xe} \simeq \tilde{T}_{ye} \simeq \tilde{T}_{ze}$) for $\theta < \theta_{ti}$. In particular, an ion ring distribution persists as long as a certain number of accelerated ions succeed in being trapped and detrapped at later times within $90° > \theta > \theta_{ti}$.

It is of interest that the dissipation mechanisms discussed here apply to a broad range of frequencies, when again no particular resonance is involved ($\omega_{ci} \ll \omega < \omega_\theta$, where $\omega_\theta = \omega_{lh}$ and ω_{ce} for the particular cases $\theta = 90°$ and $\theta = 0°$, respectively). These mechanisms can be involved in the heating of many space plasmas (or in fusion systems); either ions or electrons or both can be heated.

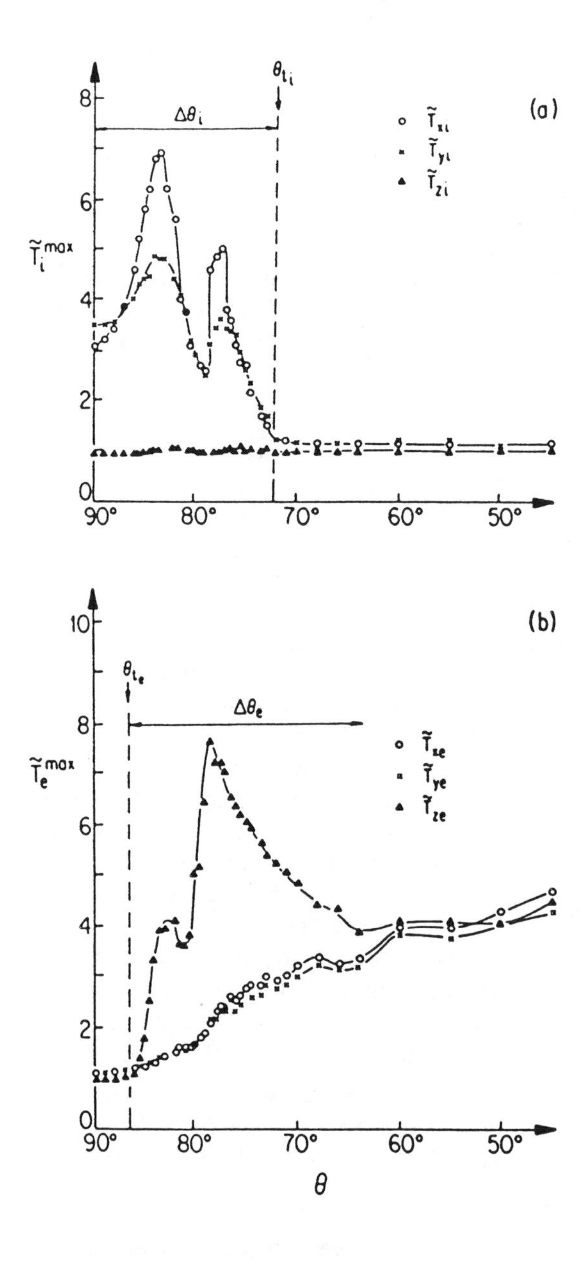

FIGURE 7 Plots of the maxima values of temperatures measured for ions (a) and electrons (b) along x, y, and z directions. These values are deduced from plots of ion temperature versus time obtained for each given angle θ; $\vec{B}_0$ is along z for $\theta = 90°$ and is defined by $\vec{B}_0 = (B_{0x}, 0, B_{0x})$, for $\theta \neq 90°$.

Magnetosonic shock

A considerable amount of both experimental and numerical studies have been devoted to magnetospheric shocks, in particular to the terrestrial bow shock located upstream of the magnetopause, as evidenced in previous reviews and monographs[59, 5,53,25,18,50,47,9]. This considerable effort has been greatly motivated by the large amount of rich experimental data gathered on board ISEE satellites. Since previous analysis has shown that shock behavior was mainly dominated by ion dynamics, a large part of the numerical effort has been based on the use of hybrid codes (electrons: massless fluid; ion: particles) and has led to a net improvement in the understanding of shock physics. Such codes are characterized by the fact that electron dynamics (i.e., the corresponding space and time scales) are neglected and the anomalous resistivity (generated by cross-field current instabilities) is included by a phenomenological ad hoc parameter, coupling the electron fluid equations and the ion particles pusher (Lorentz force). Such a procedure allows us to follow the shock dynamics over long ion time scales, including large ion space scales, without following the electrons' dynamics in detail [42,43,60,51,32]; this leads to important savings in computational cost.

Professor Dawson was particularly interested in studying the shock dynamics but by using a fully particle code. In collaboration with Lembège, he started to study shocks in conditions where the electrostatic field is mainly dictated by space charge effects ($\omega_{pe}/\omega_{ce} \leq 1$), which is in contrast with the use of hybrid codes including the quasi-neutrality condition (approximation valid for $\omega_{pe}/\omega_{ce} \gg 1$). This point was of importance, since subcritical and supercritical shocks, respectively defined by $\mathrm{M_A} < \mathrm{M_A^*}$ and $\mathrm{M_A} > \mathrm{M_A}*$, (where $\mathrm{M_A^*}$ is the first critical Mach number) can be mainly characterized by important resistive and viscous dissipation, respectively. How viscous dissipation varies from high to low $\mathrm{M_A}$ was not clarified until the mid 80s; in other words, since viscous dissipation is strongly associated with ion reflection, one wonders how the amplitude of the electrostatic field at the shock front (which is responsible for this reflection) evolves for varying $\mathrm{M_A}$. Such a question also leads to the necessity for comparing the shock features and the associated particle dynamics, in cases where the electrostatic field is strictly self consistent (particle code) or is imposed by $N_e = N_i$ condition (hybrid code). Another motivation was to determine carefully the characteristics of an oblique magnetosonic shock ($90° > \theta > 45°$), where the electron dynamics (in particular parallel to $\vec{B}_0$) are expected to play an important role[15].

Numerical studies of shocks, based on the use of a 1-2/2D fully electromagnetic particle code started at UCLA in 1984[35]. As a first step, dynamics of magnetosonic shocks and of both associated particle species (electrons and ions) were studied in detail for a direction of propagation perpendicular to $\vec{B}_0$. The main numerical results obtained for various regimes of the shock may be summarized as follows:

i. For a supercritical shock ($\mathrm{M_A} > \mathrm{M_A^*}$), an important number of upstream ions behave in a way similar to that quoted in the previous section, and suffer a "3-step mechanism" (acceleration-trapping-detrapping) at the shock front, which plays the role of an electrostatic as well as a magnetic barrier; analogous ion dynamics are also observed at the overshoots of the downstream trailing wave-train, where the local electric field can be large enough to trigger some ion trapping. This interpretation of the ion dynamics through the shock front is not only a semantic difference with respect to the ion reflection mechanism commonly referred to in previous studies; rather, it offers the possibility for determining carefully the resonant interaction range $\Delta\tilde{t} = \tilde{t}_{dt,i} - \tilde{t}_{tr}$ (between ion detrapping and trapping times), and then can be used as a support for determining the corresponding local wave damping and for a possible confrontation with analytical calculations.

Reflected ions describe very large Larmor orbits and form a ring distribution; a large rapid non-stochastic ion heating results from this ion gyration. This heating is the main source of viscous dissipation and is responsible for large field damping. Reflected ions accumulate in front of the shock during their large gyration before they get enough energy to be able to cross the shock front, and propagate in the downstream region. This continuous accumulation leads to the formation of a foot which, in the present case, can grow large in time and reach the level of the first magnetic overshoot; then, it plays the role of a new front (Figure 8). While propagating, the shock front develops a self-reformation similar to that already observed by Biskamp and Welter[6], This self-reformation mechanism not evidenced in the hybrid simulations is still a subject of controversy. In summary, three ion distributions can be distinguished (Figure 9): a Maxwellian distribution of the upstream ions (peak P_1 of the unperturbed plasma), a drifted Maxwellian distribution of the directly transmitted ions (which succeeds in passing freely through the shock, and form the peak P_2), and a ring distribution of the reflected ions. At the end of the self-reformation cycle, a new ion ring starts building up, as evidenced in Figure 9. In contrast, electrons poorly interact with the electrostatic field, since $v_{thi} \ll v_{the} < v_\phi$ at time $\tilde{t} = 0$; they only suffer a weak adiabatic compression effect mainly at the locations where the ion density becomes large, i.e., mainly at the foot (arrow f), at the shock ramp, and at the overshoots of the downstream wavetrain (Figure 10) .

ii. For a subcritical shock ($M_A < M_A^*$), the situation differs completely: the density of trapped-reflected ions drops to a low (but finite) value, leading to a broad ion distribution function with a weak tail (instead of an ion ring). A weak adiabatic bulk ion heating results. The self-reformation of the shock does not take place clearly. Indeed, since the amplitude of the foot is much weaker (less reflected ions), its velocity is also weaker. Then, the foot cannot separate clearly in time from the front itself, which overtakes it. In summary, the transition between high and low Mach regimes takes place through a certain range of M_A (instead of one critical value M_A^*), through which viscous dissipation (and associated perpendicular ion heating) drops from strong to weak values.

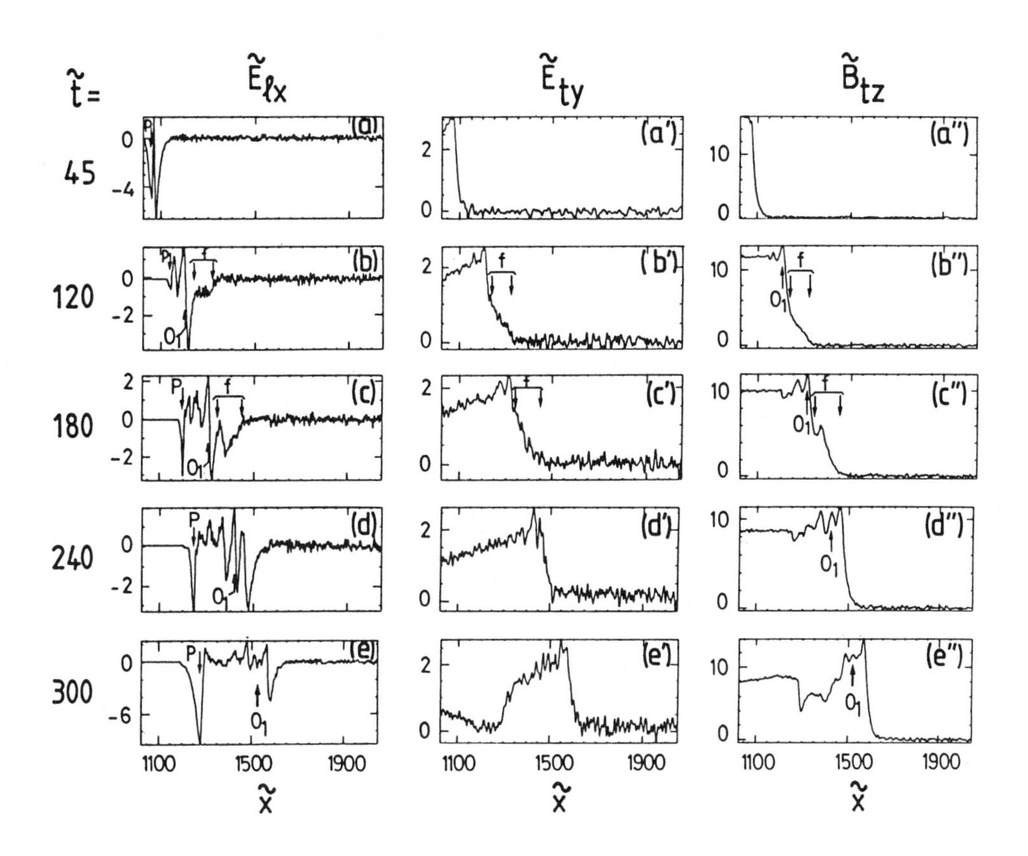

FIGURE 8 Space profiles of the longitudinal field $\tilde{E}_{ty}$ and $\tilde{B}_{tz}$ at different times; P, O_1, and f indicate the locations of the piston, the first overshoot, and of magnetic foot, respectively.

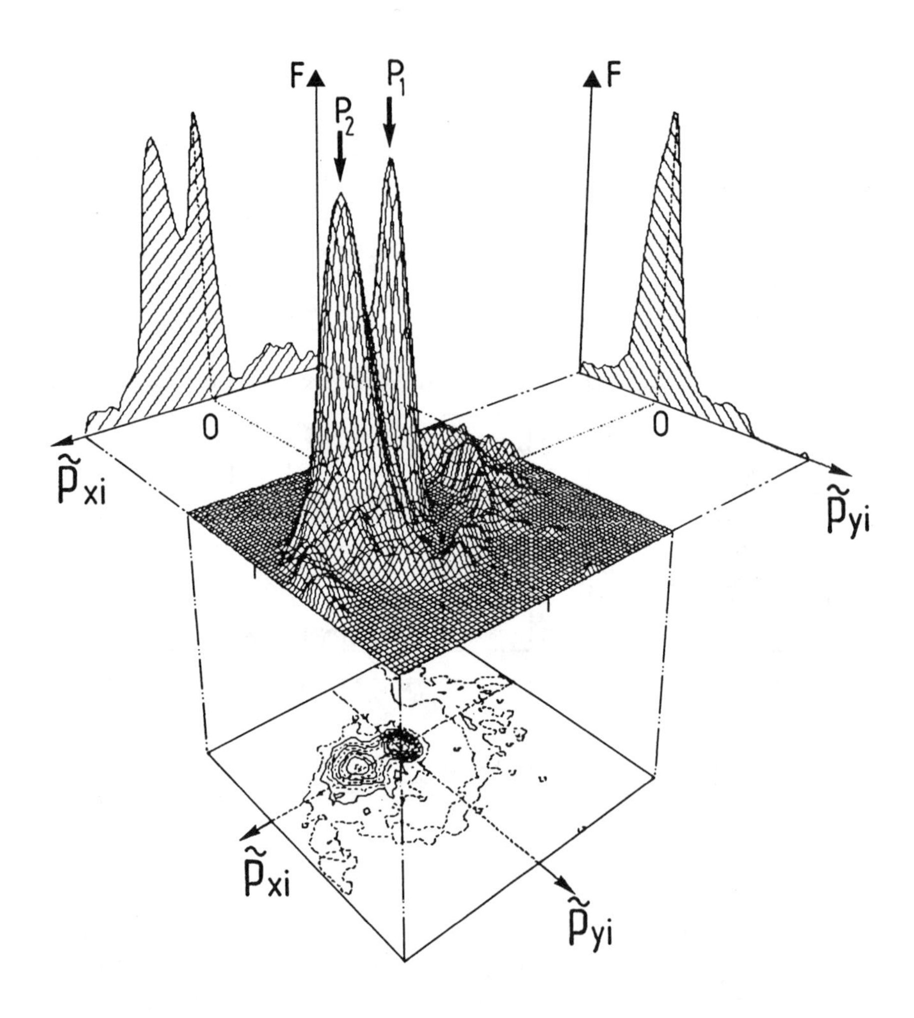

FIGURE 9 Three dimensional plot of Log (f), where $f(\tilde{p}_{xi}, \tilde{p}_{yi})$ is the space-integrated ion distribution, and associated 2-D plot projections. Peaks P_1 and P_2 correspond to the unperturbed upstream ions and to the compressed downstream (directly transmitted) ions, respectively. Ion ring is clearly apparent; a second ring starts forming in the front part of the figure.

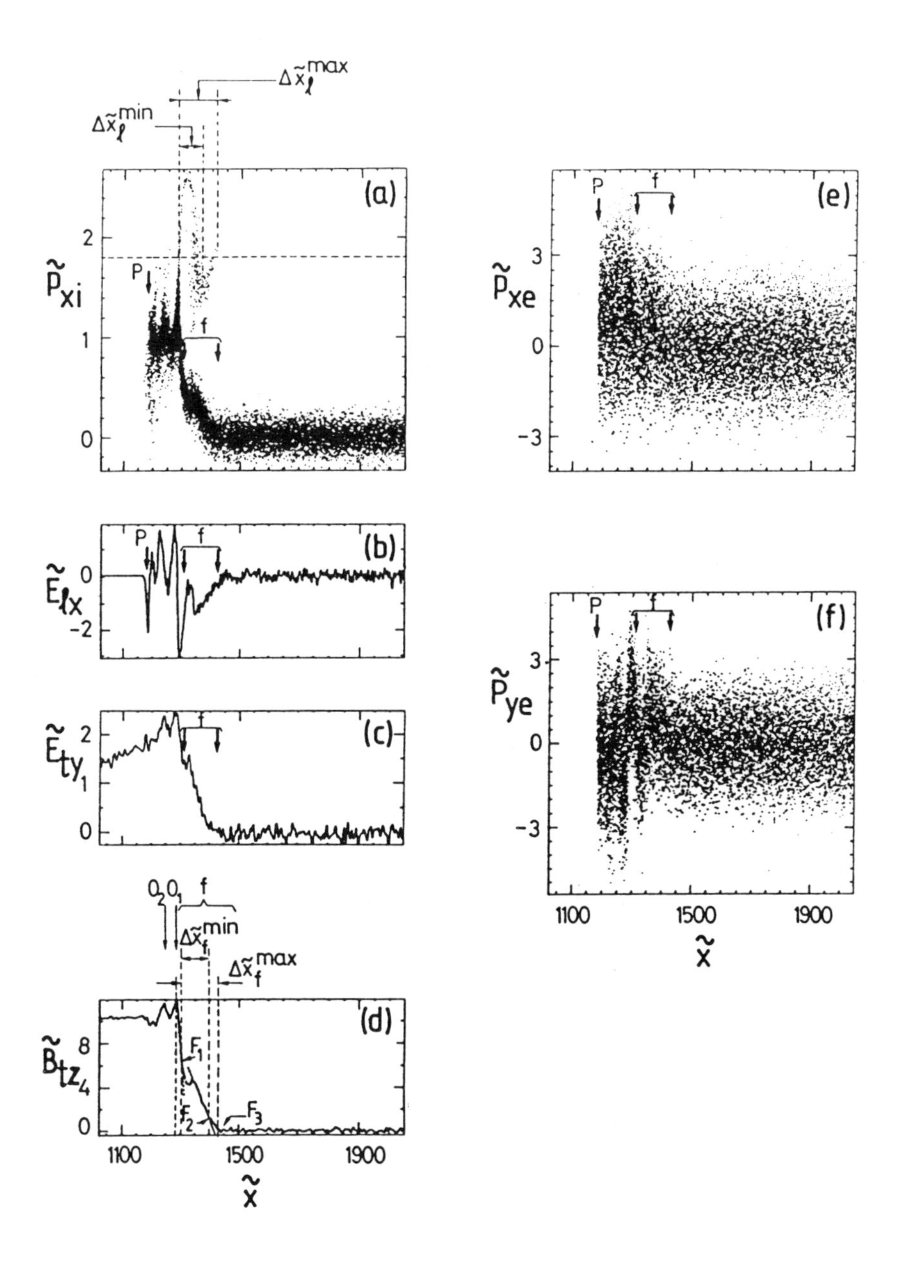

FIGURE 10 Plots of: (a) the ion phase space $(\tilde{x}, \tilde{p}_{xi})$; (b) the electrostatic field $\tilde{E}_{lx}$; (c) the transverse electric field $\tilde{E}_{ty}$; (d) the magnetic field $\tilde{B}_{tz}$; and (e),(f) the electron phase space $(\tilde{x}, \tilde{p}_{xe})$ and $(\tilde{x}, \tilde{p}_{ye})$. The dashed line in (a) represents the phase momentum of the wave $\tilde{p}_\phi \simeq 1.8$; the arrows f, P, O_1, and O_2 indicate, respectively, locations of the foot, piston, the first and the second overshoot.

All the shocks characteristics quoted above have been obtained for strictly perpendicular propagation. An extensive study has been performed for oblique cases within the range $(90° > \theta > 45°)$. In such cases, the electron dynamics may be important, since the shock electric fields in the direction of propagation (along the normal $\vec{n}$ to the wave front) have a component along the magnetic field $E_{\|} = E_n \cos\theta$, allowing the electrons as well as the ions to be accelerated by these large electric fields[15,38]. As a consequence, a fluid treatment of electrons may be inadequate. Main simulation results including a kinetic treatment of electrons may be summarized as follows:

a. Critical angles θ_{ce} and θ_{ci}, similar to those defined for the interaction of magnetosonic waves with an ambient plasma (see previous subsection), have been recovered. Electron heating and acceleration, particularly large along $\vec{B}_0$ are observed for $\theta < \theta_{ce}$, while those for the ions are strong within the plane $(\perp_1, \perp_2)$ perpendicular to $\vec{B}_0$. Perpendicular ion trapping, and consequent heating, dominates when θ is around 90° $(90° > \theta > \theta_{ti})$, and is mainly located at the shock front; in contrast, electron trapping is dominant for smaller angles and takes place both at the shock front and in the downstream trailing wavetrain (Figure 11). Detrapping of ions reflected by the shock front limits their resonant interaction with the electrostatic field $\vec{E}_{lx}$ and restricts their acceleration along the direction parallel to the shock front during their trapping. In contrast, electrons primarily move parallel to $\vec{B}_0$ and do not suffer any detrapping; they largely overshoot the parallel momentum of the shock. Then, an important electron current $\tilde{j}_{ez}$ results along the z direction and can be large enough to induce new electromagnetic components $\tilde{B}_{ty}$ and $\tilde{E}_{tz}$ (which are negligible for $\theta = 90°$); as a reference, $\vec{B}_0$ is along the z direction when $\theta = 90°$ and $\vec{B}_0 = (B_{0x}, 0, B_{0z})$ for $\theta \neq 90°$.

b. The electron acceleration along $\vec{B}_0$ (and the resulting plasma current) can be large enough to generate a whistler precursor, which escapes ahead of the shock front provided that θ deviates from 90° below a certain value. In other words, the shock front plays the role of an antenna, and evacuates the forward propagating waves by radiation from the location where they are generated (Figure 12). The dispersion properties show that low and high wavenumbers defined for a given angle $\theta < \theta_{ce}$ correspond respectively to forward and backward propagating waves; they are respectively at the origin of a precursor and a trailing wavetrain[59].

c. When $\theta < \theta_{ce}$, our numerical results have shown, for the first time (to our knowledge), that a double layer pattern forms at the shock front and plays an important role in the shock dynamics[39]. Indeed, narrow trapping loops of energetic electrons take place parallel to $\vec{B}_0$. In contrast with ions, these energetic electrons are very sensitive to the field fluctuations, due to their light mass. Hence, the narrow width $\Delta\tilde{x}_{le}$ of their trapping loops. Numerical results have indicated that $\Delta\tilde{x}_{le} \ll \Delta\tilde{x}_{li}$, which is the origin of a double layer parallel to $\vec{B}_0$, as seen in the profile of $\tilde{E}_{lx}$ (Figure 12); $\Delta\tilde{x}_{li}$ is the wide trapping loop of ions also accelerated parallel to $\vec{B}_0$. Moreover, electrons are strongly accelerated through this "self-maintained" double layer, and are scattered off by ion and magnetic fluctuations behind the shock. The potential drop at the front is required to slow down the ions, and the electrons must also respond to this jump. In summary, for ions the processes are more or less limited to the direction perpendicular to $\vec{B}_0$, while for electrons acceleration parallel to $\vec{B}_0$ and scattering perpendicular take place.

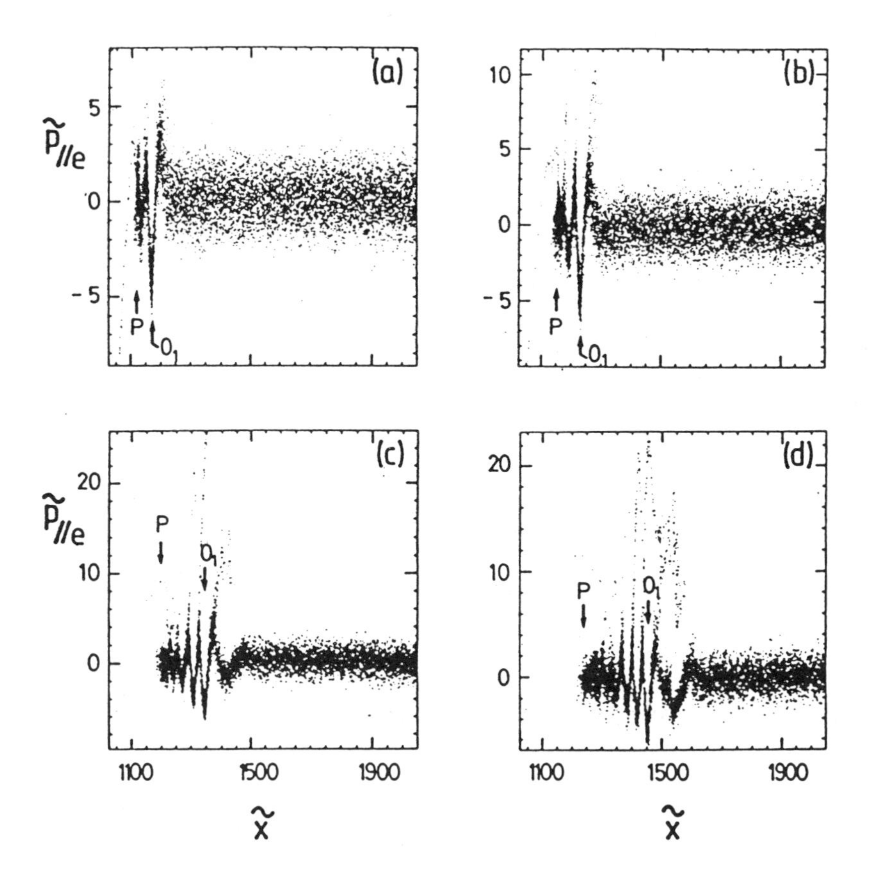

FIGURE 11 Electron phase space $(\tilde{x}, \tilde{p}_{\parallel e})$ for $\theta = 70°$ at various times (a) $\tilde{t} = 0.14\tilde{\tau}_{ci}$, (b) $0.19\tilde{\tau}_{ci}$, (c) $0.28\tilde{\tau}_{ci}$, and (d) $0.38\tilde{\tau}_{ci}$. Here, P and O_1 represent the locations of the magnetic piston (where the particle density drops to zero), and the first overshoot (maximum value of the magnetic field $\tilde{B}_z$ at the ramp), respectively.

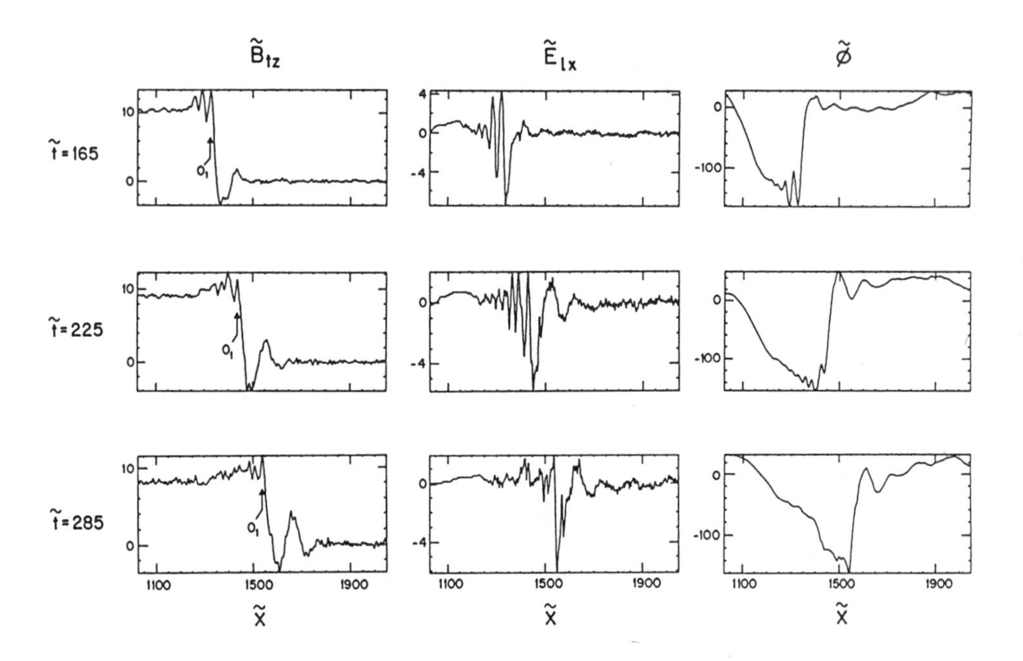

FIGURE 12 Space profiles of field components and electrostatic potential for $\theta = 65°$ at various times of the run; $\tilde{t} = 165$, 225, and 285 correspond to $0.26\tilde{\tau}_{ci}$, $0.36\tilde{\tau}_{ci}$ and $0.45\tilde{\tau}_{ci}$, respectively.

When detailed dynamics of both electrons and ions need to be included, long numerical runs covering a few ω_{ci}^{-1} can become very costly on "standard" supercomputers. Very early, Professor Dawson pointed out this problem and encouraged a few of us to engage some efforts in numerical applications on the new promising architectures of highly parallel supercomputers. The topic of oblique shocks represents a typical example for large computational needs and has been recently selected for this purpose, as described in a future section.

Simulations of Relativistic Particles in a Strongly Magnetized Plasma

Acceleration of both electrons and ions to relativistic energy by large amplitude magnetosonic waves has been investigated in the case of a strongly magnetized plasma. Such a study may have various applications in the domain of solar physics as well as in astrophysics. Indeed, high amplitude electromagnetic waves may occur during solar flares events or in the outer magnetospheres of pulsars. In the case of solar physics, direct observations in soft-rays[22] of interconnecting coronal loops suggest that loop coalescence may be a very important process for energy release in the solar corona. It has been suggested[56] that the most likely instability for impulsive energy release in solar flares is the coalescence instability[56,57]. Tajima et al. (private communication) have examined the existing observational and theoretical results, together with a global energy transfer model, and have concluded that the merging of two currents carrying solar loops can explain many of the known characteristics of solar flares, such as their impulsive nature, heating and high energy particle acceleration, and amplitude oscillations of electromagnetic emission, as well as the characteristics of microwave emission obtained during a flare. In particular, it was noted that the presence of strong electric fields and super-Alfvenic flows during the course of the instability play an important role in the production of non-thermal particles; however, the relationship between both events and the mechanisms involved in the wave-particle interactions are still sources of open questions. As mentioned earlier, numerical simulation reveals to be a very powerful tool to approach this problem.

Complementing the studies quoted in previous sections of this paper, an extensive study has been performed on the acceleration and heating of a very low β plasma by a strongly nonlinear magnetosonic wave[40]. Such a model has two advantages: (i) to follow how a nonlinear wave (sinusoidal at time $t = 0$, i.e., with a very narrow spectrum centered around a wave number k_0) may steepen into a shock profile (very large k-spectrum) in a strongly magnetized plasma; and (ii) to follow in detail how both electrons and ions reach a relativistic regime for high values of ω_{ce}/ω_{pe}. In so doing, a 1-2/2D fully electromagnetic, relativistic particle code similar to that described earlier has been used, where the magnetosonic wave is excited by an external current with a phase velocity $\tilde{v}_\phi = \tilde{v}_Z \simeq 0.5\tilde{c}$, where $\tilde{v}_A$ and $\tilde{c}$ are, respectively, the Alfven velocity and the light velocity; the wave frequency $\tilde{\omega} = 0.14$ while gyrofrequencies are $\tilde{\omega}_{ce} = 3$ and $\tilde{\omega}_{ci} = 0.166$, respectively, for electrons and ions.

In a first approach, the study has been restricted to a perpendicular propagation ($\vec{k} \perp \vec{B}_o$). In such a case, both particle species and fields exhibit a very rich dynamics and an highly nonlinear interaction which may be summarized as follows:

a. Since we have $v_\phi \gg v_{the}$, electrons are strongly accelerated by the large electrostatic field and can reach relativistic velocity relatively shortly because of their light mass. Indeed, at large amplitude the wave requires very large transverse currents which are carried by the electrons ($\tilde{E} \times \tilde{B}$ drift); these can require electron velocities greater than $\tilde{v}_\phi$, in which case the electrons become relativistic. When this happens, a strong electron heating results. This feature greatly differs from the non-relativistic case ($\tilde{v}_\phi \simeq \tilde{v}_{the}$), where electrons only exhibited a weak adiabatic compression heating.

b. Both the fields $\tilde{E}_{lx}$ and $\tilde{E}_{ty}$ contribute to the ions acceleration until they become trapped, and detrapped at later times by the $\tilde{E} \times \tilde{B}$ drift. Since we are in a strongly magnetized plasma, the width of the trapping loop is very narrow;

ions are very energized by the $\tilde{E} \times \tilde{B}$ drift and succeed in reaching a relativistic regime. Instead of forming a well collimated ring within the plane perpendicular to B_0, as in the nonrelativistic case, detrapped ions largely diffuse within a large area of the "light" circle ($\tilde{c} = 10$). In addition, at early times the relation $\tilde{\omega} \simeq \tilde{\omega}_{ci}$ is verified; but during the build-up phase of the wave the $\tilde{B}_z$ field increases and $\tilde{\omega}$ becomes locally less than $\tilde{\omega}_{ci}$ at the wave maxima. At later times, relativistic effects become important (the relativistic factor $\tilde{\gamma}_i$ increases), and the corrected ion gyrofrequency ($\omega_{ci} = qB/m_i\gamma_i$) decreases and is forced to recover the resonance condition $\tilde{\omega} \simeq \tilde{\omega}_{ci}$. Then, a very large ion heating results perpendicular to $\vec{B}_0$ by these two mechanisms: a large ion "diffuse" gyromotion and a "corrected" gyroresonance. Both effects lead to an important wave damping.

c. Nonlinear effects and a phase shift between $\tilde{E}_{lx}$ and $\tilde{E}_{ty}$ fields are of central importance; amplitude variations of both these fields strongly affect the dynamics of both electrons and ions, but in a different way. Wave steepening appears with particle trapping and becomes noticeable after it; then, the initial sinusoidal waveform evolves into a sawtooth. A ramp builds up in the wave profile similar to a shock, and a large number of higher harmonics are generated in both electromagnetic and electrostatic components. For the first time, to our knowledge, a new event has been evidenced at later times of the run: solitary electron wavelets are emitted from the ramp, and may be explained as resulting from a balance between the wave steepening and dispersion effects. Indeed, the relativistic regime observed at later times is mainly characterized by spiky profiles of particle densities and of plasma currents; ion trapping observed in the relativistic regime drastically increases the wave steepening and hence the number of higher harmonics. Low k-modes mark each other by propagating at a velocity which is dependent on its amplitude. When triggered with a large enough amplitude, they can easily pass the phase velocity $\tilde{v}_\phi$ of the input wave. Indeed, the location of the ramp is associated with a very large local $\tilde{B}_z$ (much larger than $\tilde{B}_0$), and the local Alfven velocity may overcome $\tilde{v}_\phi$ ($\simeq \tilde{v}_A$ at $\tilde{t} = 0$). The high k-modes propagating more slowly (since $d\omega/dk < 0$ for $\theta = 90°$), lie on the main waveform and promote the wavelets' separation. Different solitary wavelets pass each other according to their relative velocity; during their propagation, they locally accelerate electrons which largely diffuse within the "light" circle (Figure 14). This overall mechanism is the origin of a noticeable electron heating, which largely increases but stays less than ion heating ($\tilde{T}_e \ll \tilde{T}_i$).

d. An extended parametric study was necessary in order to determine the efficiency of this mechanism according to the wave amplitude. Numerical results have stressed two possible regimes: for "small" wave amplitude, $\tilde{T}_e \ll \tilde{T}_i$ while for "large" wave amplitude, $\tilde{T}_e \gg \tilde{T}_i$. The second regime corresponds to the formation of highly relativistic particles and may be explained as follows: the wave steepening is reinforced because of the initial larger amplitude, i.e., higher frequencies ($\tilde{\omega}_{st}$) modes are excited. On the other hand, the electron relativistic factor $\tilde{\gamma}_e$ becomes so large that the corrected gyrofrequency $\tilde{\omega}_{ce}$ drastically decreases until becoming of the order of the maximum frequency $\tilde{\omega}_{st}^{\max}$, where the subscript "st" holds as steepening. When this happens, a very important electron heating sets in by this "corrected" gyroresonance and leads to a strong increase of the temperature $\tilde{T}_e$.

Such a result has been obtained for a finite "unrealistic" mass ratio ($m_i/m_e = 50$) in order to save computer time and to follow the dynamics of both particle species over a few ion gyroperiods. The striking feature is the particular efficiency of the mechanism as soon as it takes place. For large wave amplitude, the dynamics of ions nearly approaches that of electrons (some of them gyrate and accumulate very near the "light" circle itself). These results illustrate how a sinusoidal waveform (here a magnetosonic wave) propagating within an electron-ion highly magnetized plasma tends towards the features of a relativistic wave propagating within an electron-positron plasma.

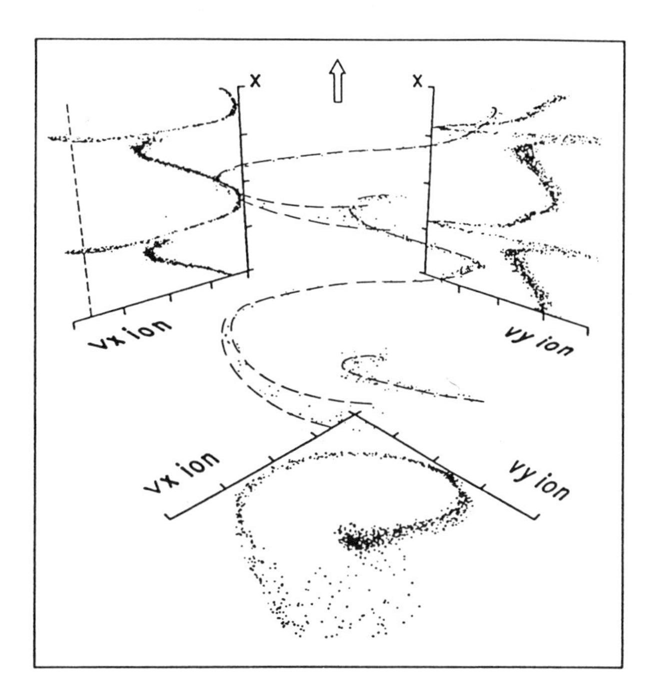

FIGURE 13 Three dimensional plot of the ion phase space $(\tilde{x}_i, \tilde{v}_{xi}, \tilde{v}_{yi})$ (similar to that of Figure 6) in the case of relativistic regime; $\omega_{ce}/\omega_{pe} = 3$, $\tilde{c} = 10$, and $\vec{B}_0$ is along the z axis.

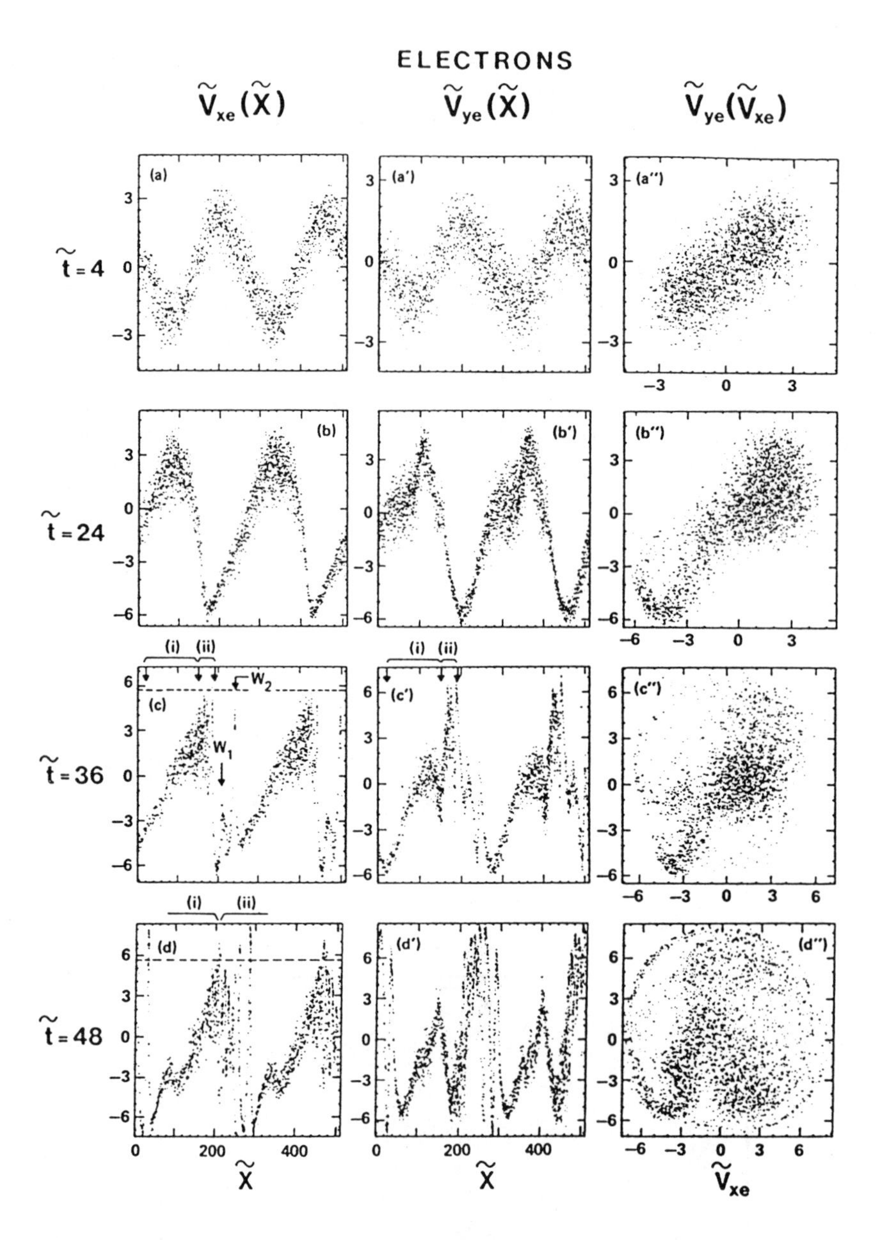

FIGURE 14 Plots of the electron phase space $(\tilde{x}, \tilde{v}_{xe})$ and $(\tilde{x}, \tilde{v}_{ye})$ and velocity space $(\tilde{v}_{xe}, \tilde{v}_{ye})$ at different times $\tilde{t}$; the dashed line indicates the phase velocity of the wave $(\tilde{v}_\phi = 5.67)$; the light velocity is $\tilde{c} = 10$ and $\vec{B}_0$ is along z axis.

Particle Simulations on New Architectures of Highly Parallel Computers: an Application to a Magnetospheric Shock

In the near future, parallel computers with terraflop performance will become available. The use of these machines will revolutionize computer simulations in many domains, particularly in studies of space plasma phenomena. Recently, Professor Dawson, in collaboration with Liewer, Decyk and Lembège, has used the 64-node Mark IIIfp hypercube parallel supercomputer to study electron dynamics in oblique magnetosonic shocks. The use of the parallel computers has allowed these shock studies to be extended into new parameter regimes which were previously computationally prohibitive[45].

The foundation for these shocks studies on parallel computers was laid by Professor Dawson in about 1987 when he encouraged V. Decyk and P. Liewer to implement a 1-D electrostatic code on the 32-node Mark III hypercube parallel computer, then under construction at JPL. The 32-node Mark III, a predecessor of the Mark IIIfp, was significantly below Cray speed. However, Professor Dawson recognized the potential of these parallel computers at a time when most computational physicists were of the opinion that parallel computers were too difficult to program to ever be useful.

The recent concurrent simulations focused on the study of electron dynamics in quasi-perpendicular oblique shocks, e.g., shocks with a propagation angle relative to the shock normal in the range $\pi/2 - \sqrt{m_e/m_i} > \theta_{Bn} > \pi/4$. Such shocks are characterized by a standing whistler wavetrain preceding the shock[59]. It was found that linear and nonlinear interaction of the electrons with the precursor wave played an important role in these shocks. As mentioned earlier, to study electron kinetic effects in shocks over long time scales it is necessary to follow both the ion and electron dynamics, making such codes much more computationally intensive than hybrid codes (fluid electrons-kinetic ions). The studies were carried out on the 64-processor Caltech/JPL Mark IIIfp hypercube concurrent supercomputer, which provided the computational power to run this full particle code with realistic mass ratios ($m_i/m_e = 1600$) and extend the earlier studies of electron kinetic effects on shocks[38,39] into new parameter regimes.

Weak quasi-perpendicular oblique shocks have been observed in the earth's bow shock under low β conditions[19,14,46]. Detailed studies of such shocks by Mellott and Greenstadt[46] using data from the ISEE 1 and 2 dual spacecraft mission, showed that the widths of these low Mach number laminar shocks scaled with the ion inertial length (c/ω_{pi}) as expected theoretically for dispersive shocks; no additional dissipation from cross-field streaming instabilities was necessary to explain the measured widths.

Both experimental[15] and numerical studies[38,39] have emphasized that electrons may be accelerated along $\vec{B}_0$ in quasiperpendicular oblique shocks under the effect of the parallel component of large electric fields. Analysis of ISEE 1 and 2 weak oblique shocks by Gary and Mellott[16] indicated that electron damping of the whistler precursor wavetrain was important; a self-consistent study of such effects also requires a kinetic treatment of electrons.

In the recent simulations of low Mach number ($M_A \leq 3$) oblique quasi-perpendicular shocks[45], we found that electron dynamics play an important role in the shock structure. Specifically, we observed a strong interaction between the upstream electrons and the whistler precursor, leading to a damping of the precursor and a heating of the electrons. In some parameter regimes, the electrons are seen to be trapped along the field lines in the electrostatic potential of the whistler, indicating significant nonlinear damping of the precursor. This can lead to strong electron heating in front of the shock when the parallel phase velocity of the whistler exceeds the electron thermal velocity.

A one-dimensional electromagnetic particle-in-cell code with kinetic electrons and ions was used for these studies, similar to that described in Lembège and Dawson[38]. The code uses standard PIC techniques[4] to solve the coupled equations for the particle orbits and the electromagnetic fields as an initial value problem. The parallel electromagnetic code was developed from a parallel electrostatic code, using the General Concurrent PIC algorithm described in Liewer and Decyk[44]. Run times for the studies in this paper range from four to fourteen hours.

Figure 15 shows results at two times in a simulation for a shock with $M_A \approx 2.8$ (determined from the observed propagation speed), $\beta_e = 0.02$, $\omega_{pe} = 2$, $m_i/m_e = 1600$, $T_i/T_e = 4$, and $\theta_{Bn} = 70°$ (Case 1). Figures 15a-d show $B_z(x)$, electron v_x and v_z phase space, and $\phi(x)$, respectively, at $t\omega_{pe} = 600$,, and Figures 15e-h the same at $t\omega_{pe} = 1560$. Several features of the precursor wavetrain formation and electrons-precursor interaction in low Mach number oblique quasi-perpendicular shocks are illustrated in these series of plots.

The phase front of the wavetrain propagates faster than the shock, consistent with the observation of an increase in wave number in the packet with distance from the shock. Throughout this run, the wavetrain continued to extend farther from the shock, with new wave crests appearing, until the run was terminated when the system boundary was reached. Although the precursor wavetrain did not reach a "steady-state" shape in this run, the amplitudes of the wavecrests nearer the shock front have reached steady-state amplitudes by the later time ($t\omega_{pe} = 1560$).

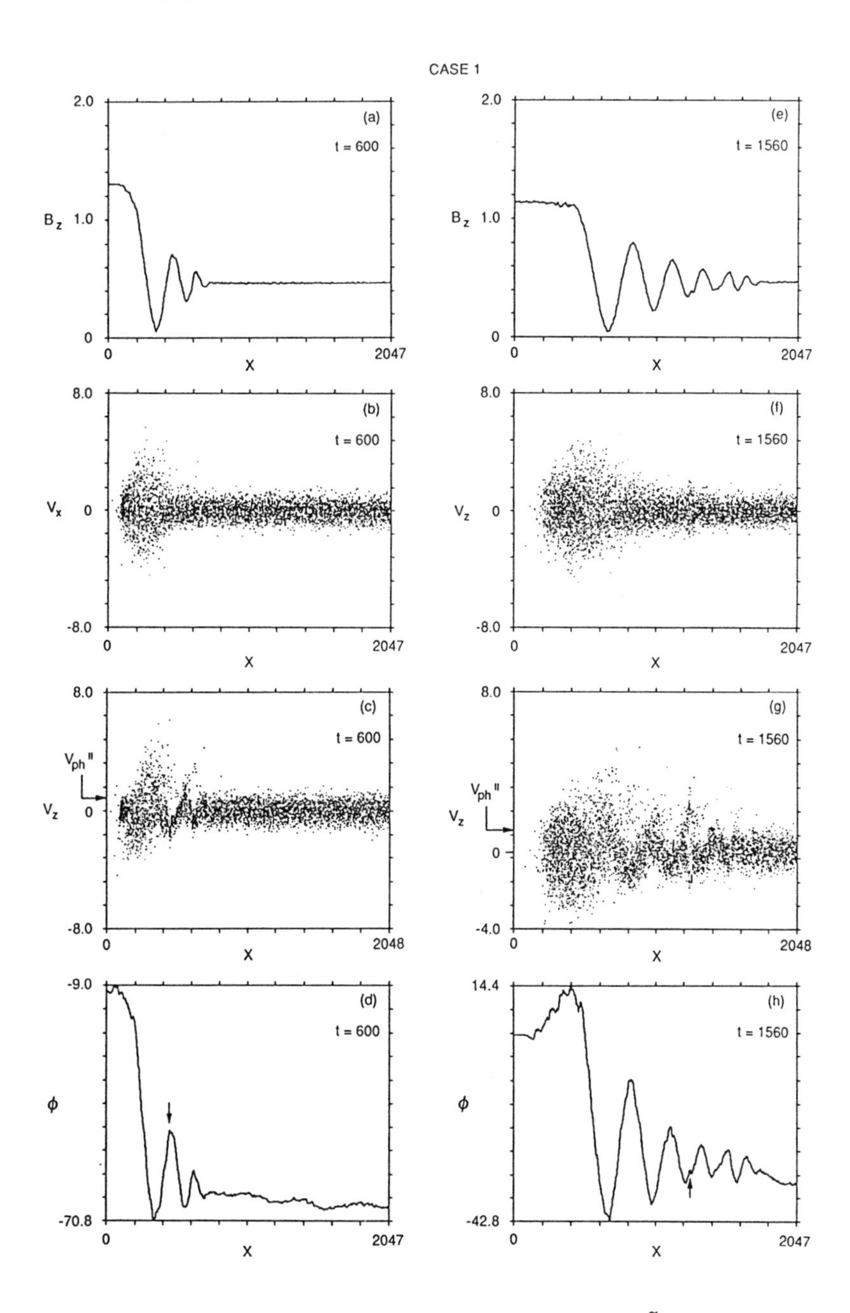

FIGURE 15 Simulation results at two times from Case 1 for $\tilde{B}_z$ (normalized to $\sqrt{4\pi n_0 m_e c^2}$), electron $(\tilde{x}, \tilde{v}_x)$ and $(\tilde{x}, \tilde{v}_z)$ phase space, and the electrostatic potential $\tilde{\phi}$ (normalized to $T_e^0/3$). This effect of the precursor wavetrain on the electrons can be seen in the phase space plots. for this case, $v_{\phi\|}/v_{te} \approx 2.0$. In (d), the arrow shows the electron potential well of the first precursor. This precursor, as well as the smaller precursor, is large enough to trap the entire electron distribution. In (h), the arrow shows a finer scale secondary wave, which appears to be associated with the trapped electrons.

The decrease in amplitude of the precursors with distance from the shock front indicated the presence of some convective damping mechanism. From the electron v_z versus x phase space plots in Figures 15c and 15g, it can be seen that the electron are interacting strongly with the precursor wavetrain (trapping in the electric potential), suggesting that electron damping may be present. Since v_z phase space is dominated by the electron motion parallel to the field for this $\theta_{Bn} = 70°$ case, the electrons are apparently interacting with the precursor via the parallel electric field of the precursor. Because the shock is propagating at an angle θ_{Bn} to the upstream magnetic field, there is a component of the shock and precursor electric fields parallel to the field:

$$E_{\|} = \nabla_{\|}\phi = \cos\theta_{Bn}\,\frac{d\phi}{dx}$$

To determine whether the potential is large enough to trap the electrons, the observed value of the potential well in the simulations can be compared to that required for trapping.

The phase velocity of the precursor parallel to the field is much higher than the velocity in the direction of propagation x, $v_{\phi\|} = v_{\phi x}/\cos\theta_{bn}$. Using $v_{\phi x} = \mathrm{M_A}\, v_A$ for the standing precursor, the expression for the parallel phase velocity becomes $v_{\phi\|}/v_{te} = \mathrm{M_A} v_A/v_{te}\cos\theta_{Bn}$. For the run in Figure 15, the electron temperature was $T_e^0 = 0.5\,T_e$, and Alfven speed relative to the initial electron thermal speed was $v_A/v_{te} = 0.25$, yielding $v_{\phi\|}/v_{te} \approx 2$.

For this simulation, the damping appears to be dominated by nonlinear (trapping) effects. However, if the run could be carried out for a longer time on a larger system so that a region of lower amplitude whistler were reached, a linear (Landau) damping region might also be observed.

If electron interaction with the parallel electric field of the precursor whistler is the cause of the electron heating and precursor damping, as suggested by the simulation in Figure 15 (Case 1), then results should depend on the ratio of the precursor parallel phase velocity relative to the electron thermal velocity,

$$\frac{v_{\phi\|}}{v_{te}} = \frac{\mathrm{M_A} v_A}{\cos\theta_{Bn} v_{te}} = \frac{\mathrm{M_A}}{\cos\theta_{Bn}}\sqrt{\frac{2}{\beta_e}\frac{m_e}{m_i}} \qquad (15)$$

Figure 16, along with Figure 15(e), shows results for the magnetic field at $t\omega_{pe} = 1560$ from three simulations with varying values of electron temperature, with other parameters the same as in Case 1 (Figure 15). Thus, for Case 2, $v_{\phi\|}/v_{te} \simeq 1$, and for Case 3, $v_{\phi\|}/v_{te} \simeq 0.5$, compared to $v_{\phi\|}/v_{te} \simeq 2$ for Case 1.

Comparison of the precursor in the magnetic fields in Figures 15 and 16 (Cases 1-3) shows that the decrease in $v_{\phi\|}/v_{te}$ (from the increase in T_e), has lead to an increased convective damping of the whistler precursor. From these simulation results, we find that for $v_{\phi\|}/v_{te} < 1$, the phase velocity lies well within the distribution and the electrons damp the precursor (linear damping) more than when $v_{\phi\|}/v_{te} > 1$. For $v_{\phi\|}/v_{te} > 1$, the precursor causes more heating of the electrons as long as the precursor potential is large enough to trap a significant portion of the distribution function; this leads to a nonlinear damping of the precursor.

The simulations here have shown that significant electron heating is observed when $v_{\phi\|}/v_{te} > 1$. Using eq. (15), the condition for a significant amount of electron heating can be written as:

$$\beta_e < 2 \frac{\mathrm{M}_{\mathrm{A}}^2}{\cos\theta_{Bn}} \frac{m_e}{m_i}$$

Thus for $m_i/m_e = 1836$ and $\cos\theta_{Bn} \simeq 0.3$, $\beta_e < 0.01\ \mathrm{M}_{\mathrm{A}}^2$ is required. Thus the 1-D numerical simulations indicated that heating by the precursor whistler will be important in planetary and interplanetary shocks under low β_e conditions.

It is thus clear that details of electron dynamics are important for some shock structures, and simple fluid approximations are likely to miss important aspects of the physics. The calculations presented here have simply pointed up the richness of the physical phenomena that can exist in plasma shocks. They cast light on only a small part of this, and there are many questions left unanswered by these simple 1D electromagnetic calculations. In the future, as faster parallel supercomputers become available and two-dimensional shock simulations on the electron time scale become practical, many more question about magnetosonic shocks will be answered.

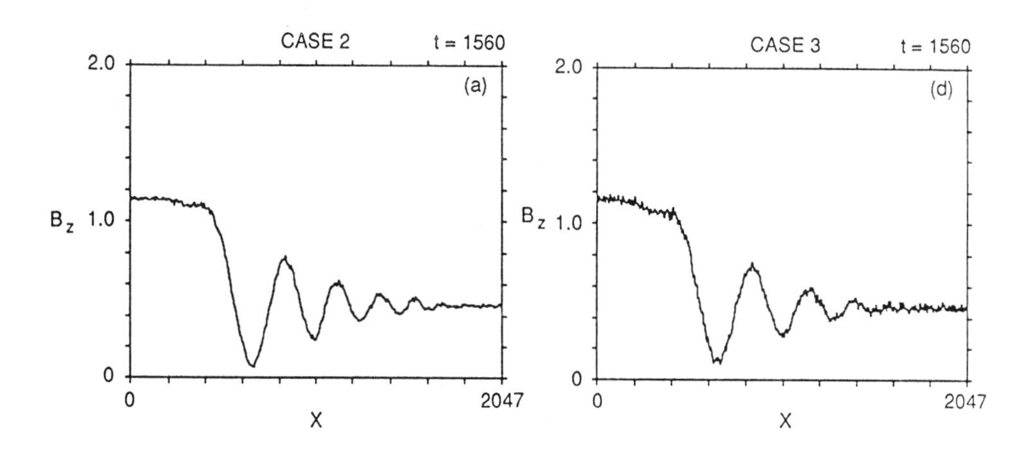

FIGURE 16 Simulation results for $\tilde{B}_z$ for Cases 2 and 3, all at the same time as the later pictures in Figure 15 ($t\omega_{pe} = 1560$). Parameters are the same as Case 1 (Figure 15), except $(v_{\phi\|}/v_{te} \approx 1.0)$ for Case 2 and $(v_{\phi\|}/v_{te} \approx 0.4)$ for Case 3. More convective damping of the precursor wavetrain is evident in the warmer case (Case 3). $\mathrm{M_A} \approx 2.8$ for Cases 1 - 3.

Conclusions

We have given a synoptical review of various topics and innovative studies that Professor Dawson has carried out in numerical simulations of space plasma phenomena. Today's computers are able to follow the time evolution of systems containing many millions of degrees of freedom, all of which are simultaneously interacting with each other. One can often test theoretical predictions and the assumptions and approximations which go into them in ways inaccessible to laboratory or space experiments. For example, it is possible to turn effects on and off in such models and see how the results change. Needless to say, this is something which is often impossible in the real world.

On the other hand, such numerical tools are far more complete and realistic than we can hope to handle analytically. They reproduce both linear and nonlinear behavior. The results can be used to predict the behavior, test predictions and to gain an understanding of the phenomena involved. This aspect is particularly helpful for performing a comparative study with space experimental data and/or for the preparation of a new space project.

As mentioned by Professor Dawson[10],

> The real power of numerical simulation does not lie [only] in reproducing complex physical phenomena. The results of such calculations often show us which are the important effects among the many possibilities and thus let us construct useful simple physical models which we would be hard pressed to justify a priori. Such calculations do not reduce the amount of physics we are called on to do, but rather increase the time spent on physics and put a premium on physical intuition...
>
> On the experimental side, one is limited to measurements of only a small fraction of these quantities of interest in a process and even these may be only sampled at a few times and positions and with a limited degree of accuracy. This is particularly true for observations of natural phenomena such as are encountered with space plasmas.

We can add that, in many cases of large-scale and long-time natural phenomena where observational studies are difficult or bring incomplete information, such a numerical approach represents an unique way for analyzing in detail the complexity of the phenomena and the simultaneous interaction of many effects.

Finally, let us remember that Professor Dawson also made an important contribution in educating new researchers in the field of plasma simulation; in particular, his active participation in successive International Simulation Schools of Space Plasmas illustrates quite well not only his deep interest in studying space plasma phenomena, but also his care for establishing a link between fusion, space and the astrophysics communities.

References

1. Ashour-Abdalla, M., J.N. Leboeuf, J.M. Dawson, and C.F. Kennel, "A Simulation Study of Cold Electron Heating By Loss Cone Instabilities", *Geophys. Res. Lett.* 7, 889, 1980.

2. Ashour-Abdalla, M., J.N. Leboeuf, T. Tajima, C.F. Kennel, and J.M. Dawson, "Ultrarelativistic Electromagnetic Pulses in Plasmas", *Phys. Rev.* A23, 1906, 1981.

3. Ashour-Abdalla, M., J.N. Leboeuf, J.M. Dawson, and C.F. Kennel, "A Simulation Study of Cold Electron Heating by Loss Cone Instabilities", *Geophys. Res. Lett.* 7, 889, 1980.

4. Birdsall, C.K., and Langdon, A.B., *Plasma Physics Via Computer Simulation*, McGraw-Hill, New York, 1985.

5. Biskamp, D., "Collisionless Shock Waves in Plasma", *Nuclear Fus.*, 13, 719-740, 1973.

6. Biskamp, D. and H. Welter, "Numerical Studies of Magnetosonic Collisionless Shock Waves", *Nuclear Fus.*, 12, 663 , 1973.

7. Brunel, F., J.N. Leboeuf, T. Tajima, J.M. Dawson, M. Makino, and T. Kamimura, "Magnetohydrodynamic Particle code: Lax-Wendroff Algorithm With Finer Grid Interpolations", *Journ. Comp. Phys.* 43, 268, 1981.

8. Brunel F., T. Tajima and J.M. Dawson, *Phys. Rev. Lett.*, 49, 323, 1982.

9. Burgess, D., "Numerical Simulation of Collisionless Shocks", in *Proceedings of International Conference on Collisionless Shocks, Bulatonfured, Hungary*, pp. 89-111, ed. by Szego, 1987.

10. Dawson, J.M., "Simulations of Space Plasma Phenomena", in *Physics of Auroral Arc Formation*, Geophys. Monograph Series, 25, 1981.

11. Dawson, J.M., V.K. Decyk, R.W. Huff, I. Jechart, T. Katsouleas, J.N. Leboeuf, B. Lembège, R.M. Martinez, Y. Ohsawa, and S.T. Ratliff, "Damping of Large Amplitude Plasma Waves Propagating Perpendicular to the Magnetic Field", *Phys. Rev. Lett.* 50, 1455, 1983.

12. Décréau, P.M.E., C. Béghin and M. Parrot, "Global Characteristics of the Cold Plasma in the Equatorial Plasmapause Region As Deduced From the GEOS 1 Mutual Impedance Probe", *J. Geophys. Res.*, 87, 695, 1982.

13. Dungey, J.W., "Interplanetary Magnetic Field and the Auroral Zones", *Phys. Rev. Lett.* 6, 47, 1961.

14. Fairfield, D.H. and W.C. Feldman, "Standing Waves At Low Mach Number Laminar Bow Shocks", *J. Geophys. Res.*, 80, 515, 1975.

15. Feldman, W.C., "Electron Velocity Distributions Near Collisionless Shocks", in *Collissionless Shocks in the Heliosphere: Reviews of Current Research* (B. Tsurutani and R.G. Stone, eds, AGU, Washington D.C.) p. 195, 1985.

16. Gary, S.P. and M.M. Mellott, "Whistler Damping At Oblique Propagation: Laminar Shock Precursors", *J. Geophys. Res.*, 90, 99, 1985.

17. Gendrin, R., "Consequences of Hydromagnetic Waves on Magnetospheric Particle Dynamics", *Space Science Rev.* 42, 515, 1985.

18. Goodrich, C.C., "Numerical Simulations of Quasi-Perpendicular Collisionless Shocks", in *Collisionless Schocks in the Heliosphere: Review of Current Research*, Geophys. Monograph. Ser., Vol. 35, pp. 153-168, ed. by B.T. Tsurutani and R.G. Stone, AGU Washington, D.C., 1985.

19. Greenstadt, E.W., C.T. Russel, F.L. Scarf, V. Formisano, and M. Neugebauer, "Structure of the Quasi-Perpendicular Laminar Bow Shock", *J. Geophys. Res.*, 80, 502, 1975.

20. Gurnett, D.A., F.L. Scarf, R.W. Fredricks and E.J. Smith, "The ISEE-1 and ISEE-2 Plasma Investigation", *Geoscience Electronics* GE-16, 225, 1978.

21. Hones, E.W., Jr., D.N. Baker, S.J. Bame, W.C. Feldman, J.T. Gosling, D.J. McComas, R.D. Zwickl, J.A. Slavin, E.J. Smith, and B. Tsurutani, "Structure of the Magnetotail at 220 RE and Its Response to Geomagnetic Activity", *Geophys. Res. Lett.* 11, 5, 1984.

22. Howard, R. and Z. Svestka, *Solar Phys.*, 54, 65, 1977.

23. Katsouleas, T. and J.M. Dawson, "Unlimited Electron Acceleration in Laser Driven Plasma Waves", *Phys. Rev. Lett.* 51, 392, 1983.

24. Kennel, C.F., F.S. Fujimura, and R. Pellat, *Space Sci. Rev.* 24, 407, 1979.

25. Kennel, C.F., Edminston, J.P. and T. Hada, "A Quarter Century of Collionless Shock Research", in *Collisionless Schocks in the Heliosphere: Reviews of Current Research*, Geophys. Monograph. Ser., Vol. 35, pp. 1-36, ed. by B.T. Tsurutani and R.G. Stone, AGU Washington, D.C., 1985.

26. Korth, A., G. Kremser, A. Roux, S. Perraut, J.A. Sauvaud, J.M. Bosqued, A. Pedersen and B. Aparicio, "Drift Boundaries and ULF Wave Generation Near Noon at Geostationary Orbit", *Geophys. Res. Lett.*, 10, 639, 1983.

27. Korth, A., G. Kremser, S. Perraut, and A. Roux, "Interaction of Particles with Ion Cyclotron Waves and Magnetosonic Waves. Observations from GEOS 1 and 2", *Planet Space Sci.*, 32, 1393, 1984.

28. Kurth, W.S., L.A. Frank, M. Ashour-Abdalla, D.A. Gurnett, and B.G. Burek, "A Free Energy Source for Intense Electrostatic Waves", *Geophys. Res. Lett.* 7, 293, 1980.

29. Leboeuf., J.N., T. Tajima, C.F. Kennel, and J.M. Dawson, "Global Simulation of the Time-Dependent Magnetosphere", *Geophys. Res. Lett.* 5, 609, 1978.

30. Leboeuf, J.N., T. Tajima, and J.M. Dawson, "A Magnetohydrodynamic Particle Code for Fluid Simulations of Plasmas", *Journ. Comp. Phys.* 31, 379, 1979.

31. Leboeuf, J.N., T. Tajima, C.F. Kennel, and J.M. Dawson, "Global Magneto-Hydrodynamic Simulation of the Two-Dimensional Magnetosphere", in *Quantitative Modeling of Magnetospheric Processes*, Geophysical Monograph 21, edited by W.P. Olson, American Goephysical Union, Washington, D.C., 1979.

32. Leboeuf, J.N., T. Tajima, C.F. Kennel, and J.M. Dawson, "Global Simulations of the Three-Dimensional Magnetosphere", *Geophys. Res. Lett.* 8, 257, 1981.

33. Leboeuf, J.N., M. Ashour-Abdalla, T. Tajima, C.F. Kennel, F.V. Coroniti, and J.M. Dawson, "Ultrarelativistic Waves in Overdense Electron-Positron Plasmas", *Phys. Rev.* A, 25, 1023, 1982.

34. Lembège, B., S.T. Ratliff, J.M. Dawson, and Y. Ohsawa, "Ion Heating and Acceleration by Strong Magnetosonic Waves", *Phys. Rev. Lett.* 51, 264, 1983.

35. Lembège, B., and J.M. Dawson, "Plasma Heating and Acceleration by Strong Magnetosonic Waves Propagating Obliquely to a Magnetostatic Field", *Phys. Rev. Lett.* 53, 11, 1053, 1984a.

36. Lembège , B., and J.M. Dawson , "Kinetic Perpendicular Collisionless Shocks", UCLA Report, PPG-832, 1984b.

37. Lembège, B., and J.M. Dawson, "Self-Consistent Study of a Perpendicular Collisionless and Nonresistive Shocks", *Phys. Fluids* 30(6), 1767, 1987a.

38. Lembège, B., and J.M. Dawson, "Plasma Heating Through a Supercritical Oblique Collisionless Shock", *Phys. Fluids* 30 (4), 1110, 1987b.

39. Lembège, B., and J.M. Dawson, "Formation of Double Layers Within an Oblique Collisionless Shock", *Phys. Rev. Lett.* 62 (23), 2683, 1989a.

40. Lembège, B. and J.M. Dawson, "Relativistic Particle Dynamics in a Steepening Magnetosonic Wave", *Phys. Fluids* B1, 1001, 1989b.

41. Lembège, B. and J.M. Dawson, "Self-Consistent Plasma Heating and Acceleration by Strong Magnetosonic Waves for $\theta = 90°$. Part I: Basic Mechanisms", *Phys. Fluids* 29, 3, 821, 1986.

42. Leroy, M.M., C.C. Goodrich, D. Winske, C.S. Wu and K. Papadopulos, "Simulation of a Perpendicular Bow Shock", *Geophys. Res. Lett.* 8, 1269 ,1981.

43. Leroy, M.M., D. Winscke, C.C. Goodrich, C.S. Wu and K. Papadopoulos, "The Structure of Perpendicular Bow Shocks", *J. Geophys. Res. Lett.*, 87, 5081, 1982.

44. Liewer, P.C., and V.K. Decyk, "A General Conurrent Algorithm for Plasma Particle-In-Cell Simulation Codes", *J. Comp. Physics*, 85, 302, 1989.

45. Liewer P.C., V.K. Decyk, J.M. Dawson and B. Lembège, "Numerical Studies of Electron Dynamics in Oblique Quasi-Perpendicular Collisionless Shock Waves", *J. Geophys. Res.*, (in press) 1991.

46. Mellott, M.M., and E.W. Greenstadt, "The Structure of Oblique Subcritical Bow Shocks: ISEE1 and 2 Observations", *J. Geophys. Res.* 89, 2151, 1984.

47. Papadopoulos, K., "Microinstabilities and Anomalous Transport", in *Collisionless Shocks in the Heliosphere: Review of Current Research*, Geophys. Monograph. Ser., Vol. 35, pp. 59-88, ed. by B.T. Tsurutani and R.G. Stone, AGU Washingtn, D.C., 1985.

48. Perraut, S., A. Roux, P. Robert, R. Gendrin, J.A. Sauvaud, J.M. Bosqued, G. Kremser and A. Korth, "A Systematic Study of ULF Waves Above fH+ from GEOS 1 and 2 Measurements and Their Relationships With Proton Ring Distributions", *J. Geophys. Res.* 87, 6219, 1982.

49. Podgorny, I.M., E.M. Dubinin, P.L. Izrailevich, Y.N. Potanin, "Plasma Dynamics in Laboratory Models of the Magnetosphere of the Earth and Uranus", *Izv. An. Ser. Fiz.* 41, 1870, 1977.

50. Quest, K.B., "Simulations of High Mach Number Collisionless Perpendicular Shocks in Astrophysical Plasmas", *Phys. Rev. Lett.*, 54, 1872-1874, 1985.

51. Quest, K.B., "Simulations of High Mach Number Perpendicular Shocks With Resistive Electrons", *J. Geophys. Res.* 91, 8805, 1986.

52. Quest, K.B., "Hybrid Simulation", in Numerical Simulation of Space Plasmas, Proceedings of ISSS-3, Part 1, pp. 177-182, Cepadues Editions, Toulouse France (1989).

53. Sagdeev, R.Z., "The 1976 Oppenheimer Lectures: Critical Problems in Plasma Astrophysics. Part II: Singular Layers and Reconnection", *Rev. Mod. Phys.*, 51, 11, 1979.

54. Sentman, D.D., J.N. Leboeuf, T. Katsouleas, R.W. Huff, and J.M. Dawson, "Electrostatic Instabilities of Velocity-Space-Shell Distributions in Magnetized Plasmas", *Phys. Fluids* 29, 2569, 1986.

55. Tajima, T. and J.M. Dawson, Phys. Rev. Lett. 43, 267, 1979.

56. Tajima, T., F. Brunel and J.I. Sakai, "Loop Coalescence in Flares and Coronal X-Ray Brightening", *The Astrophys. J.*, 258, L45, 1982.

57. Tajima, T., in *Fusion Energy*, International Atomic Energy Agency, ICTP, Trieste, 403, 1982.

58. Tajima, T., F. Brunel, J.I. Sakai, Toyama, L. Vlahos and M.R. Kundu, "The Coalescence Instability of Solar Flares", (private communication).

59. Tidman, D.A. and N.A. Krall, "Shock Waves in Collisionless Plasma", Wiley, Interscience, New York, 1971.

60. Winske, D. and M.M. Leroy, "Hybrid Simulation Techniques Applied to the Earth's Low Shock", in *Computer Simulation of Space Plasmas*, ISSS-1, ed. by Matsumoto and T. Sato, pp. 255-278, Kluwer Academic, Hingham, Mass 1984.

W. L. Kruer
Lawrence Livermore National Laboratory, Livermore, CA 94550

Suprathermal Particle Generation and Other Plasma Effects in Laser Fusion

Abstract

Two basic plasma topics important to the understanding of laser plasma coupling are suprathermal electron generation and nonlinear ion waves. Some old and new results are discussed, including hot electron generation in hohlraum targets.

Introduction

This symposium is a fitting tribute to a remarkable scientist and teacher: John M. Dawson. As this symposium emphasizes, he has made pioneering contributions to so many different areas of plasma physics and to widespread applications. As I can attest, Professor Dawson has also been a source of inspiration and guidance to numerous students and post-docs.

This discussion will focus on two basic topics in plasma physics: suprathermal electron generation and nonlinear ion waves. Early papers by Dawson and colleagues serve as key references on both these topics. Both topics are very important for laser fusion applications. Indeed, suprathermal electron generation has had a critical influence on the choice of laser light wavelength.

Suprathermal Electrons

Suprathermal electron generation is a key feature of collisionless plasma heating via excitation of electron plasma waves. This important result was emphasized by early work on anomalous high frequency resistivity.[1] In the simplest model, a pump field with frequency near the electron plasma frequency excites electron plasma waves and ion fluctuations via the oscillating two stream and ion acoustic decay instabilities. In the nonlinear state, energy transferred into plasma waves is in turn passed on to the electrons and ions. Hence the plasma heats at a rate corresponding to an enhanced collision frequency. However, the heating is very different from collisional heating, in which electrons in the main body of the distribution function are heated. In contrast, electron plasma waves heat the faster, more nearly resonant electrons. The heated velocity distribution function is characterized by a tail of suprathermal electrons, as indicated by the example in Fig. 1.

The mechanisms for suprathermal electron generation depend on the size of the irradiated plasma. In the early 1970s, the laser energy was small (tens of joules). Short pulse lengths and small focal spots were used to achieve high intensity irradiation and drive exploding pusher targets. These experiments were characterized by a small region of underdense plasma. The density scalelength L of this plasma can be crudely estimated as a characteristic expansion velocity ($\sim 3 \times 10^7$ cm/sec) times the pulse length. As an example, for a 30 ps pulse of 1.06 μm laser light, $L \sim 10\lambda_0$, where λ_0 is the laser light wavelength.

For laser fusion applications, suprathermal electron generation is a deleterious effect to be avoided. High energy electrons have a long mean free path (proportional to the square of their energy). Hence they can preheat the fuel inside a capsule, preventing the compression needed to achieve high gain implosions.

In short scalelength plasmas, the coupling is principally determined by interaction with plasma near the critical density. Resonance absorption then plays a central role as does a self-consistent steepening of the density profile near the critical density. As illustrated in Fig. 2, resonance absorption of intense laser light generates suprathermal electrons.[2,3] In short pulse experiments, suprathermal or hot electron generation was inferred from the high energy x rays generated when they bremsstrahlung. It was typically found[4] that about 20-30% of the laser energy was absorbed into hot electrons with a characteristic temperature which depends on the intensity and wavelength of the laser light. Indeed, the inferred hot electron temperatures were in reasonable agreement with those found in simulations.

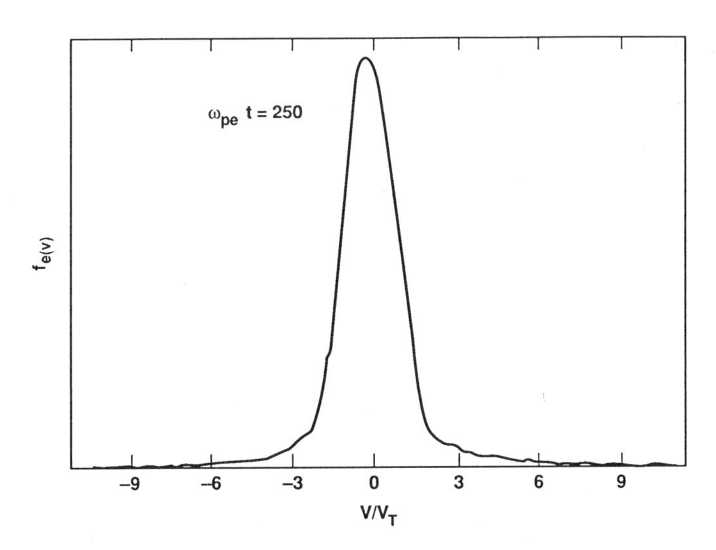

FIGURE 1 A heated electron velocity distribution function from a particle simulation[1] of a uniform plasma driven by a pump field oscillating near the electron plasma frequency.

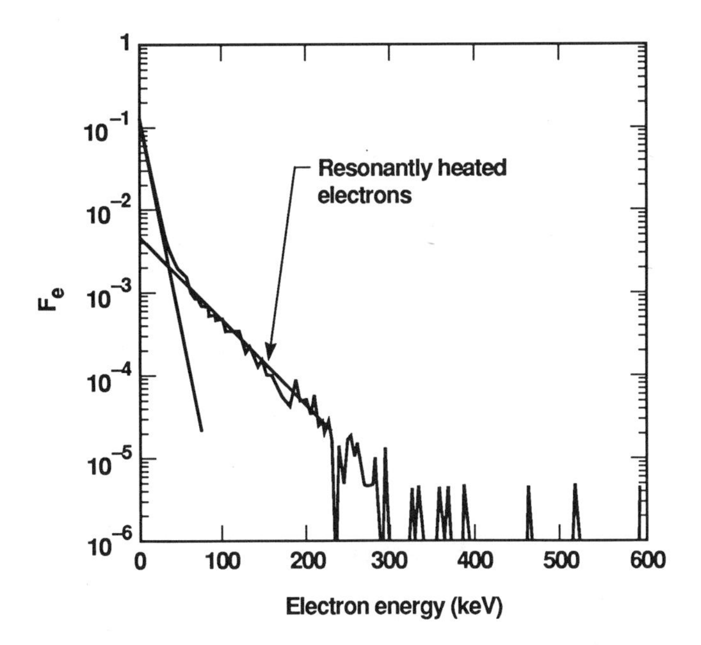

FIGURE 2 A heated electron distribution function from a two-dimensional simulation[2] of resonance absorption.

In current experiments, laser light with a pulse length $\gtrsim 1$ ns is used for ablatively driven compressions. Such experiments are characterized by a large region of underdense plasma. Typical density scalelengths are of order 1 mm, which corresponds to thousands of λ_0.

In large underdense plasmas, the Raman and $2\omega_{pe}$ instabilities can lead to strong hot electron generation.[5–7] This so-called tail heating is well-illustrated by a one-dimensional simulation[7] of the Raman instability in a uniform plasma with density equal to $0.1\, n_{cr}$ (where n_{cr} is the critical density). Figure 3 shows an electron distribution function in the nonlinear state. Note the heated tail due to the electron plasma wave driven up in the Raman process. In very strongly driven plasmas, this tail heating can stabilize[8] the instability. In general,[4] ion dynamics play an important role. The $2\omega_{pe}$ instability also leads to suprathermal tails.

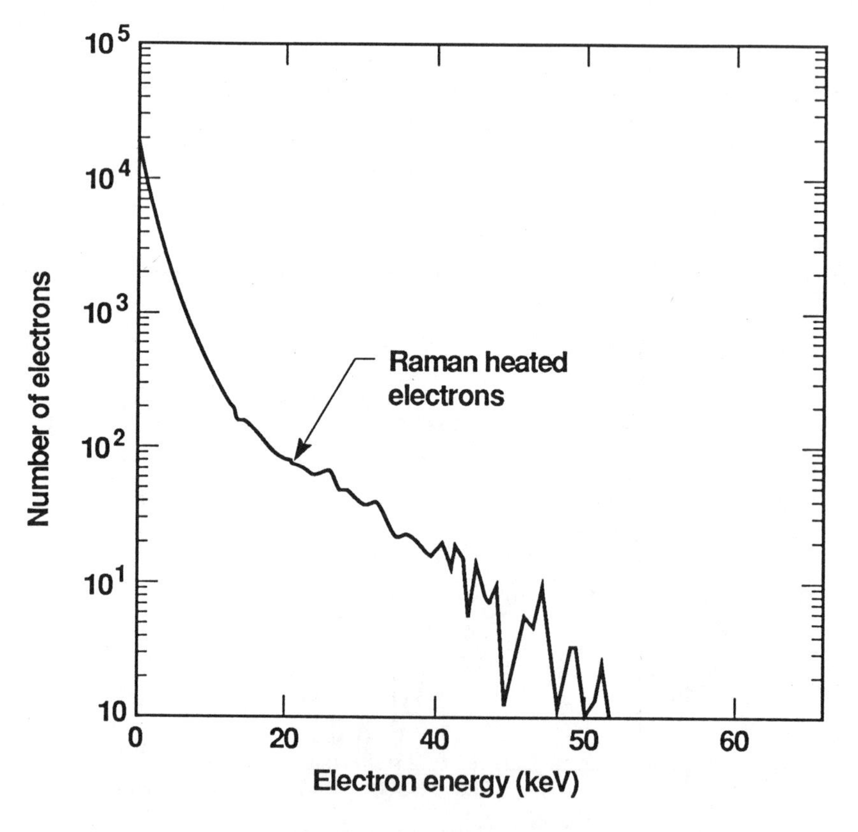

FIGURE 3 A heated electron distribution function from a one-dimensional simulation[7] of the Raman instability.

A correlation of hot electron generation with the Raman instability has been found in experiments with long scalelength plasmas. Figure 4 shows the fraction of the laser energy in hot electrons (as inferred from the hard x rays) versus the measured fraction of the energy which is Raman-scattered. In these experiments,[9] Au disks were irradiated with 1 ns pulse of 0.53 μm light at intensities ranging from 10^{14} - 10^{16} W/cm^2. The correlation predicted using the Manley-Rowe relations is shown by the solid line. Within the error bars, other processes such as the $2\omega_{pe}$ instability may also be contributing.

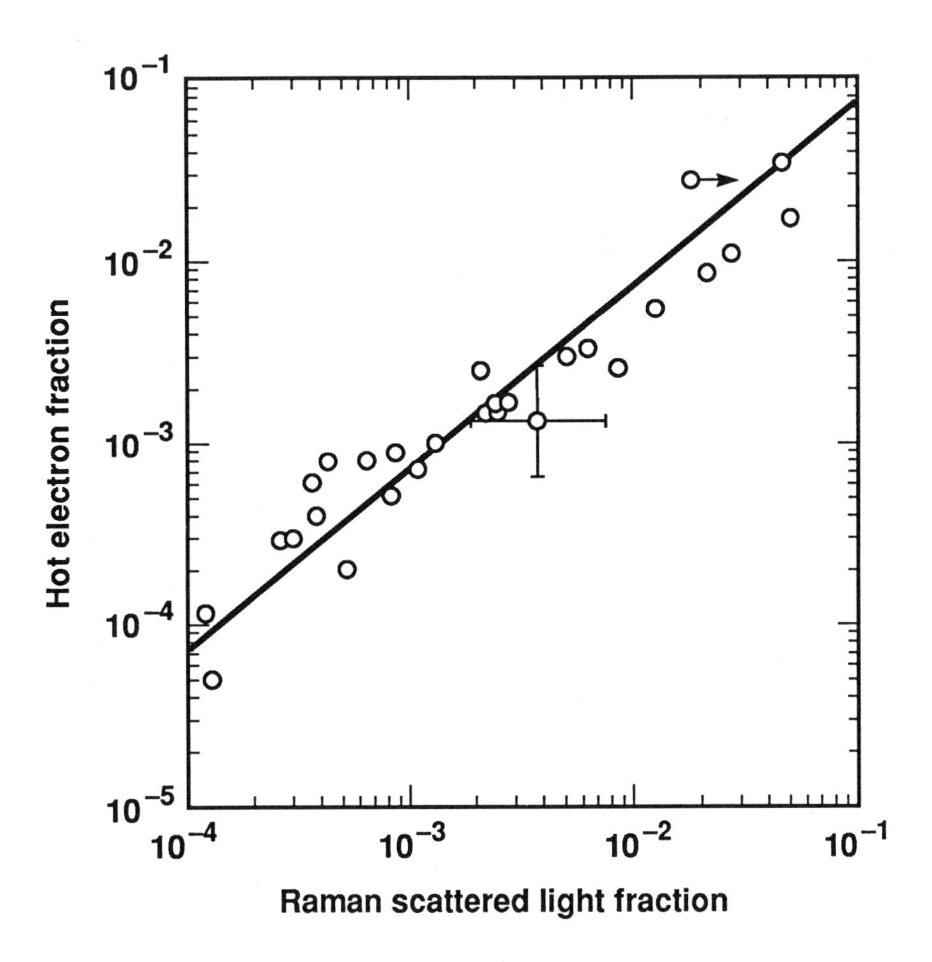

FIGURE 4 The fraction of the incident laser energy absorbed into hot electrons versus the measured fraction in Raman-scattered light.[9]

Significant levels of stimulated Raman scattering have been observed in large, strongly driven plasmas. Measurements of the fraction of the light which is Raman-scattered are shown in Fig. 5 versus the estimated size of the underdense plasma. The boxes denote the range of values in various experiments using[9,10] low Z targets irradiated with 0.53 μm light at intensities $> 10^{15}$ W/cm^2. In disk experiments with the Argus laser, the scalelength was about $10^2\lambda_0$, and the Raman reflectivity was rather small. Raman reflectivities up to several percent were measured for the plasmas accessed with disks on the Novette laser. Even larger plasmas were obtained by using thin disks which burn through and become underdense near the peak of the irradiation pulse. Plasma sizes of roughly $10^3\lambda_0$ were then obtained, giving a peak Raman reflectivity in excess of 10%. For such totally underdense plasmas, the size is estimated as the distance over which the density falls to about half the peak density. Finally, even larger plasmas were accessed by Chris Darrow in Nova experiments[10] with burn-through disks and 0.35 μm light. The measured Raman reflectivities in these experiments with $L \sim 5000\lambda_0$ were up to about 20%. As indicated by the arrow in Fig. 5, the peak reflectivity is probably even higher. Simple estimates show that in these large plasmas irradiated with 0.35 μm light, half or more of the scattered light would be collisionally absorbed before reaching the detector. Finally, the dashed curve shows a theoretical estimate based on the tail heating model mentioned earlier.

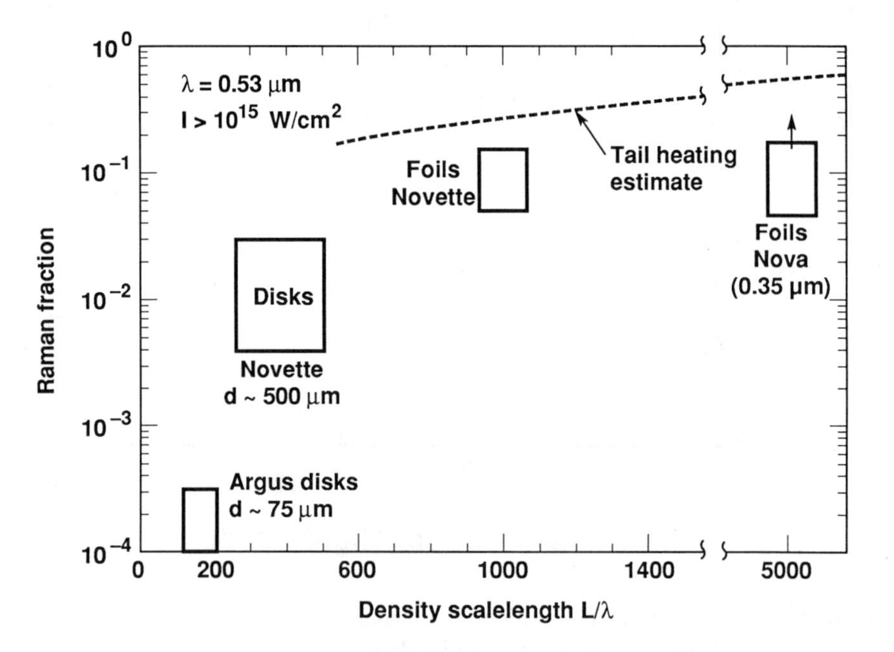

FIGURE 5 The measured fraction of the laser energy in Raman-scattered light versus the estimated size of the plasma.

Significant levels[11] of stimulated Raman scattering and concomitant hot electron generation were found in early hohlraum experiments using the Shiva laser. In these experiments, the interior of a gold hohlraum with characteristic dimensions of order 1 mm was irradiated with 1.06 μm laser light with a pulse length of order 1 ns. A large region of underdense plasma with density about 1-2.5 $\times$ 10^{20} cm^{-3} accumulated within the can. This plasma provided an ideal environment for excitation of stimulated Raman scattering.

Indeed, measurements of the high energy x-ray fluence showed that up to 50% of the laser energy was deposited into high energy electrons with a characteristic temperature of about 50 keV. Such electrons are deleterious since they preheat the capsules and degrade the implosions. Measurements also confirmed that intense stimulated Raman scattering was occuring in these experiments. In fact, up to 1/3 of the incident laser light was observed[12] to be Raman-scattered. The wavelength spectrum of the scattered light showed scattering from a plasma with density about 1-2.5 $\times$ 10^{20} cm^{-3}.

The hot electron problem was solved by using shorter wavelength laser light, which strongly reduces the excitation of underdense plasma instabilities. The plasma in which stimulated Raman scattering is efficient has a density of about 0.1-0.25 of the critical density. Since the critical density increases inversely as the square of the wavelength, there is less plasma at relevant densities as the wavelength is decreased. In addition, the more weakly driven instability can even be collisionally quenched by using sufficiently short wavelength light. For example, if 0.35 μm laser light is used, the Raman instability would be below collisional threshold in a Au plasma for intensities $\lesssim 5 \times 10^{14}$ W/cm^2.

Strong reductions in high energy electron generation with shorter wavelength light was first confirmed[13] in hohlraum experiments with the Argus laser. The wavelength of the light was changed by doubling or tripling the frequency of the Nd laser light. The measured fluence of high energy x rays decreased dramatically as the wavelength was changed from 1.06 μm to 0.53 μm to 0.35 μm. The Nova laser was then designed to operate with an output wavelength of 0.35 μm. Indeed, high energy electron generation has been no problem[14] in Nova hohlraum experiments. The fraction of the laser energy deposited into hot electrons is typically $\lesssim 0.1 - 1\%$.

Nonlinear ion waves

The second topic is the nonlinear behavior of ion waves.[15] The level to which ion acoustic waves are nonlinearly driven by the beat between two electromagnetic waves is a very important question for laser plasma interactions. For example, ion waves from the Brillouin instability can both affect the location of the absorption and significantly modify various processes in which electron plasma waves are generated. In microwave plasma experiments at UCLA, detailed measurements[16] have been made of the amplitude and harmonic generation of ion waves driven by the beat between opposed microwave beams. In addition, the ion tail formation has been measured. These measurements have both extended our quantitative understanding and provided a detailed test-bed for simulations.

Simulations[17] of these experiments have been carried out. The saturation amplitudes of both the driven ion wave and its second harmonic agree with experiments to within $\leq 30\%$. As in the experiments, an ion tail is formed when the ion fluctuation reaches a modest amplitude of $\delta n/n \simeq 6\%$, where δn is the density fluctuation in the ion wave and n the background plasma density. The observed density of the ion tail is obtained within a factor of less than 2.

The simulations use a one-dimensional particle ion, fluid electron code. This description includes ion kinetic effects in order to model the crucial phenomena of ion trapping and tail formation. The code also allows us to follow the evolution of the ion wave on the experimental time scale and to at least crudely model some important real world effects such as ion neutral collisions and ion loss from the interaction region.

In the code ions are treated as particles and electrons as a warm fluid with density n_e and temperature θ_e. Neglecting electron inertia, we obtain

$$n_e = n_{oe}\exp\left(\frac{e\phi}{\theta_e} + \alpha\right), \tag{1}$$

where n_{oe} is the initial uniform plasma density, ϕ is the low frequency electrical potential, and α is proportional to the so-called ponderomotive potential. We neglect pump depletion and calculate the ponderomotive potential due to the beat between two counter-streaming light waves with constant amplitudes. In particular,

$$\alpha = \frac{mv_1v_2}{2\theta_e}\cos[(k_1 - k_2)x - (\omega_1 - \omega_2)t], \tag{2}$$

where $v_{1,2} = eE_{1,2}/m\omega_{1,2}$. Here $E_{1,2}$ is the electrical field amplitude, $\omega_{1,2}$ the frequency and $k_{1,2}$ the wave number of the light waves. An ion acoustic wave is resonantly driven when $\omega_1 - \omega_2 = (k_1 - k_2)v_s$, where v_s is the ion sound velocity.

Ion neutral collisions are significant in the experiments. An effective collision frequency ν describing these collisions plus the damping effect due to ion escape from the interaction region is estimated to be $\nu/kv_s \simeq 0.07$ in the experiments discussed. This damping rate is measured by observing the decay of a low amplitude ion wave after the microwave beams are turned off. Collisions are modeled in the code in two different ways: (1) by adding a collisional drag term to the force equation for the ions and (2) by statistically replacing ions at an appropriate rate. In either case, the results were checked by observing the decay of an excited ion wave. No significant differences were noted depending on which scheme was used.

In the experiment,[16] two opposed microwave beams were propagated through a plasma with rather uniform density. The frequencies were adjusted to resonantly drive an ion wave, and the saturated amplitude of both the ion wave and its second harmonic was measured via Thomson scattering (and Langmuir probes). Let's consider a typical experiment with a plasma density of $0.15n_{cr}$ and $(P_1P_2)^{1/2} = 85$ kW, where P_1 and P_2 are the powers in the microwave beams. As expected, the results were observed to depend only on the product of the powers; i.e., on P_1P_2. In addition, pump depletion was observed to be a negligible effect for this density.

The measurements of the saturated amplitudes of the driven ion wave and its second harmonic as a function of distance along the plasma column is shown in Fig. 6. Note that the average driven wave is $\delta n/n \simeq 5\%$ and the average amplitude of the second harmonic wave is about 0.2-0.3 as large. The analogous simulation result is shown in Fig. 7. In the simulation, we treat the temporal evolution, ignoring boundary effects. Note that the driven ion wave grows and saturates at an amplitude of $\delta n/n \simeq 5.8\%$ and the amplitude of the second harmonic saturates at $\delta n/n \simeq 1.7\%$. Both saturation values are in quite reasonable agreement with the experiment.

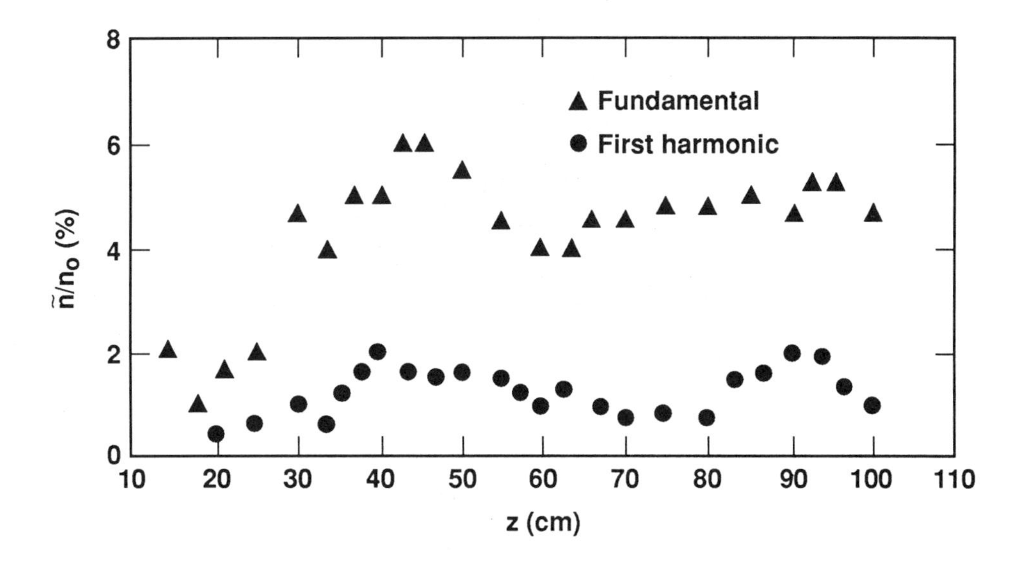

FIGURE 6 The measured amplitudes of the driven ion wave and its second harmonic as a function of distance along the plasma column in an experiment with $(P_1P_2)^{1/2} = 85\text{kW}$.

(a)

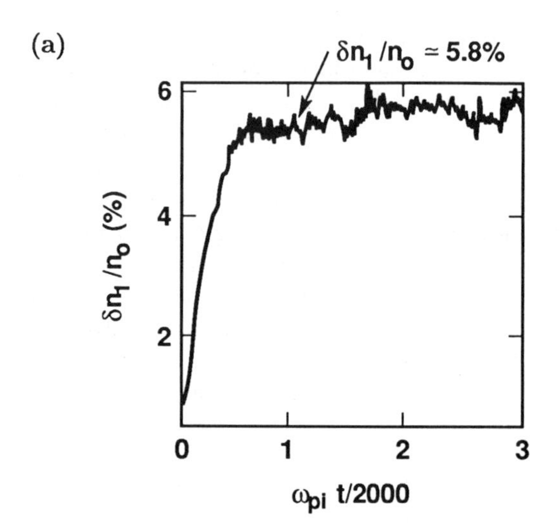

(b)

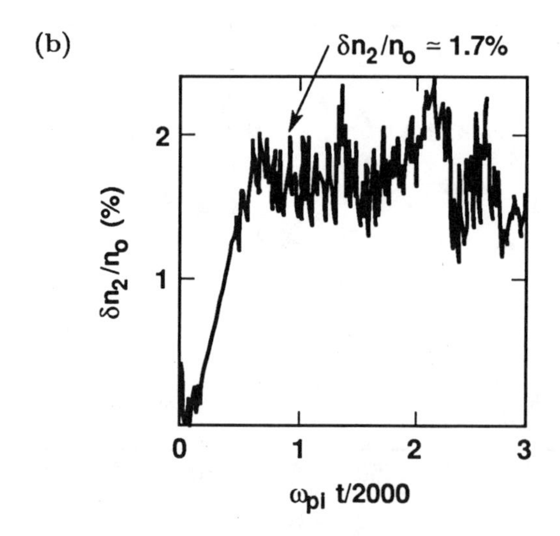

FIGURE 7 The evolution of the amplitudes of (a) the driven ion wave and (b) its second harmonic in a simulation with $(P_1 P_2)^{1/2} = 85\text{kW}$, $T_i/Te = 0.04$, $\nu/kv_s \simeq 0.07$, and $k\lambda_{\text{De}} \simeq 0.03$.

As shown by the ion distribution function exhibited in Fig. 8, a suprathermal ion tail is formed in the nonlinear state. This happens even though the amplitude of the driven ion wave is quite modest, $\delta n/n \lesssim 6\%$. Steepening of the wave form by the harmonic generation helps to allow this trapping to onset at a lower than originally expected wave amplitude. Nonlinear damping associated with the tail

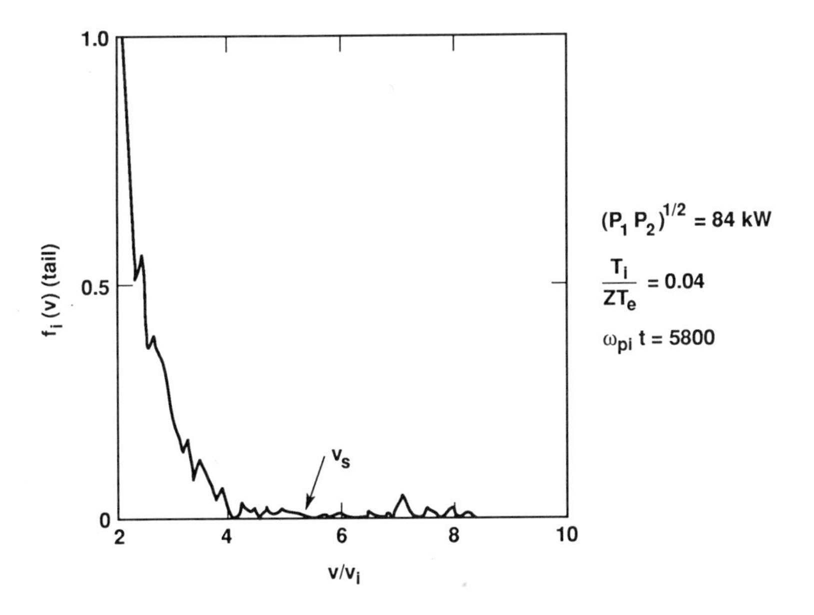

FIGURE 8 A plot of the tail of the ion distribution function calculated in the simulation.

formation appears to be the most potent nonlinear saturation effect in both the experiments and the simulations. The density of the tail is roughly a factor of two less in the simulations than in the measurements. The resulting discrepancy is most noticeable at the largest value of $P_1 P_2$ used in the experiment. Then the measured amplitude of the driven wave is about 30% less than in the simulation.

A Note on Quasi-Resonant Decay of Ion Waves

Quasi-resonant decay of finite-amplitude ion-acoustic waves has been proposed[18] as a saturation mechanism for the Brillouin instability. This decay of a driven ion wave into lower-wave-number ion waves has been examined in theory and in simulations with a two-fluid code. We found that it is essential to allow for harmonic generation, which is also quasi-resonant and dominates the mode coupling. In general, harmonic generation suppresses decay into lower wave numbers.

The decay of a finite-amplitude ion wave into two-lower-frequency ion waves has been predicted.[18] For ion waves,

$$\omega = \frac{k v_s}{\sqrt{1 + k^2 \lambda_{\mathrm{De}}^2}} \tag{3}$$

where ω is the frequency, k is the wave number, λ_{De} is the electron Debye length, and v_s is the ion sound velocity. (For simplicity, corrections due to ion temperature and damping are neglected.) Since ω is not simply proportional to k, a small frequency mismatch must be overcome, i.e., the decay is quasi-resonant.

The theory of quasi-resonant decay is readily obtained from the two-fluid equations. An equation for the low-frequency density fluctuation is obtained and then linearized by assuming for the density n:

$$n = n_0 + \delta n + \tilde{n}, \tag{4}$$

where n_0 is the uniform background-plasma density, δn is the density fluctuation associated with the large-amplitude ion wave with frequency ω_0 and wave-number k_0, and $\tilde{n}$ is the fluctuation associated with the perturbation.

A dispersion relation is obtained by truncating to two modes with lower wave numbers. This dispersion relation predicts that the maximum growth rate occurs for decay into the half-harmonic $(k_0/2)$ and is

$$\gamma_0 = \frac{k_0 v_s}{4}\frac{\delta n}{n_0}. \tag{5}$$

If we allow for the frequency mismatch $\Delta\omega$, growth requires $\gamma_0 > \Delta\omega$, or

$$\frac{\delta n}{n_0} > \frac{3}{4} k_0^2 \lambda_{\text{De}}^2 . \tag{6}$$

However, derivation of this decay instability omits a very important effect, harmonic generation, which is also quasi-resonant and which leads to a steepening of the large-amplitude ion wave. We carried out simulations to include these higher wavenumber waves. In our code, we used a two-fluid description and drove up the large amplitude ion wave by a ponderomotive potential. An ion-wave damping proportional to wave number was also used, as is appropriate for Landau damping. For more details on the code, see Ref. 19.

Two examples illustrate our results. In the first, only two modes are included in the simulations–the driven mode and its half-harmonic. As shown in Fig. 9, the driven mode ($k\lambda_{\text{De}} = 0.05$) is excited to an amplitude of $\delta n/n = 0.10$. The amplitude of the half-harmonic mode ($k\lambda_{\text{De}} = 0.025$) then exponentiates with a growth rate in very good agreement with the two-mode theory. Indeed, this half-harmonic mode efficiently takes energy from the driven mode and subsequently dominates.

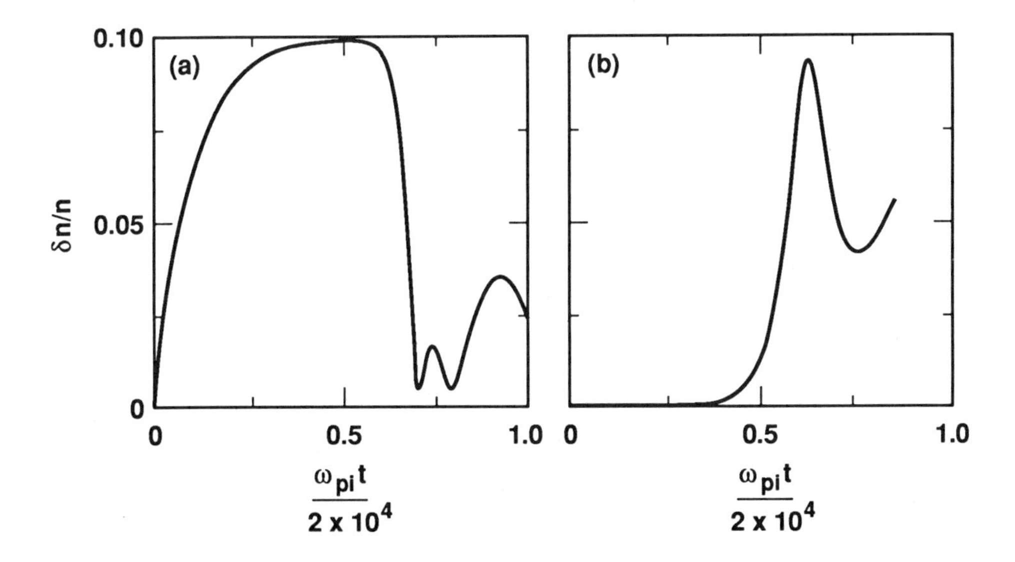

FIGURE 9 Evolution of (a) driven mode ($k\lambda_{De} = 0.05$) and (b) half-harmonic mode ($k\lambda_{De} = 0.025$) in a two-mode simulation. The energy damping rate is 0.02 of the ion-wave frequency. Here ω_{pi} is the ion plasma frequency.

The behavior is qualitatively different if ion waves with higher numbers are included in the simulation. In this second example, the same driven mode is again excited, but now mode numbers up to 20 are retained. Hence this simulation allows the competitive effect of harmonic generation. Evolution of the amplitude of the driven mode, its half-harmonic, and its second harmonic is shown in Fig. 10. The amplitude of the driven mode now saturates at $\delta n/n \simeq 0.048$, concomitant with significant excitation of the second harmonic. In contrast to predictions of the two-mode theory, the half-harmonic wave does not grow, even though ion-wave damping was reduced by a factor of 2.

Many different simulations were carried out, varying the $k\lambda_{De}$ of the driven mode and the strength of the ponderomotive driver. The above results were typical. No growth of the half-harmonic was observed when higher-wave-number modes were included. Growth of the half-harmonic mode was observed in only one case in which the zero-order state became extremely distorted. Finally, we note that half-harmonic generation was not observed in recent detailed experiments on ion-wave excitation. However, the ion-wave damping would have been sufficient to suppress the half-harmonic generation in these experiments, even if the truncated mode analysis were valid.

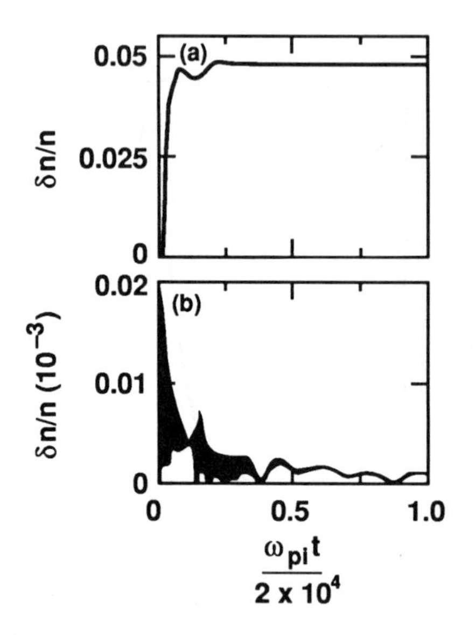

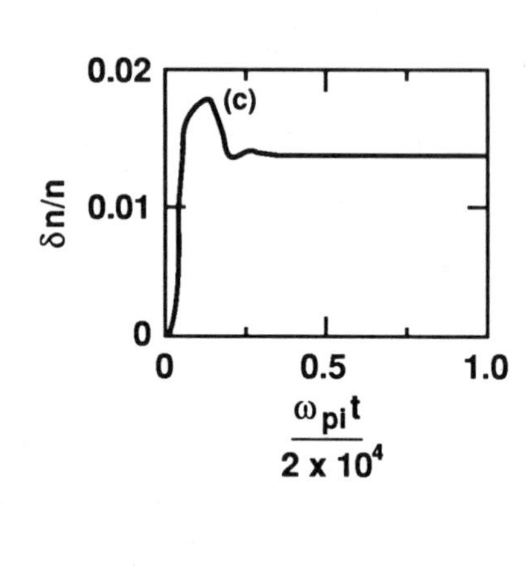

FIGURE 10 Evolution of (a) driven mode ($k\lambda_{De} = 0.05$), (b) half-harmonic mode ($k\lambda_{De} = 0.025$), and (c) second harmonic mode ($k\lambda_{De} = 0.1$). In this simulation, 20 modes were allowed, and the energy damping rate was 0.01 of the ion-wave frequency.

Acknowledgements

I am very grateful for the collaborations discussed here and cited in the references. Special acknowledgement is due to D.W. Phillion, E.M. Campbell, and R.E. Turner for permission to cite their hohlraum data. Work performed under the auspices of the United States Department of Energy by the Lawrence Livermore National Laboratory under contract number W-7405-ENG-48.

References

1. W.L. Kruer, P.K. Kaw, J.M. Dawson, and C. Oberman, *Phys. Rev. Lett.* **24**, 987 (1970).

2. Kent Estabrook and W.L. Kruer, *Phys. Rev. Lett.* **40**, 42 (1978).

3. D.W. Forslund, J.M. Kindel, and K. Lee, *Phys. Rev. Lett.* **39**, 284 (1977).

4. H.A. Baldis, E.M. Campbell, and W.L. Kruer, in *Handbook of Plasma Physics*, vol. 3, edited by S. Witkowski and A.M. Rubenchik, Chapter 9, North-Holland, Amsterdam, (in press).

5. D.W. Forslund, J.M. Kindel, and E.L. Lindman, *Phys. Rev. Lett.* **30**, 739 (1973);
P W.L. Kruer, K.G. Estabrook, and K.H. Sinz, *Nuclear Fusion* **13**, 952 (1973); H.H. Klein, W.M. Manheimer, E. Ott, and W.L. Kruer *Phys. Rev. Lett.* **31**, 1187 (1973).

6. A.B. Langdon, B.F. Lasinski, and W.L. Kruer, *Phys. Rev. Lett.* **43**, 133 - (1979).

7. Kent Estabrook, W.L. Kruer, and B.F. Lasinski, *Phys. Rev. Lett.* **45**, 1399 (1980).

8. W.L. Kruer and Kent Estabrook, in *Laser Interaction and Related Plasma Phenomena*, vol. 5, pp. 783-800, Plenum Press, New York (1981).

9. R.P. Drake, et al., *Phys. Rev. Lett.* **53**, 1739 (1984); R.P. Drake, *Laser and Part. Beams* **6**, p. 235 (1988).

10. C.B. Darrow, et al., *Phys. Fluids* in press.

11. W.L. Kruer, Lawrence Livermore National Laboratory, UCRL-JC-104618 (1990).

12. D.W. Phillion, E.M. Campbell, and R.E. Turner, "Raman Scattering from Hohlraum Targets," Lawrence Livermore National Laboratory UCRL-53372 (1982).

13. E.M. Campbell, et al. (to be published).

14. J.D. Kilkenny, et al., in *Plasma Physics and Controlled Fusion Research*, vol. 3, p. 29, IAEA, Vienna (1989); J.D. Lindl, this symposium.

15. J.M. Dawson, W.L. Kruer, and B. Rosen, *Dynamics of Io nized Gases*, edited by M. Lighthill, I. Ima, and H. Sato, pp. 47-61, Univ. of Tokyo Press, Tokyo (1973).

16. C.W. Pawley, "Linear and Nonlinear Response of Ion Acoustic Waves Driven by Optical Mixing," Ph.D. Thesis, UCLA, 1986.

17. C.W. Pawley, N.C. Luhmann, Jr., and W.L. Kruer, *Plasma Physics and Controlled Nuclear Fusion Research*, vol. 3, pp. 137-144, International Atomic Energy Agency, Vienna, Austria (1987).

18. S.J. Karttunen, J.N. McMullin, and A.A. Offenberger, *Phys. Fluids* **24**, 447 (1981).

19. *Laser Program Annual Report–1979*, pp. 3-43 and 3-44, Lawrence Livermore National Laboratory, Livermore, Calif., UCRL-50021-79 (1980).

John D. Lindl
Lawrence Livermore National Laboratory, Livermore, CA 94550

Progress on Ignition Physics for ICF and Plans for a Nova Upgrade to Demonstrate Ignition and Propagating Burn by the Year 2000

Inertial Confinement Fusion is an approach to fusion which relies on the inertia of the fuel mass to provide confinement. To achieve conditions under which this confinement is sufficient for efficient thermonuclear burn, high gain ICF targets designed to be imploded by laser light have features similar to those shown in Fig. 1. These capsules are generally a spherical shell which is filled with low density gas. The shell is composed of an outer region which forms the ablator and an inner region of frozen or liquid DT which forms the main fuel. Energy from the driver is delivered to the ablator which heats up and expands. As the ablator expands and blows outward, the rest of the shell is forced inward to conserve momentum. In this implosion process, several features are important. We define the in-flight-aspect-ratio (IFAR) as the ratio of the shell radius R as it implodes to its thickness ΔR. Hydrodynamic instabilities during the implosion impose limits on this ratio which results in a minimum pressure requirement for any given implosion velocity. The convergence ratio is defined as the ratio of the initial outer radius of the ablator to the final compressed radius of the hot spot. This hot spot is the central region of the compressed fuel which is required to ignite the main fuel in high gain designs. The typical configuration of the compressed fuel at ignition is shown in Fig. 2. The hot spot ρR must be equal to about 0.3 g/cm^2, for effective self-heating from a temperature of 4-5 keV achieved by PdV work during the compression process. Typical convergence ratios to the hot spot are 30-40. To maintain a nearly spherical shape during the implosion, when convergence ratios are this large, the flux delivered to the capsule must be uniform to a few percent.

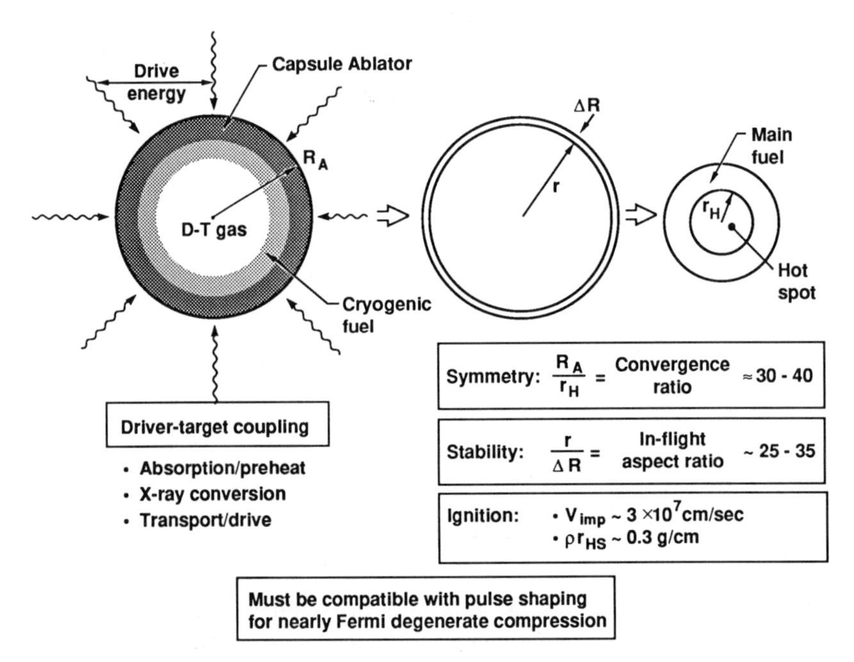

FIGURE 1 Many high gain ICF targets have common features

In general, ICF capsules rely on either electron conduction (direct drive) or x rays (indirect drive) to drive an implosion.

In direct drive, the laser beams (or charged particle beams) are aimed directly at a target. The laser energy is transferred to electrons via inverse bremsstrahlung or a variety of plasma collective processes. This absorption occurs at a density equal to or less than the plasma density $n_c = 10^{21}/\lambda^2$ where λ is the laser wavelength in microns. Electron conduction must transport the energy to the ablation front which typically has an electron density of about 10^{24} cm^3. Uniformity of the flux must be obtained by direct overlap of a large number of very uniform beams, or by lateral electron conduction smoothing.

In the x-ray or indirect drive approach to ICF, the laser beams are first absorbed in a high-Z enclosure, a hohlraum, which surrounds the capsule. A significant fraction of this absorbed energy is converted to x-rays which then drive the capsule implosion. This two step process results in a decoupling of the absorption and smoothing processes and relaxes the beam quality requirements.

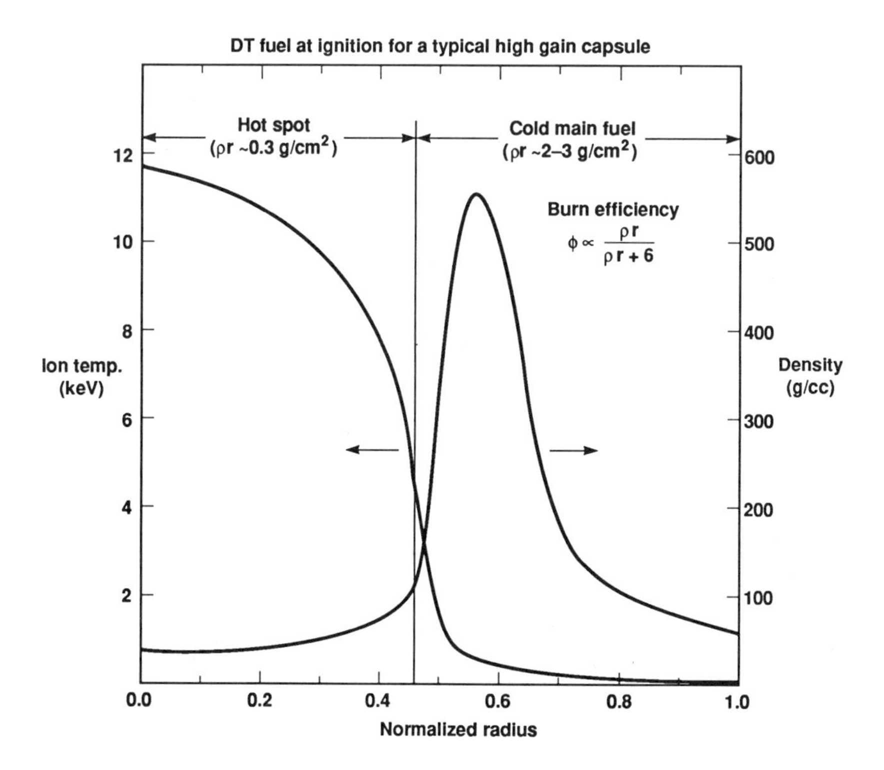

FIGURE 2 ICF capsules utilize central ignition and propagation into compressed DT to achieve high gain.

The minimum driver energy required to achieve ignition and propagating burn into the main fuel depends strongly on the achievable implosion velocity of the capsule. This velocity in turn is primarily determined by the peak pressure that can be generated consistent with efficient coupling of the driver energy to the capsule, and by the degree of hydrodynamic instability of the implosion process. The following discussion makes these relationships more quantitative.

The pressure P generated by ablation scales as a power of the incident flux I:

$$P = P_0 I^{a_1} \tag{1}$$

where a_1 is less than 1. The mass ablation rate $\dot{m}$ is also given by a power of the incident flux:

$$\dot{m} = \dot{m}_0 I^{a_2} \tag{2}$$

where a_2 is also less than 1. If the pusher is imploded adiabatically, then the pressure and density ρ are related by:

$$P = 2\beta\rho^{5/3} \text{ or } \rho = \frac{P^{3/5}}{1.5\beta^{3/5}} = P_0^{3/5} I^{3/5^{a_1}} \tag{3}$$

where β is the ratio of the pressure at a given density to the Fermi pressure. Pressure is in megabars and density is in g/cm^3. The implosion velocity V_{imp} of a shell accelerated by ablation is given by a rocket equation:

$$V_{\text{imp}} = \frac{P}{\dot{m}} ln\frac{m_0}{m} = \frac{P_0}{\dot{m}_0} I^{a_1-a_2} ln\frac{m_0}{m} = V_{\text{ex}} ln\frac{m_0}{m}; V_{\text{ex}} = \frac{P_0}{\dot{m}_0} I^{a_1-a_2} \tag{4}$$

The ablation velocity, the velocity with which the ablation front moves through the shell, V_{abl}, is given by the mass ablation rate divided by the shell density:

$$V_{\text{abl}} = \frac{\dot{m}}{\rho} = 1.5\beta^{3/5} \frac{\dot{m}_0}{P_0^{3/5}} I^{a_2-3/5a_1} = V_{\text{abl}}^0 I^{a_1-3/5a_1} \tag{5}$$

Examples of some of the analytical and numerical work which was used to derive these equations are given in Refs. 1-3.

The shell aspect ratio, $R/\Delta R$, the ratio of the shell radius R to its thickness can be related to V_{imp}, V_{abl}, and V_{ex} by integrating the rocket equation. If we assume the shell is accelerated over half its radius, we have:

$$\int_0^{t_1} \text{vdt} \sim 1/2R = \int_0^{t_1} V_{\text{ex}} ln\left(\frac{m_0}{m}\right) \text{dt where } ln\frac{m_1}{m_0} = ln\left(1 - \frac{v_{\text{abl}} t_1}{\Delta R}\right) = \frac{V_{\text{imp}}}{V_{\text{ex}}} \tag{6}$$

$$\Rightarrow \frac{R}{\Delta R} = 2\frac{V_{\text{ex}}}{V_{\text{abl}}} f_1\left(\frac{V_{\text{imp}}}{V_{\text{ex}}}\right) \text{ where } f_1(x) = [1 - (1+x)\exp(-x)] \tag{7}$$

If the implosion velocity is greater than the rocket exhaust velocity V_{ex}, then f_1 is approximately linear in x. In this case, there is a simple linear relationship between the shell aspect ratio and the ratio of implosion velocity to ablation velocity:

Case a) $$R/\Delta R = 0.56\frac{V_{\text{imp}}}{V_{\text{abl}}} \text{ for } 4 > \frac{v_{\text{imp}}}{v_{\text{ex}}} > 0.8 \tag{8a}$$

In this regime, the achievable implosion velocity is proportional to the product of the shell aspect ratio and the ablation velocity. If the implosion velocity is less than the rocket exhaust velocity, then f_1 is approximately quadratic in x and the relationship between the shell aspect ratio and the rocket parameters is somewhat more complex:

Case b) $$R/\Delta R = 0.70\frac{V_{\text{imp}}^2}{V_{\text{abl}} V_{\text{ex}}} \text{ for } 0.8 > \frac{V_{\text{imp}}}{V_{\text{ex}}} > 0.1 \tag{8b}$$

In general, hydrodynamic instability sets an upper limit to the value of the shell aspect ratio. The growth rate for Rayleigh-Taylor type instabilities in the presence of ablation and a density gradient is given by:

$$\gamma = \sqrt{\frac{ka}{1+kL}} - \alpha k V_{\rm abl} \tag{9}$$

In this equation, k is the mode wavenumber, a is the acceleration, L is the density gradient scale length in the ablation front, and α is a constant between 1 and 3. Recent analytical and numerical work which arrived at this dispersion relation is found in references 4-5. If we assume,

a = const
$R/2 = 1/2\ \mathrm{a}t_1^2$
$L = \Delta R/2$
$k = \ell/R$

where ℓ is the spherical harmonic mode number, then the number of e-foldings for a given mode is given by:

$$\begin{aligned} n \sim\ & \ell^{1/2}/\left[1+(\ell/2)(\Delta R/R)\right]^{1/2} - \alpha\frac{\ell}{R}\int_0^{t_1} V_{\rm abl}dt = \\ & \ell^{1/2}\left[1+(\ell/2)(\Delta R/R)\right]^{1/2} - \alpha\ell(\Delta R/R)\, f_2(V_{\rm imp}/V_{\rm ex}) \end{aligned} \tag{10}$$

where $f_2(x) = [1 - \exp(-x)]$. Alternatively, the shell aspect ratio can be replaced by Eq. (7) to give:

$$n = \ell^{1/2}\left[1+\frac{\ell}{4}\frac{V_{\rm abl}}{V_{\rm ex} f_1(V_{\rm imp}/V_{\rm ex})}\right]^{-1/2} - \frac{\alpha\ell}{2}\frac{V_{\rm abl}}{V_{\rm ex}}\frac{f_2(V_{\rm imp}/V_{\rm ex})}{f_1(V_{\rm imp}/V_{\rm ex})} \tag{11}$$

For the case with $V_{\rm imp} > V_{\rm ex}$, $f_2(x)$ is nearly a constant and Eq. (11) takes on the simple form:

$$\begin{aligned} n &= \ell^{1/2}/\left[1+(\ell/2)(\Delta R/R)\right]^{1/2} - 0.9\alpha\ell(\Delta R/R) \\ &= \ell^{1/2}/\left[1+(\ell/1.12)(V_{\rm abl}/V_{\rm imp})\right]^{1/2} - 1.06\alpha\ell(V_{\rm abl}/V_{\rm imp}) \end{aligned} \tag{12}$$

for $4 > V_{\rm imp}/V_{\rm abl} > 5/4$ so that the number of e-foldings of a given mode number depends only on the shell aspect ratio, or equivalently, on the ratio of $V_{\rm imp}$ to $V_{\rm abl}$. This is the case that will be followed in the rest of the analysis, but the conclusions would be very similar for the general case. The perturbation caused by the growth of all modes will depend on the spectrum of modes and the surface finish. The best estimates of these effects require the extensive use of detailed numerical calculations. These calculations predict that shells with $R/\Delta R$ <25-35 having state of the art machined or plasma vapor deposited surfaces with an RMS surface finish of about 1000Å can be imploded without breaking up. In addition, these shells can be compressed to produce a near spherical hot spot at a capsule absorbed energy about a factor of two above the minimum for a given implosion velocity. From Eq. (8a), the achievable implosion velocity will be linearly proportional to the ablation velocity.

To complete the analysis, we require a relationship between the capsule absorbed energy required for ignition and burn propagation, and the implosion velocity. This relationship is not unique. It depends on the type of capsule being imploded. The model developed below is for capsules such as that shown in Fig. 1 which ignite from a hot spot and propagate into a compressed main fuel region. It is fairly accurate for the type of capsule of greatest interest to energy applications which require high gain.

The hot spot energy E_h, for a hot spot heated to the ignition temperature T_h is given by:

$$E_h \sim (0.096)\left(\bar{Z} + 1/\bar{A}\right) T_H M_H = 1/2 M_C V_{\rm imp}^2 f \sim \eta_c f E_I \tag{13}$$

where M_c is the pusher mass of dense main fuel, f is the coupling fraction between the pusher energy and the hot spot, η_c is the coupling efficiency between the driver and the imploding fuel mass and $E_{\rm I}$ is the incident driver energy. If we assume that f, T_H and η_c are nearly constant, then we have:

$$M_H \sim M_c V_{\rm imp}^2 \sim E_I \tag{14}$$

For the hot spot mass, we have:

$$M_H \sim \left[4\pi \rho r^3/3\right] \sim \left[(\rho r)_H^3 / \rho_H^2\right] \tag{15}$$

Since $(\rho\, r)_{\rm H} \sim 0.3{\rm g/cm}^2$, if we assume that $\rho_c \propto \rho_{\rm H}$ where ρ_c is the cold fuel density and ρ_c is the hot spot density, we have:

$$E_I \sim M_H \sim \left(1/\rho_c^2\right) \tag{16}$$

For an adiabatic compression of the pusher as it decelerates, we have:

$$\epsilon_C \sim V_{\rm imp}^2 \sim \beta \rho_C^{2/3} \quad \text{or} \quad V_{\rm imp}^6 \sim \beta^3 \rho_C^2 \tag{17}$$

where ϵ_c is the specific energy per gram of the imploding cold fuel. Using this result in Eq. (16) gives:

$$E_I \sim \beta^3 V_{\rm imp}^{-6} \tag{18}$$

Very similar conclusions were obtained by Colgate and Petschek in Ref. 6.

Finally, we can use Eqs. (5) and (8a) to get:

$$E_I \propto \beta^3 \left(R/\Delta R\right)^{-6} V_{\rm abl}^{-6} \quad \text{or} \quad E_I = E_0 \beta^{-3/5} \left(R/\Delta R\right)^{-6} 1/I^{6(a_2 - 3/5 a_1} \tag{19}$$

For all implosions, the threshold energy depends strongly on the acceptable capsule aspect ratio. The tolerable aspect ratio depends in turn on the achievable capsule surface finish and the spectral distribution of perturbations as well as the extent of reduction in the growth rate for given modes can be reduced by optimization the density and ablation stabilization effects in Eq. 9. The variation of the threshold energy on the incident intensity depends on the specific values of the exponents a_1 and a_2. In general, the exponent of the intensity varies from near zero to a value greater than unity. Limitations on the achievable intensity occur primarily because of deleterious laser plasma interaction effects occurring at high intensity which can either reduce absorption, adversely affect the absorption uniformity or result in the production sources of capsule preheat which can prevent efficient compression of the fuel.

Equations can also be developed for the yield and gain as a function of the driver energy. The yield can be written approximately as:

$$Y \sim M_C \left[(\rho r)\, c/6\right] \epsilon_{TN} \tag{20}$$

where ϵ_{TN} is the fusion energy per gram for DT. From Eq. (14), we have:

$$M_C \sim \left(\eta_C E_I / V_{\mathrm{imp}}^2\right) \tag{21}$$

If we assume constant coupling efficiency η_c and use Eq. (18), we get:

$$M_C \sim E_I^{4/3} \tag{22}$$

Using

$$M_C \sim \left[(\rho r)_c^3 / \rho_c^2\right] \tag{23}$$

and Eq. (17) we get

$$(\rho r)_C \sim \left(\rho_C^2 M_C\right)^{1/3} \sim E^{1/9} \tag{24}$$

Combining Eqs. (20), (22), and (24), we get:

$$Y \sim E^{13/9} \tag{25}$$

In general, because of the need for increased preheat shielding at smaller sizes, we expect both the effective fuel mass and rr to scale more rapidly with energy than Eq. (25) and the yield more closely follows:

$$Y\,(MJ) \sim 25 E^{5/3}\,(MJ) \tag{26}$$

This equation holds when the capsule at each energy is imploded to a velocity equal to the minimum required to ignite for a given surface finish. This will require a minimum intensity or capsule aspect ratio at each energy as given by Eq. (19). At any given energy, it will be possible to meet or exceed the minimum ignition velocity for all intensities or aspect ratios above this minimum. For any given implosion velocity, it is possible to calculate capsule gain versus driver energy. Gain curves for indirect drive at three different velocities are shown in Fig. 3. These curves, for three different implosion velocities, are all calculated under the assumption of a fixed hohlraum coupling efficiency of laser energy to a capsule. The shaded band at the left of each set of curves corresponds to the uncertainty in the achievable capsule surface finish. The far left hand edge corresponds to the gain achievable for perfectly uniform implosions. The right hand edge of the band corresponds to the gain for targets with surface finishes which we expect to achieve in the next several years. As the driver energy increases, the minimum implosion velocity required to ignite a capsule decreases. If we exceed the minimum velocity at any driver size, the capsules will still ignite. However, there is a performance penalty for operating above the minimum velocity. The gain will drop because we will implode less mass and get less yield for a given energy. Hence, the optimum strategy implies operation at the minimum implosion velocity consistent with the desired yield or driver size. For a fixed capsule surface finish, this optimum is given by the dashed line through the three curves. It is this line that follows the scaling given by Eq. (26).

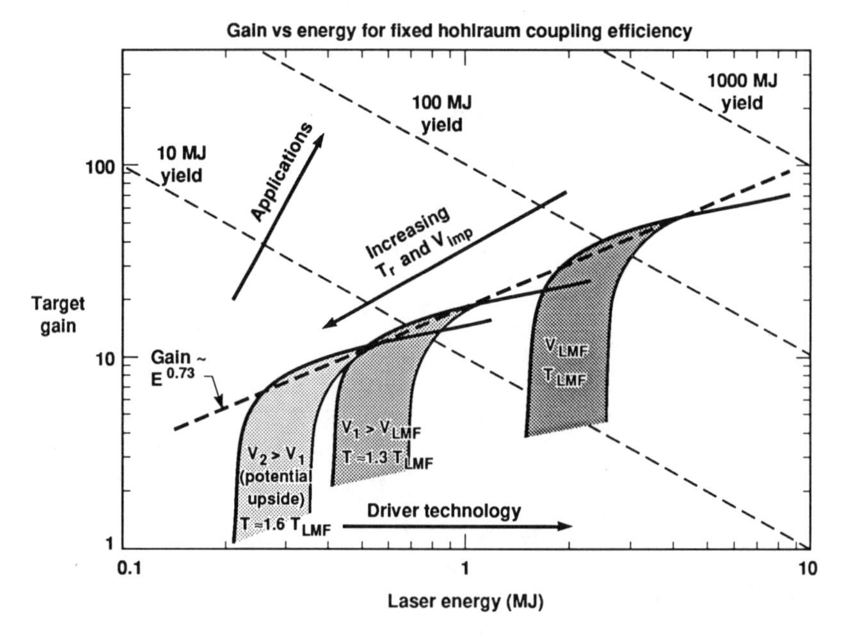

FIGURE 3 Increasing implosion velocity for a fixed hydro stability criterion allows ignition and propagating burn at low drive energy.

(a)

(b)

FIGURE 4 (a) Noval Laser Bay; (b) Nova Target Chamber

Experiments during the past five years, principally those conducted on the Nova laser at the Lawrence Livermore National Laboratory (LLNL) and in the Halite/Centurion Program utilizing a small fraction of the energy from an underground nuclear explosive, have allowed us to make dramatic progress in establishing the hydrodynamics requirements and the plasma physics limitations of laser drivers at various wavelengths and sizes. The three sets of curves in Fig. 3 summarize the conclusions of that work for a 0.35 micron laser driver. Until recently, most of the experiments on Nova had been focused on demonstrating that the hohlraum and plasma physics constraints were consistent with achieving the velocity labelled V_{lmf}. As a result of the work done with 1.06 micron light on the Shiva laser, it was our expectation that plasma effects on Nova using 0.35 micron light would allow us to achieve this velocity, but probably not much more. In addition, for yields of a few hundred megajoules or more, this is an adequate velocity. Yields of this magnitude are required for many weapon physics applications, and are also required for most ICF reactor designs. However, the minimum driver size, as indicated in Fig. 3, is in the range of 4-5 MJ. A facility of this size has been labelled the Laboratory Microfusion Facility (LMF) by DOE and has been under intense study since 1988. However, in early 1990, using the recently increased power and energy available from Nova, LLNL demonstrated hohlraum drive conditions consistent with the higher velocity required for the middle set of gain curves in Fig. 3. At this higher implosion velocity, ignition and burn propagation can be demonstrated with a 1-2 MJ laser. Such a laser, using the LLNL Athena laser technology developed for the LMF, can be placed in the existing Nova building and could utilize significant portions of the existing Nova hardware. Although the expected yields of less than about 30 MJ are not large enough for ICF energy applications or for many weapon physics applications, the demonstration of ignition and burn propagation in the laboratory would complete the basic target physics stage of the ICF program. Such a demonstration would set the stage for high confidence development of the applications of ICF. This Nova Upgrade has been endorsed by Secretary of Energy Watkins' Fusion Policy Advisory Committee (FPAC) and the Koonin Committee of the NAS. Subject to successful completion of the Nova experiments and laser technology development planned for the next 2-3 years, the FPAC and NAS recommend start of construction of this facility in FY-94. Start of construction in FY-94 would allow demonstration of ignition before the year 2000.

Shown in Fig. 4a and 4b is the Nova laser bay and experimental area. Nova can deliver about 40 kJ of 0.35 micron laser light in a 1 ns pulse. Using the much more compact Athena laser amplifier architecture, shown in Fig. 5, it is possible to fit a 1-2 MJ upgrade to Nova into the existing Shiva-Nova building. The Athena beamlines are a multipass design, fully relayed, composed of a 4×4 array of 30 cm beamlets. Except for sharing the optical cavity, the beamlets are optically independent and individually pointed at the target for maximum control of illumination uniformity. The proposed Nova Upgrade would consist of 18 such beamlines, for a total of 288 beamlets arranged as shown in Fig. 6a and 6b. The power and energy which this laser system can deliver is a function of pulse length as shown in Fig. 7. The solid curve is relationship between power and energy for a single point design optimized at 8 ns. The squares are performance which could be achieved with 288 beamlets if the beamlets were optimized for the pulse length indicated next to the box. Although the power increases as the pulse length τ decreases, the amount of energy which can be extracted drops approximately as $\tau^{-1/2}$ because of a decrease in the optical component damage threshold. The rapid drop in performance at both extremes in pulse length is due to nonlinear optical effects. Additional information on the Athena architecture can be found in Ref. 7.

In summary, the ICF Program has made major progress in establishing the physics and technology base required to demonstrate ignition and burn propagation in the laboratory. With an FY-94 start of construction of a 1-2 MJ upgrade to the Nova laser, the goal could be achieved by the year 2000.

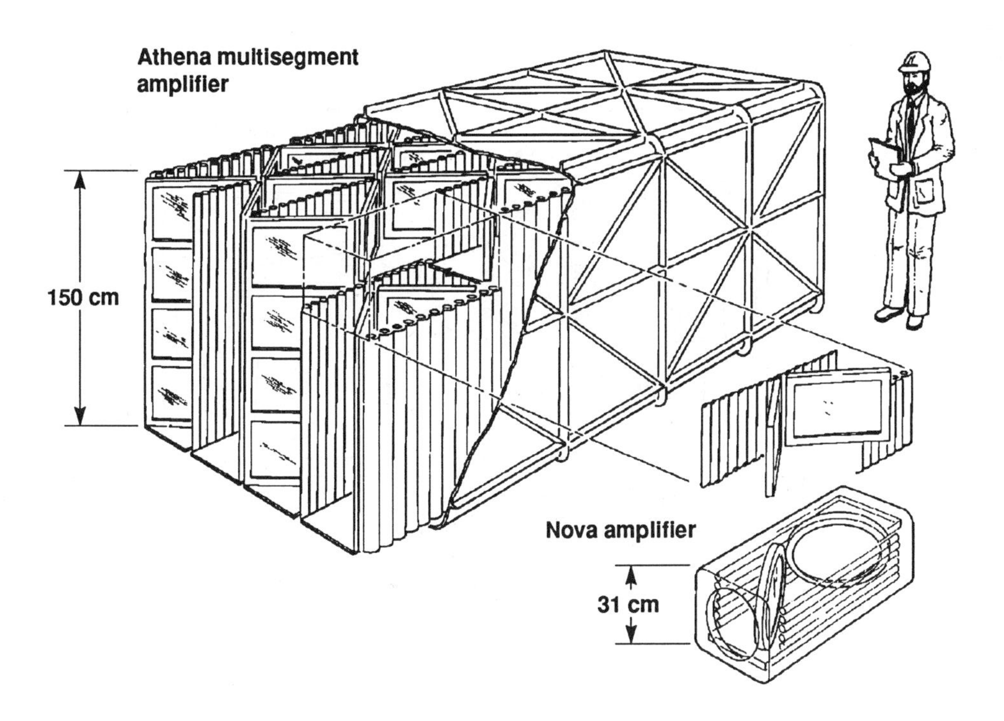

FIGURE 5 The Athena amplifier is essentially an array of Nova 31 cm amplifiers

(a)

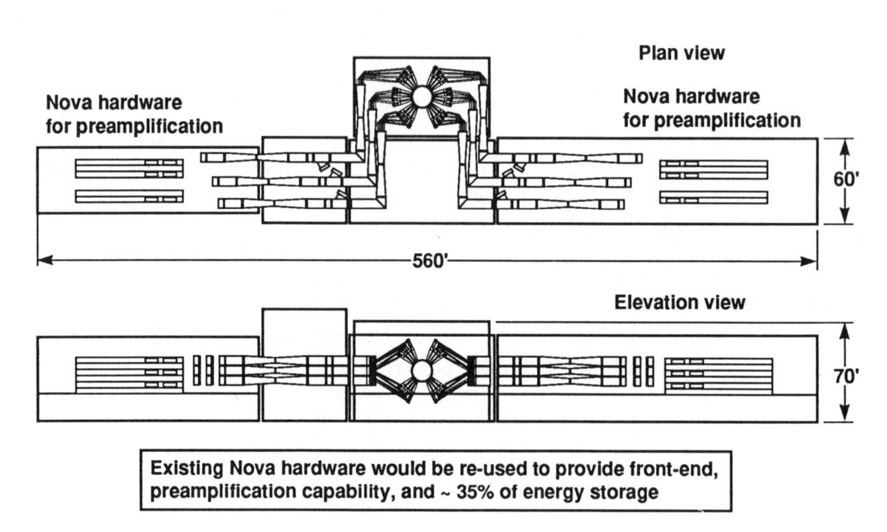

(b)

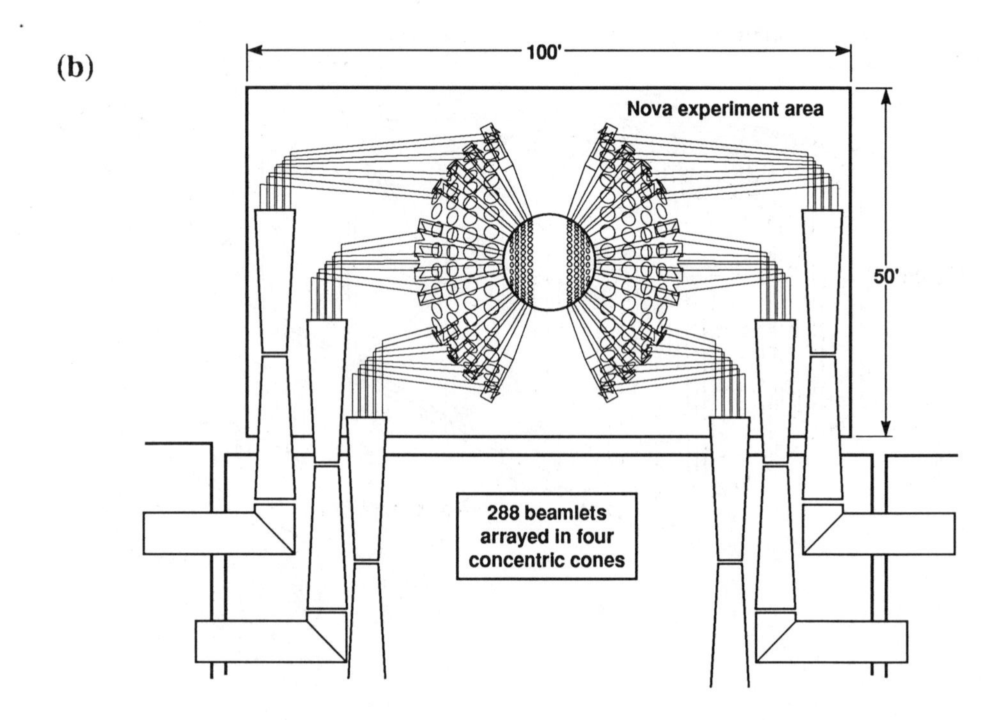

FIGURE 6 Nova Upgrade: **(a)** To demonstrate ignition and gain fits in the existing Shiva-Nova Building; **(b)** Illumination geometry provides the capsule flux uniformity required.

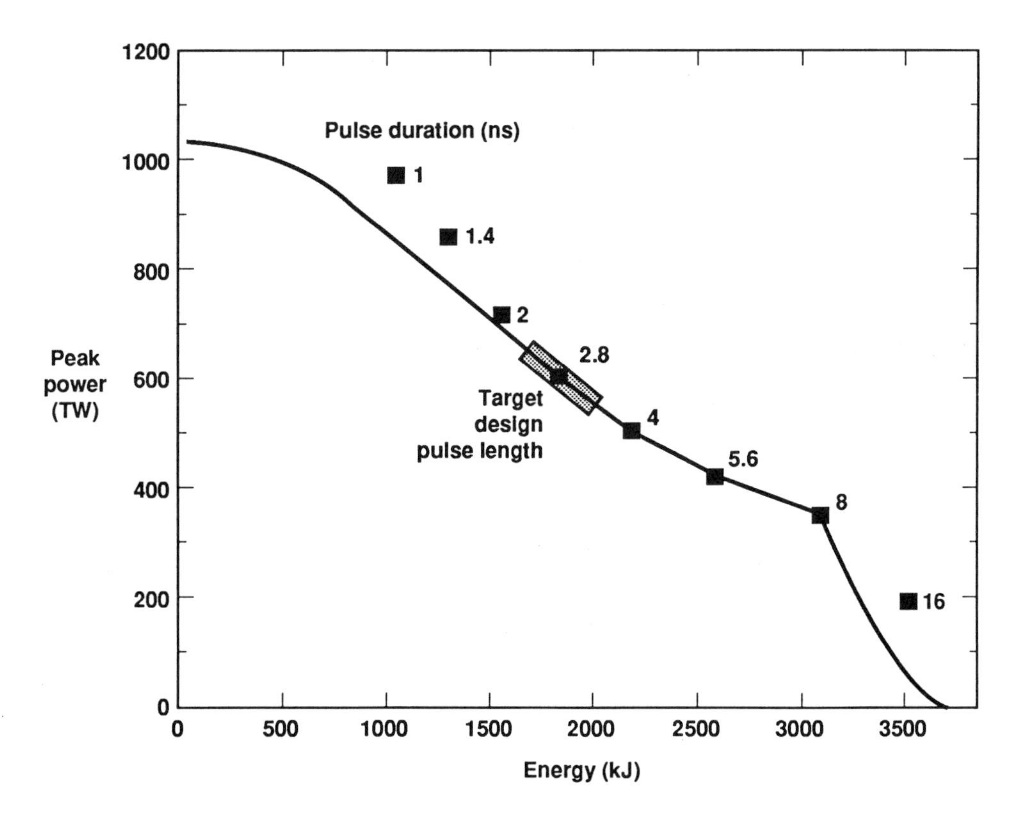

FIGURE 7 Nova Upgrade: Optimized laser has broad flexibility in performance

Acknowledgements

Work performed under the auspices of the United States Department of Energy by the Lawrence Livermore National Laboratory under contract number W-7405-ENG-48.

References

1. W.M. Manheimer and D.G. Colombant, "Steady-State Planar Ablative Flow", *Phys. Fluids* **25**, 9, pp. 1644-1652 (1982).

2. C.E. Max, J.D. Lindl, and W.C. Mead, "Effect of Symmetry Requirements on the Wavelength Scaling of Directly Driven Laser Fusion Implosions," *Nuclear Fusion*, **23**, 2, pp. 131-145 (1983).

3. S.M. Pollaine and J.D. Lindl, "Effect of Capsule Aspect Ratio on Hydrodynamic Efficiency," *Nuclear Fusion*, **26**, 12, pp. 1719-1723 (1986).

4. H. Takabe, K. Mima, L. Montierth, and R.L. Morse, "Self-consistent Growth Rate of the Rayleigh-Taylor Instability in an Ablatively Accelerating Plasma," *Phys. Fluids* **28**, 120, pp. 3676-3682 (1985).

5. M. Tabak, D.H. Munro, and J.D. Lindl, "Hydrodynamic Instability and the Direct Drive Approach to Laser Fusion," *Phys. Fluids* B**2**, 5, pp. 1007-1014 (1990).

6. Stirling A. Colgate and Albert G. Petschek, "Minimum Conditions for the Ignition of Fusion," Los Alamos National Laboratory Report LA-UR-88-1268 (1988).

7. W. Howard Lowdermilk, "Athena – An Advanced Nd:Glass Laser Driver for ICF," Lawrence Livermore National Laboratory Report UCRL-JC-103112 (1990).

Francis F. Chen
Electrical Engineering Department
University of California, Los Angeles, CA 90024-1594

Double Helix: The Dawson Separation Process

Introduction

In the 35 years during which I have had the pleasure and privilege of being associated with John Dawson, both at Princeton and at UCLA, we have worked on a large variety of subjects in plasma physics. In each case, I found that John's appreciation of experimental realities was remarkable and set him apart from all other theorists. In the 1950s and 1960s, we were concerned with Bohm diffusion, convective cells, drift waves, and minimum-B stabilization. In the 1970s, in connection with the fusion program of EPRI (Electric Power Research Institute), our attention turned to advanced fuels, synchrotron radiation, direct conversion of x-rays, multipole and surmac reactors, laser-heated solenoids, and other alternate concepts. The 1970s also saw the birth of laser fusion (which John foresaw in his Letter on *Giant Pulse Lasers*), parametric instabilities, and experiments to verify these. In the 1980s, John's ideas spawned the UCLA experiments and computations on beat-wave, surfatron, and wake-field accelerators, plasma lenses, electrostatic FEL wigglers, and photon accelerators for frequency up-conversion.

But the project which best illustrates John Dawson's ability to interact with experimentalists is the TRW program on isotope separation, which John initiated and led throughout its life in the period 1973-1986. In an internal TRW report simply entitled "Isotope Separation", dated October 15, 1973, John gave ideas on several new methods, including the ion cyclotron resonance scheme which eventually grew into a multi-million dollar project. Other physicists had also considered this method, but they never carried out experiments to the point where the real problems arose. John refrained from detailed calculations at first, knowing full well that only experiment could tell him what is important and what is not. Indeed, as the project went on, problem after problem would arise, and John would find the solution. Without his strong physical intuition on which way to go, the project would not have met with the great success that it did.

The PSP (Plasma Separation Process) project at TRW, code-named Task II (there was no such thing as Task I, as far as I know), is probably the most practical project that he and our crew of plasma physicists will ever have worked on. We had to learn the strange terminology of fission: "enrichment", "depletion", "tails", and "SWUs" (separative work units), and deal with such mundane problems as how to keep uranium from flaking off the collectors. Experiments began in small research devices of the Q-machine or filament-discharge type and then moved to M1A, a normal-magnet, ECR-heated device, and to M2A and M2B, two large, dedicated machines with 20-kG superconducting magnets. Finally, the pre-prototype device was constructed and successfully operated. This had a large superconducting magnet around a vacuum chamber with a 1-meter bore (Fig. 1). Plans were made to build a development module in full plant size (Fig. 2), and an entire uranium separation plant would consist of a dozen of these.

The PSP process was, however, in competition with two other advanced isotope separation processes aimed at reducing the cost of producing fissile isotopes below that of the standard gas centrifuge and diffusion plants. One was the MLIS (molecular laser isotope separation) scheme at Los Alamos which used an infrared laser to excite UF_6. The other was the AVLIS (atomic vapor laser isotope separation) scheme at Livermore, which uses a copper-vapor laser to separate uranium and plutonium isotopes. The Dawson process had the advantage that it could be used on any element, not just those with convenient spectral lines. Furthermore, the PSP project succeeded in producing palpable amounts of enriched uranium, while the others had not. In spite of this, a federal review of these projects came out in favor of AVLIS, and funding for the others was discontinued.

The Dawson process has a promising future nonetheless. The versatility of ICRH separation can be applied to such present uses as betavoltaic batteries for space vehicles and such future uses as tailoring of fusion wall materials, disposal of radioactive waste, and tagging with C^{13}. But the greatest potential is for producing medical isotopes. Indeed, TRW has recently provided the medical profession with 40 g of Pd^{102} (a 20,000 dose supply), which is used for radiation treatment of prostate cancer; and we can expect that the Dawson process will be of help to thousands of people requiring medical treatment in the near future.

FIGURE 1 The TRW Pre-prototype Facility [*TRW Quest*, Vol. 6, No. 1, Winter 1982/83].

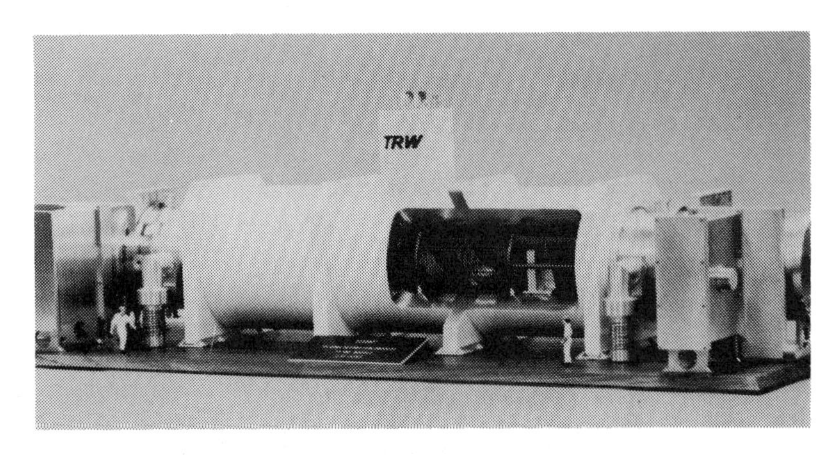

FIGURE 2 Model of full scale module [*TRW Quest*, Vol. 6, No. 1, Winter 1982/83].

The Early Stages

Natural uranium has a majority isotope U^{238} and a minority isotope U^{235} with an abundance of 0.7%. For use in reactors, the U^{235} must be enriched to above 3%. In the Dawson process (Fig. 3), a plasma containing natural uranium is produced in a strong magnetic field, and a radiofrequency field is applied at a frequency which resonates with the cyclotron motion of the minor species but is out of synch with the major species. The minor ions are then accelerated ("spun up") into large orbits and can be separated in a collector at the end away from the source. The collector, which will be described later, yields batches of depleted and enriched uranium. If necessary, the material can be run through again for further enrichment. The SWUs are computed from a formula which accounts for not only the throughput and enrichment, but also the depletion of the tails.

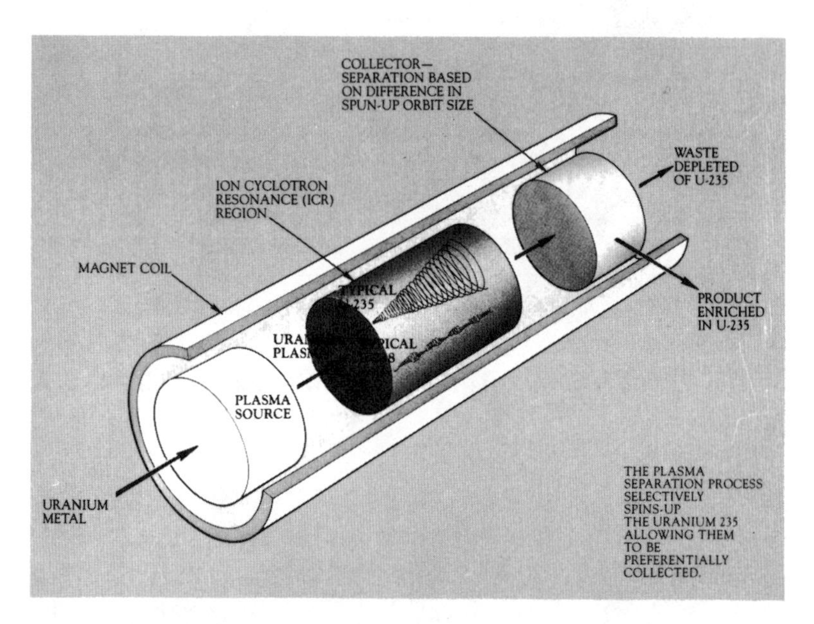

FIGURE 3 Diagram of the process [TRW brochure, *Advanced Isotope Separation Program: Plasma Separation Process for Uranium Enrichment*, August 1981].

The basic principles were outlined in a report by Fried, Dawson and Arnush[1] in 1975. In this paper, it was pointed out that the width of the cyclotron resonance had to be less than the fractional mass difference, which is only about 1% in uranium. That meant that the field uniformity $\Delta B/B$, the collision probability ν/Ω_c, and the transit time loss $v_{||}/\Omega_c L$ all had to be less than 1%. Several methods for separating the spun-up species from the cold species were proposed:

magnetic mirrors which worked on the difference in pitch angle, collectors which worked on the difference in diffusion rates, chemical reactions which depended on the temperature, and Venetian blind collectors which scraped off the large-orbit ions. Eventually the last of these was adopted. The plasma method had the advantage over the Calutron and similar devices in that space charge was automatically canceled, and the method could be used on elements other than uranium – in fact, more easily, because of the larger mass differences.

The first calculations showing the feasibility of the PSP included papers by Fried, Dawson, and Bollens[2], Wilcox[3], Coroniti and Fredricks[5], Fried[6], and Caponi[7]. Proof-of Principle experiments were done in Q-machines by A.Y. Wong et al. and in filament discharges by Stenzel et al.[8,9]. The main experimental results were summarized in a *Phys. Rev. Letter* by Dawson et al.[10], one of the few published papers by the TRW group. In this paper, data were given showing resonant acceleration of K, Ne, Cl, A, and Xe isotopes, as well as enrichment of potassium from a K^{41}/K^{39} ratio of 0.07 to a ratio of 4. The Q machine data in Fig. 4 show a clear separation of the K^{39} and K^{41} peaks. Most of the work was done in filament discharges such as that shown in Fig. 5. The rf voltage was applied across split endplates, an excitation method called "direct drive". Fig. 6 shows the separation of Xe peaks in this device.

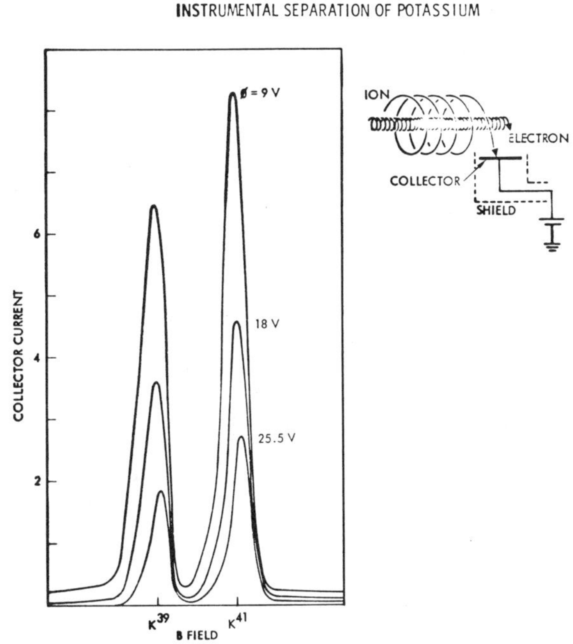

INSTRUMENTAL SEPARATION OF POTASSIUM

FIGURE 4 Cyclotron resonance peaks in potassium [Ref. 10].

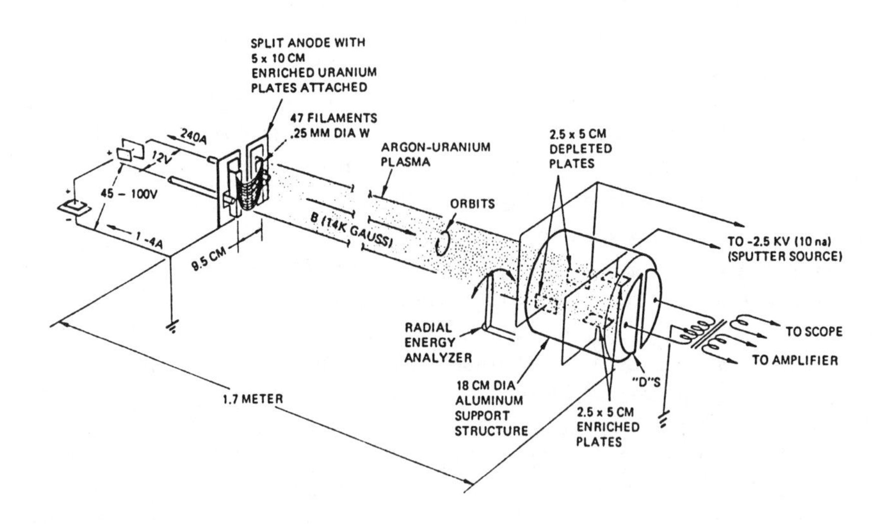

FIGURE 5 Filament discharge with direct drive "dees" [TRW Status Report DSP-192, Sept. 1976].

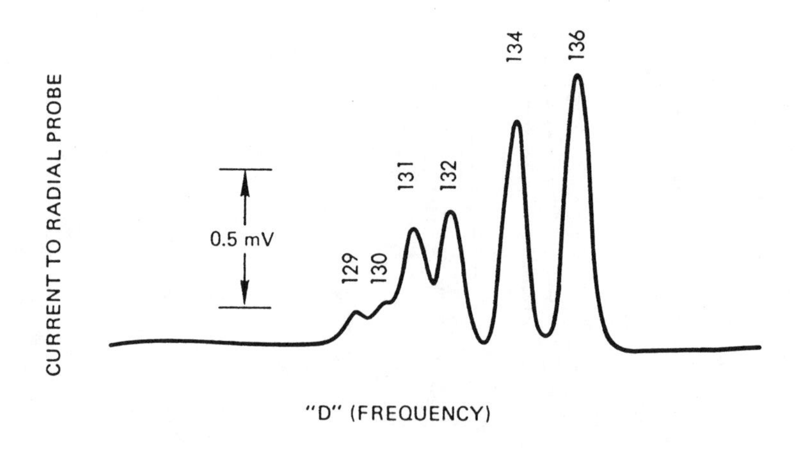

FIGURE 6 Separation of xenon cyclotron resonance peaks [Ref. 10].

Measurements of the rf field inside the plasma, shown in Fig. 7, were made in an argon plasma with a 5% krypton additive and a nitrogen impurity. It is seen that only a broad minimum of the rf E-field occurs near the argon resonance, but a sharp minimum occurs near the resonance of the minor species. The current of fast ions shows peaks near the krypton and nitrogen resonances.

These results can be understood from a simple treatment of two-ion plasmas. There is a frequency, called the two-ion hybrid frequency, at which the two ion fluids oscillate out of phase in such a way as to cancel each other's space charge. This frequency was given long ago by Buchsbaum[11]:

$$\omega^2 = \Omega_1 \Omega_2 \frac{\alpha_1 \Omega_2 + \alpha_2 \Omega_1}{\alpha_1 \Omega_1 + \alpha_2 \Omega_2} \quad . \tag{1}$$

Here Ω_1, Ω_2 and α_1, α_2 are the cyclotron frequencies and relative densities of the major and minor species, respectively. In the limit of small α_2, the two-ion hybrid frequency approaches the cyclotron frequency of the minor species. The smaller the density of the minors, the larger must their excursions be in order to cancel the space charge of the majors. To achieve large velocities for species 2, the frequency must be close to their cyclotron resonance. This explains the sharp increase in dielectric constant near Ω_2 in Fig. 7. The two-ion resonance is ideal for isotope separation because it is self-adjusting: the minor ion velocity is larger, the smaller its concentration.

Eq. (1) holds only for $k_{||} = 0$. If the propagation $\mathbf{k}$ has a sufficiently large component along $\mathbf{B}$, the electrons can move along the field to cancel the space charge, and the frequency will deviate significantly from that in Eq. (1). Fig. 8 shows ω/Ω_1 as a function of $k_{||}/k_{\perp}^{11}$. At vanishingly small $k_{\perp}$, the two roots are the lower hybrid and the two-ion hybrid oscillations. When $k_{||}/k_{\perp}$ is of order $(m/M)^{\frac{1}{2}}$, there are two electrostatic ion cyclotron waves, one near Ω_1 and one near Ω_2. For $k_{||}/k_{\perp} > 1$, these become pure cyclotron oscillations. The waves near Ω_2 are shown on an expanded scale in Fig. 9; it is seen that the frequency actually differs from Ω_2 in the three regions of $k_{||}$, depending on the density and temperature of the minor species. The aforementioned advantage of the two-ion hybrid can be utilized if $k_{||}/k_{\perp}$ can be kept below $(m/M)^{\frac{1}{2}}$ by the excitation mechanism, or else simple acceleration at Ω_2 can be applied.

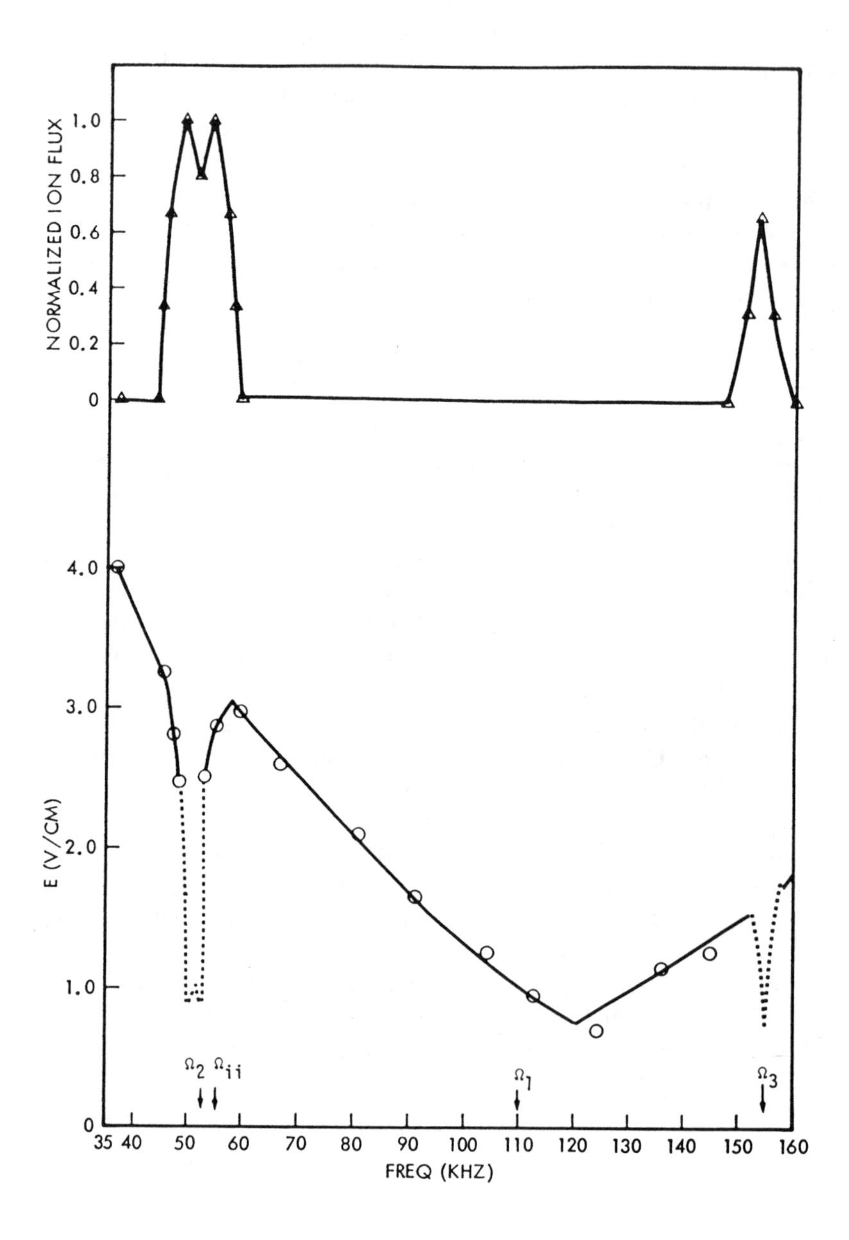

FIGURE 7 Fast ion flux (top) and E-field (bottom) in Ar-Kr plasma, as a function of frequency, in a 3-kG magnetic field. Here Ω_1, Ω_2, and Ω_3 are, respectively, the argon, krypton, and nitrogen cyclotron resonance frequencies, and Ω_{ii} is the two-ion resonance frequency [Ref. 9].

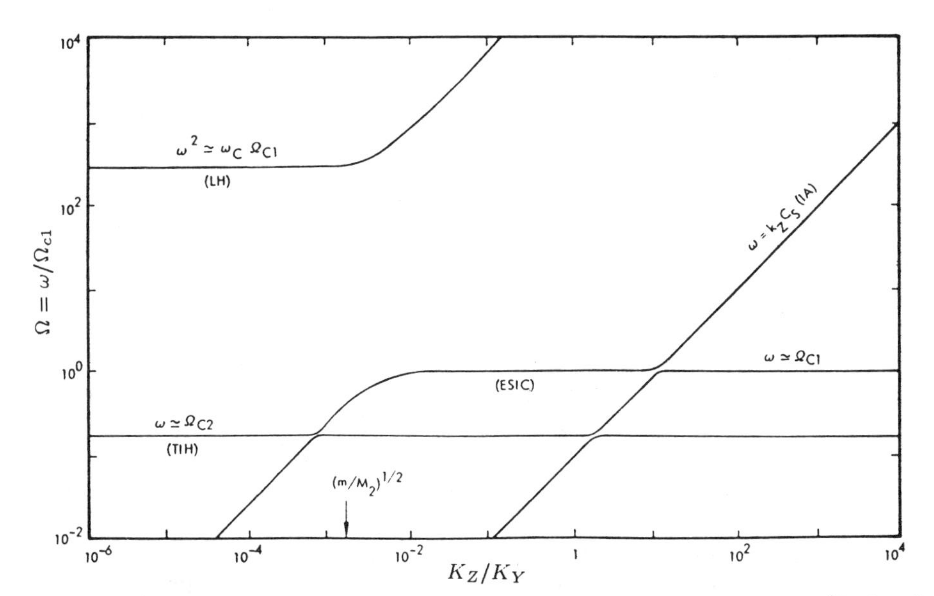

FIGURE 8 Dispersion curves for electrostatic waves in a two-ion plasma [Ref. 11].

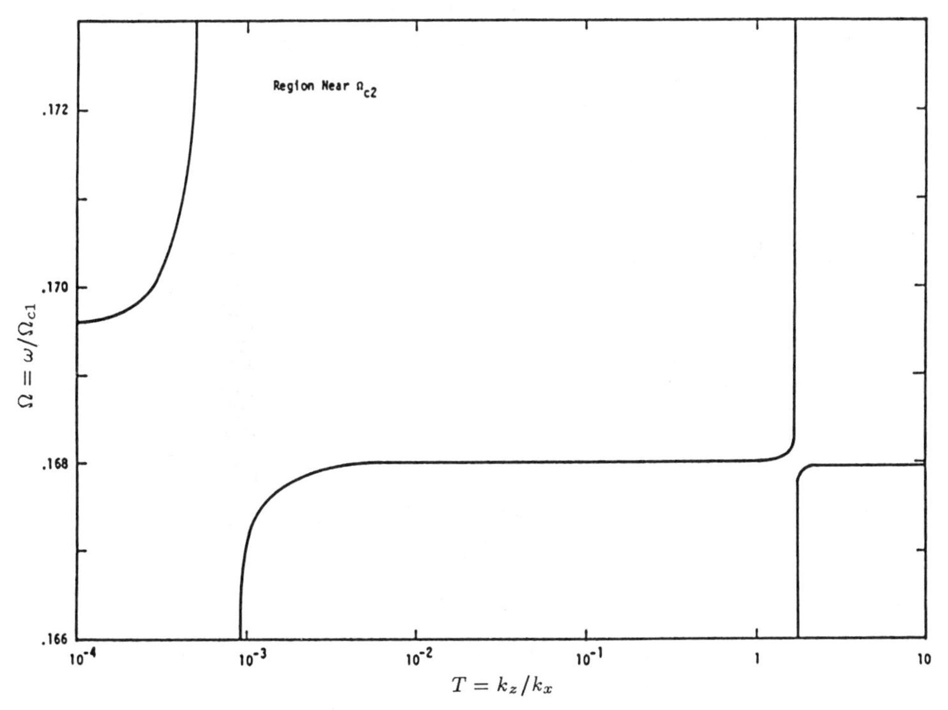

FIGURE 9 The region of Fig. 8 near the minor resonance, shown on an expanded vertical scale [Ref. 11].

Experiments with Uranium

The serious work on separating uranium was done in three large machines with uniform magnetic fields provided by superconducting magnets. One was devoted to plasma measurements, a second to uranium collections, and the third, the large pre-prototype device, was run under production conditions. The different parts involved in each apparatus are shown schematically in Fig. 10. In the source region, uranium atoms were sputtered into the discharge from a negatively biased plate faced with natural uranium. These are ionized by ECRH at 28 GHz. The discharge is started with argon, but the argon can be valved off later to obtain a pure uranium plasma. The plasma streams into the excitation region of uniform field, where a helical antenna generates a left-hand polarized field ($\sim$ 120 kHz, $\sim$ 2V/cm) resonant with U^{235}. These ions are spun up to an energy of about 500 eV. The U^{238} ions, being nonresonant, are alternately accelerated and decelerated, ending up with an effective temperature of about 100 eV. The electrons cool away from the source to a temperature of about 1 eV. With neutral U pressures of 1-50 mtorr, plasma densities up to 1.6×10^{12} cm^{-3} could be produced. The ions stream into the collector region, where a "Venetian blind" array of fences scrapes off the hot ions, which have an average Larmor radius of 2.7 cm, and allow the cold ions ($r_L \sim 1.2$ cm) to pass through to the tails plate. The separation is further aided by a bias voltage. The solid products and tails are removed periodically in batches.

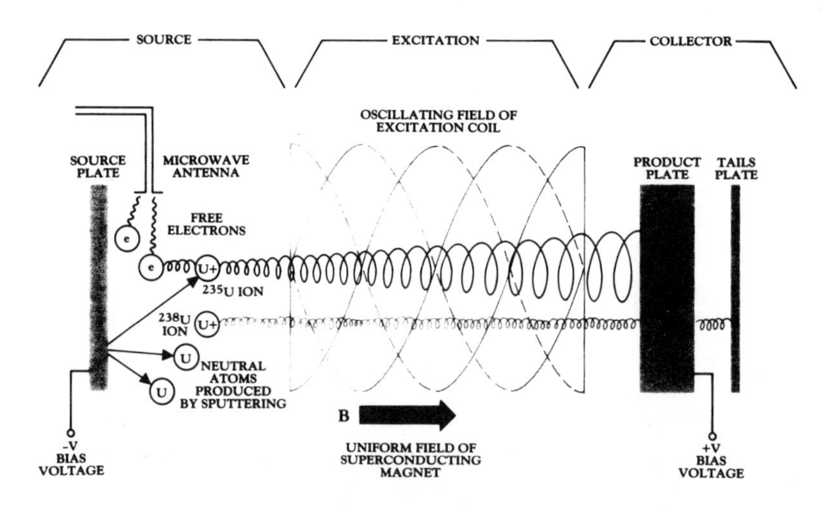

FIGURE 10 Schematic of the apparatus used for separation of uranium [*TRW Quest*, Vol. 6, No. 1, Winter 1982/83].

Each part of the process had problems which had to be solved, and we discuss each of these in turn.

Field uniformity. By purchasing large-diameter superconducting solenoids, it was easy to keep the magnetic field sufficiently constant and uniform. The price that had to be paid was accessibility. Diagnostics had to be introduced from the ends, and an array of long probe shafts protruded from the collector end of the machines.

Vapor source. Sputtering was found to be the most convenient way to introduce uranium atoms. The mounting and cooling of the uranium plates were not severe problems, and control over the neutral pressure and electron temperature could be achieved by varying the bias voltage.

Plasma production. After experimentation with various sources, it was decided that the best method was electron cyclotron resonance heating using the 28 GHz sources developed for ECRH in magnetic fusion. The magnetic field was expanded into a conducting cavity containing the typical curved surface where $\omega = \omega_c$, and the neutrals from the sputter plate were effectively ionized by the microwave power. There were, however, hotspots in the cavity which created ripples in density and potential which propagated along the field lines and could be seen with probes. These inhomogeneities caused broadening of the ion resonance and could have been removed by stirring the microwaves or modulating the frequency. However, this was never done because this was not the limiting effect. In high-power operation, the usual waveguide problems of windows and arc detection had to be overcome.

Wave excitation. Direct drive using split endplates causes the potential on the field lines terminating in each segment to fluctuate at the rf frequency. However, the rf field is not uniformly distributed over the plasma volume. Therefore, inductive drive was used, with an antenna inside the vacuum chamber. The antenna consisted of two intertwined helices (hence the title of this paper) of water-cooled copper, fed 90° apart in phase so as to excite a field in the ion gyration direction (Fig. 11). Extensive calculations of the field pattern inside the plasma produced by antennas of different pitch angles, numbers of wavelengths, and end rings were carried out by McVey[12,13]. Measurements confirmed these field patterns.

Species control. Doubly ionized uranium could have been spun up by the second harmonic of the rf field, but it would not have been collected properly and would have decreased the enrichment factor. The ionization potentials 6.2 and 11.8 eV for U^+ and U^{++} are sufficiently far apart that the population of U^{++} could be held down by running at the low T_e of 1 eV.

Ion acceleration. Aside from transit time, the factor limiting the energy to which the minority ions could be spun up was collisions, both Coulomb and hard-body. For this very practical application, the hard-body cross section had to be known very accurately. Throughout the project, extensive computer modeling accompanied the experiments, and the enrichment ratios and yields were predicted for the actual operating conditions. The success of differential ion heating was reported to the public by Romesser et al.[14].

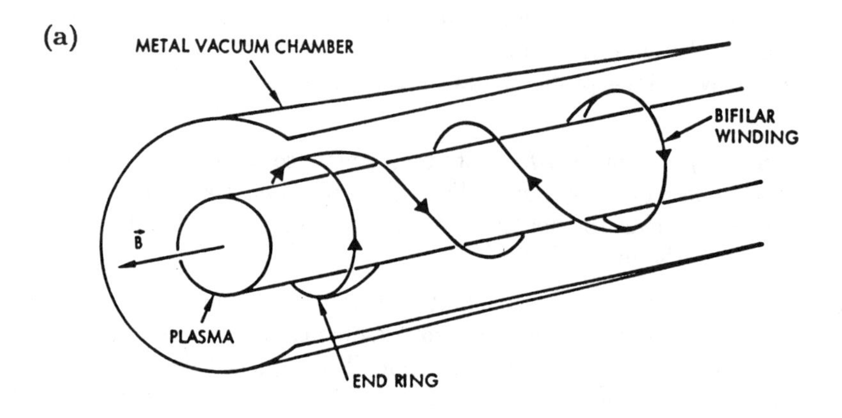

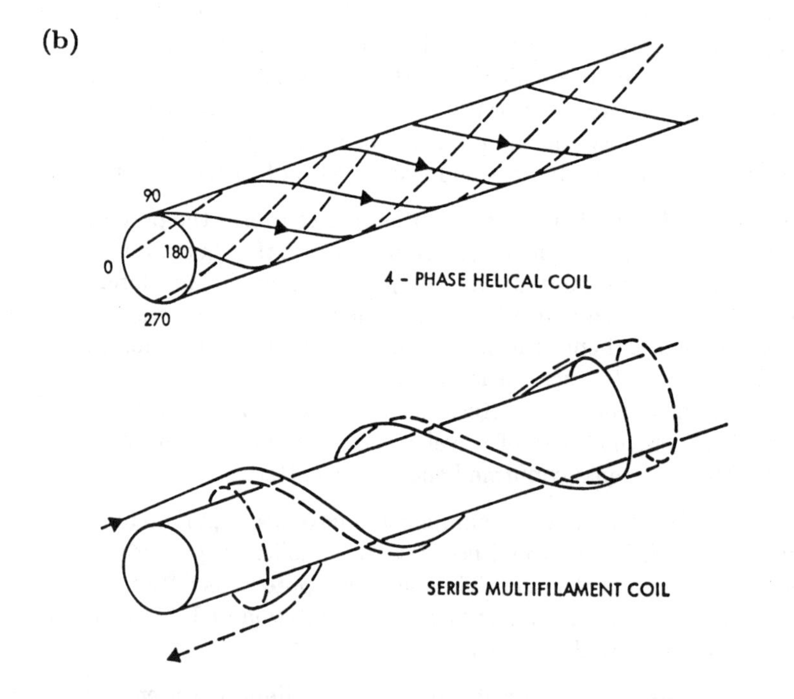

FIGURE 11 Examples of helical windings [Ref. 13]: **(a)** filamentary current model; **(b)** helical coil configurations

Product collection. The Venetian-blind collectors consisted of plates shielded by fences which prevented small Larmor radius ions from striking the plate (Fig. 12). The length of the plates and the height of the fences were adjusted to optimize the enrichment.

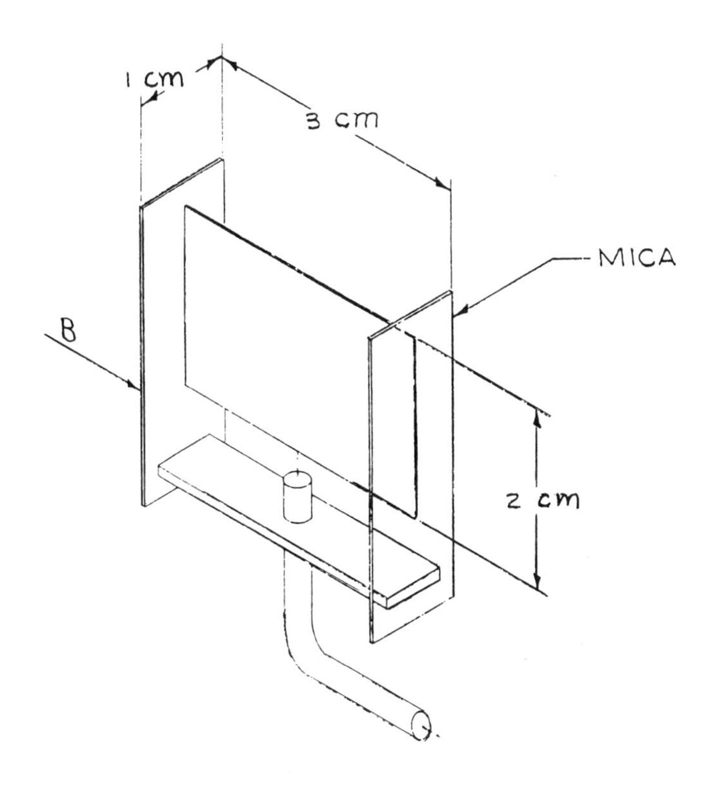

FIGURE 12 Schematic of a collector [TRW Monthly Report DSP-113, July 1976].

Diagnostics. An impressive array of diagnostics was used to characterize the plasma parameters. Standard single and double Langmuir probes, microwave interferometers, and optical spectrometers gave information on density and density fluctuations in space and time, electron temperature, and ion species. In the later stages, laser induced fluorescence was used to measure ion velocities. To measure dc and rf space potential, and hence electric field, an emissive probe was developed[15], which is heated by electron bombardment and switched electronically to a high-impedance load for measurement of the floating potential while the tip is still hot.

The main diagnostic, however, was the radial energy analyzer (REA), shown in Fig. 13, consisting of a biased collector recessed in a cylindrical tube. The tube scraped off the electrons so that only ions were collected, and the ion temperature could be obtained by varying the bias. The space-charge limit to the operation of REAs was found[16] to be $n = 2.5\,\mathrm{E(keV)}/d(mm)^2 \times 10^{10}\ \mathrm{cm}^{-3}$, where n is the density of minority ions inside the tube. This limit was not serious, and the REA was used daily to check the efficiency of acceleration. The space charge surrounding scrape-off devices like REAs and mass spectrometer probes[17] is an interesting problem (Fig. 14).

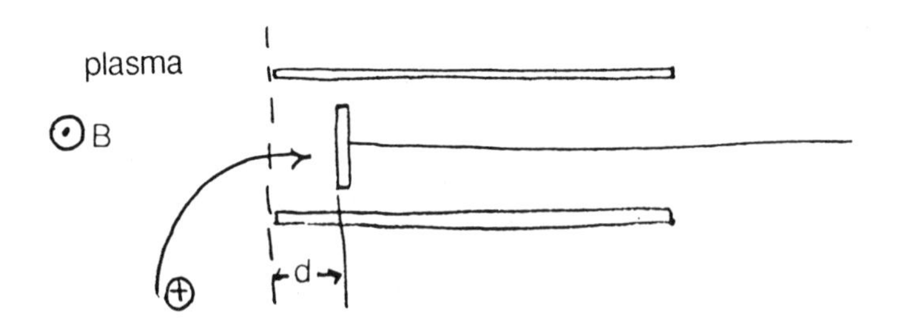

FIGURE 13 A radial energy analyzer

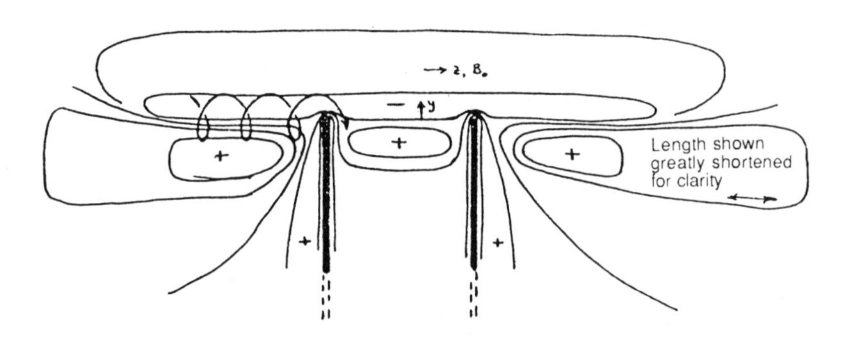

FIGURE 14 Space charge around a radial energy analyzer [Ref. 16]

Interesting Physics Questions

Research on this project turned up some rather interesting problems in basic plasma physics.

Space charge. From the beginning, those who had had experience with the Calutron expressed great skepticism over whether or not this scheme could circumvent the problems of space charge that had plagued previous attempts at plasma isotope separation. Of course, ordinary space charge had been fully accounted for in solving for the fields using standard plasma theory, but what about the edges? Would the large-orbit ions stick out radially past the well confined electron column so that they could not be neutralized? Would the column the rotate like a crankshaft? This question was answered in two ways. Kindel et al.[18] did a numerical simulation with boundaries, and Chen[19] included drift wave terms in treating the two-ion hybrid wave in a plasma with a density gradient. In each case, the effect of space charge was found to be negligible.

Endplate boundary conditions in direct drive. We see from Fig. 8 that the two-ion hybrid can be excited only if $k_{||}$ is extremely small, usually smaller than can be accommodated within a machine with a length of several meters. From Q-machine work, it is known[20] that the sheaths on the endplates can act as insulators, so that waves can have parallel wavelengths much longer than the plasma length. However, the solution for the collisional plasmas of Q-machines would not work for the more collisionless plasmas produced by ECRH. The electron parallel velocity is then limited by both resistivity and electron inertia, terms which are 90° out of phase. Is there a similar effect in this case, making it possible for $k_{||}$ to take on small values? We found[21] that, indeed, this was possible, but that the standing waves in this case are not pure cosines but are strange-looking cosines with a phase shift that varies with axial position.

Diagnostics with the two-ion hybrid. Early experiments with direct drive by the TRW group[22] showed that the two-ion resonance, as evidenced by the peak in the REA current, occurred at a frequency that varied with the density ratio α_2/α_1, as predicted by Eq. (1). This was difficult to understand, because the plasma was not long enough to let $k_{||}/k_{\perp}$ be less than $(m/M)^{\frac{1}{2}}$, and the frequency should have been much higher than that given by Eq. (1). Applying the sheath theory of Ref. 21 did not bring the theoretical curves within the experimental error (Fig. 15). We found, however, that agreement could be achieved if one accounted for the damping of the wave between the two endplates (Fig. 16). Once this is understood, detection of the two-ion hybrid excited by split endplates becomes a simple and useful technique to measuring the density of impurities in a plasma[22].

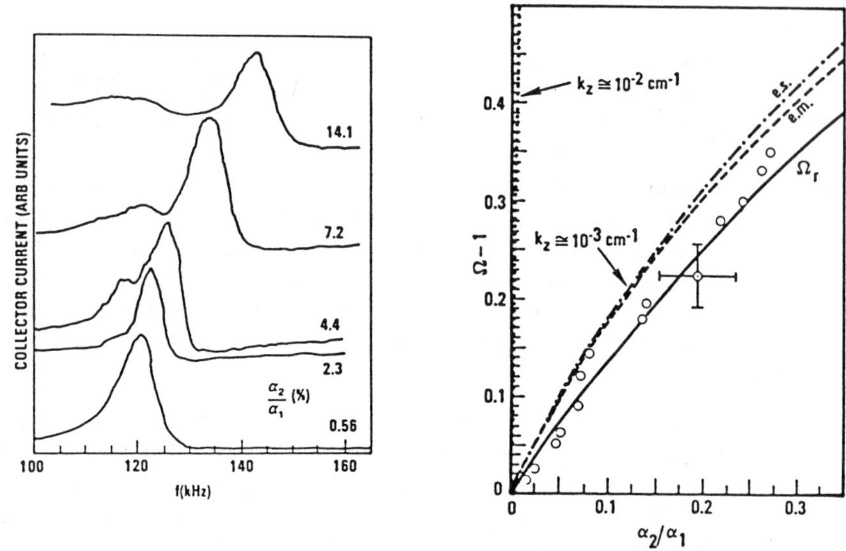

FIGURE 15 Measurements of the frequency shift from the minor species cyclotron resonance, as a function of impurity concentration. The solid line is the prediction of the Buchsbaum formula. The other theoretical curves correspond to cases (A) and (B) in Fig. 16.

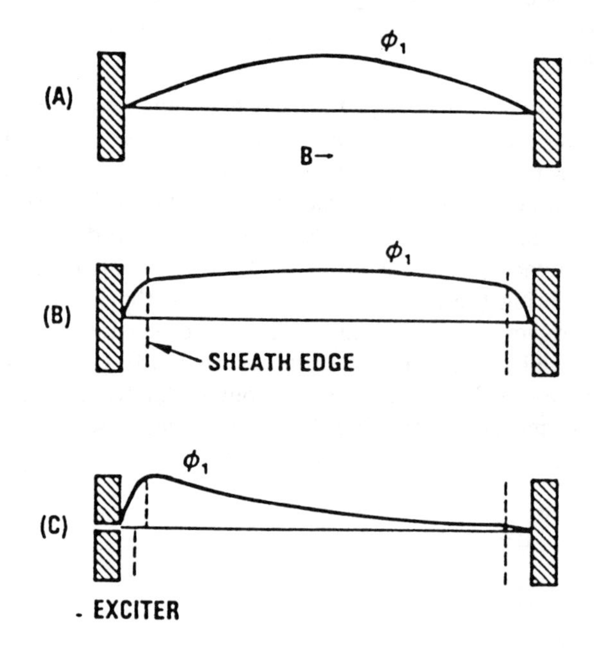

FIGURE 16 Behavior of the axial amplitude variation in (A) a standing wave with no sheath, (B) a standing wave with insulating sheaths, and (C) a damped wave excited by a split plate at one end.

Field enhancement with Type III antennas. The helical antennas used in this project were a variation of the Nagoya Type III antenna[23]. John Dawson was the first to point out the physical reason for the superior performance of such antennas. Fig. 17 shows the essential elements of the antenna, which are finite $k_{||}$ and $k_{\perp}$ and current elements along the direction of $\mathbf{B_0}$. As $\mathbf{J}$ increases, a field $E_{||}$ is induced in the plasma by the field B_1. This causes electrons to flow along each field line until they meet other electrons coming from the adjoining half-wavelength.

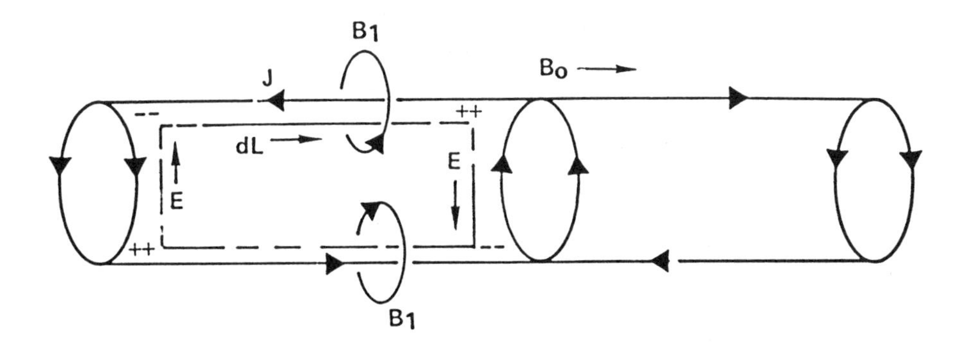

FIGURE 17 Mechanism of the Nagoya Type III antenna [Ref. 24].

Electrostatic charges then build up, giving rise to an electrostatic E-field. Since the plasma is a good conductor, the electrostatic $E_{||}$ field builds up until it cancels the electromagnetically induced $E_{||}$ field, so that the total $E_{||}$ is zero. The same space charge creates a field $E_{\perp}$, since $k_{\perp}$ is finite. This $E_{\perp}$ field is in the same direction as that induced by the transverse legs of the antenna but can be much larger. Electrons cannot move across $\mathbf{B}$ to short out this transverse field, so it can penetrate as far as an ion skin depth c/Ω_{pi}. Thus, the Type III mechanism not only converts the electromagnetic signal into an internal electrostatic field but also amplifies the induced field. In slab geometry (Fig. 18), the enhancement factor is found[24] to be approximately $(k_{\perp}/k_{||})^2$; furthermore, this factor is valid right through the ion cyclotron resonance. Detailed treatment of the cylindrical case was given by McVey[13] with essentially the same result, and experimental confirmation of field enhancement was reported by Tang et al.[25].

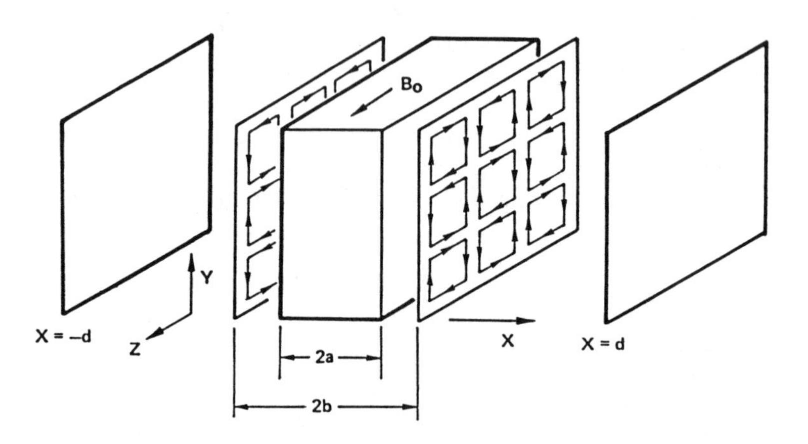

FIGURE 18 Geometry used for calculating the enhancement factor [Ref. 24].

Bandwidth broadening. Though the enrichment of U^{235} in uranium was successfully achieved in macroscopic amounts, the temperature ratio of the resonant and non-resonant species was not as high as theory predicted. The primary cause for this was traced to broadening of the resonance by random pulses in plasma potential. Typical pulses and the broadening of the rf spectrum that they produced are shown in Fig. 19. The pulses are field aligned, with transverse dimensions of the order of the ion gyroradius, and occur at random intervals averaging 1 msec when the rf excitation is strong. The pulses are always positive and have amplitudes up to 40V, much higher than the electron temperature. These observations could be explained qualitatively by a model in which the rf ponderomotive force caused unstable growth of density filaments[26]. This model was criticized, however, because it didn't fit all the data, and another mechanism based on phase-bunching of the ion orbits was proposed by M. Mussetto. John Dawson had a third idea connected with ion viscosity. Unfortunately, the project was stopped before this interesting effect was resolved. The resolution could have had ramifications in other applications of ICRF.

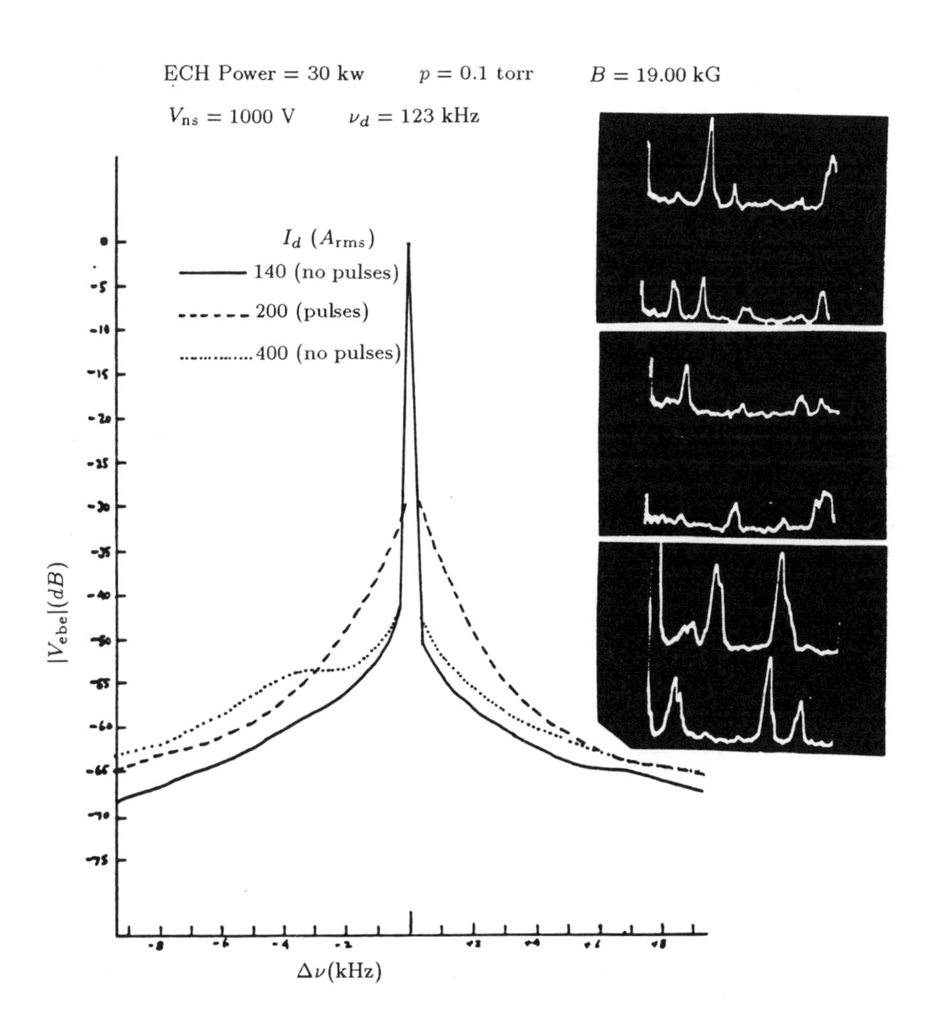

FIGURE 19 Frequency spectrum of the rf field showing the broadening caused by random pulses in plasma potential (inset).

Acknowledgments

Because this was a classified project, the references in this paper are skewed toward the work that saw the light of day in either published papers or unclassified reports. The success of the project was due to a large number of dedicated physicists and engineers, only some of whose names can be found in the references quoted here. Tom Romesser was the chief scientist, and Larry Harnett the chief engineer. Steve Korn was the program manager, and Sol Rocklin his deputy. Umbrellas were provided by Don Arnush, who headed the plasma efforts at TRW, and his boss, Pete Staudhammer. But it was John Dawson who thought up the idea in the first place and carried it over the many obstacles.

References

1. B.D. Fried, J.M. Dawson, and D. Arnush, "The Dawson Separation Process", TRW Report 29160-6001-RU-00 (Aug. 1975).
2. B. Fried, J. Dawson, and R. Bollens, "Magnetic Pumping and Ion Cyclotron Resonance Heating for a Multi-species Plasma", TRW Report (unnumbered) (Oct. 1974).
3. T.J. Wilcox, "Selective Ion Heating by Cyclotron Resonant Pumping in a Multicomponent Plasma", TRW Report 99900-7781-RU-00 (Dec. 1974).
4. F.V. Coroniti and R.W. Fredricks, "Charge-exchange Collisions in the Dawson Separation Process", TRW Report 29160-6029-RU-00 (Mar. 1976).
5. R.W. Fredricks, "Effect of Collisions on Line Broadening in Electromagnetic Cyclotron Resonance Device", TRW Report 99994-6291-RU-00 (Feb. 1975).
6. B.D. Fried, "Transit-time Effects in Ion Cyclotron Resonance Heating of Multi-Species Plasmas", TRW Report Task II-THA-01 (Dec. 1977).
7. M. Zales Caponi, "Single Particle Resonances in a Nonuniform Electric Field Produced by Direct Excitation of End Electrodes", TRW Report (unnumbered) (Feb. 1977).
8. R.L. Stenzel and J.T. Tang, "Electrostatic Excitation of Ion Cyclotron Resonances", TRW Report 99994-6324-RU-00 (Mar. 1976).
9. J.T. Tang, R.L. Stenzel, and H.C. Kim, "RF Electric Fields in Ion Cyclotron Excitation of Multispecies Plasma", TRW Report Task II-2120 (Jan. 1979); *Phys. Fluids* **22**, 1907 (1979).
10. J.M. Dawson, H.C. Kim, D. Arnush, B.D. Fried, R.W. Gould, L.O. Heflinger, C.F. Kennel, T.E. Romesser, R.L. Stenzel, A.Y. Wong, and R.F. Wuerker,

"Isotope Separation in Plasmas by Use of Ion Cyclotron Resonance", *Phys. Rev. Letters* **37** 1547 (1976).

11. F.F. Chen, "DSP Theory for Experimentalists", TRW Report Task II-1359 (Mar. 1978).
12. B.D. McVey, "Electric Field Patterns for Multiphase Helical Coils", TRW Report Task II-2584 (Sept. 1979).
13. B.D. McVey, "Excitation Theory of the Inductive Drive", TRW Report Task II-2740 (Jan. 1980).
14. T.E. Romesser, *Bull. Amer. Phys. Soc.* **24** 1052 (1979), papers 7B2, 4F8-11.
15. L.L. Higgins, *Bull. Amer. Phys. soc.* **23**, 803 (1978), paper 4F13.
16. F.F. Chen, "Space Charge in Radial Energy Analyzers", TRW Report Task II-1712 (1976).
17. F.F. Chen, "Operation of Mass Spectrometer Probes", TRW Report Task II-2235 (Mar. 1979).
18. J.M. Kindel, A.T. Lin, J.M. Dawson, and R.M. Martinez, "Nonlinear Effects of Ion Cyclotron Heating of Bounded Plasmas", *Phys. Fluids* **24**, 498 (1981).
19. F.F. Chen, "The Two-Ion Hybrid Wave in an Inhomogeneous Plasma", TRW Report PSP-5056 (1982).
20. F.F. Chen, *Plasma Physics* **7**, 399 (1965).
21. F.F. Chen, "Axial Eigenmodes of Long-$\lambda_{||}$ Waves in Plasmas Bounded by Sheaths", *Phys. Fluids* **22**, 2346 (1979).
22. F.F. Chen, G. Dimonte, T. Christensen, G.R. Neil, and T.E. Romesser, "Use of the Two-Ion Hybrid as an Impurity Diagnostic", *Phys. Fluids* **29**, 1651 (1986).
23. T. Watari et al., *Phys. Fluids* **21**, 2076 (1978).
24. F.F. Chen, "Radiofrequency Field Enhancement Near Ion GyroResonance", TRW Report Task II-3552 (Jan. 1981).
25. J.T. Tang, M. Mussetto, B. McVey, D. Dixon, and T. Romesser, *Bull. Amer. Phys. Soc.* **24**, 992 (1979), paper 4F11.
26. F.F. Chen, "An ICRF-driven Inverse Filamentation Instability", TRW Report PSP-TW-348 (1984).

Bruce Langdon
Lawrence Livermore National Laboratory, Livermore, CA 94550

30+ Years of Plasma Simulation

It's a great pleasure to be here, as part of this Symposium to honor John Dawson, with so many friends and colleagues.

Plasma physics was introduced to many of us through a course that John and Mel Gottlieb taught entering graduate students at Princeton in the '60s. This course emphasized a pictorial as well as quantitative understanding of physics, rather than mathematical formalism. John is unexcelled at drawing some sketches, invoking a conservation law or two, and finding an answer.

One day John showed me some work he was doing with the computer, in which he integrated in time the motion of a thousand or so sheets of charge in their self-consistent fields, and was able to model a lot of aspects of plasma behavior. Well, this was *very intriguing.* Computer simulation had become one of John's prime tools for learning about plasmas. He had already published a few papers, and some had appeared from elsewhere, notably Buneman's 1959 paper on dissipation in collisionless plasmas.

In 1965, I was ready for a thesis project. John returned from his very significant stay in Nagoya, and I had the good fortune to begin work with him. He and a colleague at Nagoya had tried out an idea for an *electromagnetic* particle code, and I implemented a version of this. Characteristically, John's idea was elegantly motivated by a picture, of the contributions of particle currents to left- and right-propagating electromagnetic waves (for this reason certain numerical problems affecting later codes were avoided). I think these were the first-ever electromagnetic simulation codes.

The following "ramble" illustrates how scientific computing proceeded in those days:

To set the initial particle velocities, you need some way to generate a finite set of velocities that populate the velocity range according to a Maxwellian or whatever, and don't show excessive ordering. Ways to generate so-called pseudo-random numbers are still discussed today in the numerical mathematics literature.

A wad of cards was the kit for the sort of random number generator John used then. Here's how it works. In those days, most information created for computer input was rendered as holes in cards using a keyboard machine about the size of a small desk. So you could punch up a deck of cards, each punched with a velocity. Then you could randomize them the same way you do in other card games (i.e., shuffle them).

Then you put the problem parameters, punched on other cards, with these velocity cards, and some other stuff, wrapped an elastic band around it, and put it in a box with other people's decks of cards. That evening someone took the box right to the computer, where an operator put them in a "card reader", a machine that would about fit in your modern mini-van. This card reader had a claw that would, or should, grab just one card at a time and drag it past a set of fingers that detected the holes.

So what you had punched finally went into the computer and did whatever, the computer printed on paper using another machine that made a deafening racket. The paper and your cards were brought back out to the plasma laboratory for the next morning, and you'd learn what had been accomplished.

Progress took time.

In the '60s, two dimensional simulations began in earnest. The introduction of a spatial mesh to mediate the field, and the development of fast Fourier and cyclic reduction methods at Stanford, made this possible. Around 1970, electromagnetic codes came into heavy use, for example in laser plasma studies. Former students and colleagues of John Dawson's are *very* well represented in this work. Some of this is discussed at this Symposium and in Bill Kruer's Maxwell Prize address to the APS Division of Plasma Physics in November.

Actually, particle simulation methods were first attempted nearly 50 years ago. Simulations of microwave devices were early instances of parallel computing, as Oscar Buneman can explain. Some call these systems "non-neutral" plasmas. Nowadays some of the adaptations of plasma simulation that are most interesting to me are in charged particle devices, such as accelerators.

As Professor Orbach outlined, John and his colleagues have found occasion to adapt particle simulation methods to a remarkable variety of plasma topics. Indeed, these methods are in use all over the world, in large part due to the example of Dawson and his colleagues.

Acknowledgement

Work performed under the auspices of the United States Department of Energy by the Lawrence Livermore National Laboratory under contract number W-7405-ENG-48.

Viktor K. Decyk
Physics Department and
Institute for Plasma and Fusion Research, UCLA

Future Directions in Simulation

The subject of my talk is future directions in simulation. Predicting the future is always tricky business. It is safer to predict the present. Therefore I will talk about some of the activities going on now that I think will have a significant impact on the way simulation is done in the future. In particular, I will be concentrating on those trends which will impact the average scientist, not just the computational physicist. In contrast to other talks in this symposium, which emphasize the contribution John Dawson has made to plasma physics research, this talk will be about his concern for students and teaching, which most of us at this symposium know well, but do not often talk about. The trends I will focus on are the development of user friendly interfaces, interactive simulations, animation and 3D graphics, and parallel processing. Most of my examples will be from particle simulations, but the same concepts apply to other types of numerical modeling.

The first trend I want to discuss is the development of user friendly interfaces for scientific simulations. What I mean by an interface is a program and command file which sits between the original scientific program and the human user and allows a person who is scientifically literate, but perhaps computer illiterate, to perform numerical simulations.

The significance of this for simulation is that it suddenly allows creative scientists who do not have the time or inclination to learn about computers to use simulation as a tool in their scientific work. Theorists will continue to do theory and experimentalists to do experiments, but they will both have the advantage of new insights gained from simulations. Simulation is no longer the domain of experts in the art who talk mostly with each other.

Under the guidance of John Dawson, we have begun to develop such user friendly interfaces for our simulation codes at UCLA. Other examples also exist at other places that I will mention later. But let me mention here some simple things we did at UCLA that had a substantial impact on the more widespread use of particle simulation[1]. The original goal was to have an interface that would enable students in a regular physics class to run simulations without any knowledge of computers or even editors. Only a knowledge of the physical parameters would be required. This was accomplished by adding two pieces to the original particle simulation code. One was an interactive input program. The other was a command file which took care of system specific details.

The interactive input program is in essence a menu driven preprocessor which creates an output file of parameter statements. These parameter statements contain all the information the FORTRAN program needs to run, including the input data and array sizes. Under the control of the command file, the FORTRAN program is then compiled, linked and executed. The only change to the original FORTRAN program was to replace the data statements and dimensions with parameters, and to add an include statement. The interactive input program displays a list of variables and their current values, and users change the values by entering data in the form: VARIABLE = VALUE. A simple parser evaluates expressions of this form, and sets the variable to the appropriate value. If VALUE is omitted, the meaning of the VARIABLE is displayed. In some cases the program rejects unreasonable VARIABLE values to prevent students from running nonsense simulations.

The interactive input program is itself managed by a command file, whose purpose is to hide the computer operating system from the user. This command file, or shell, will ask the user when some information is needed, such as where to send graphical output, and will act accordingly. Versions of the command file exist for various mainframe computers such as IBM, Cray, VAX, and UNIX workstations. The user does not even need to know what kind of computer he is using.

This user friendly particle simulation has been used for several years in teaching both graduate and undergraduate plasma physics courses. The most common use has been to test students' understanding of lecture material. After some concepts have been derived in lecture, students perform computer experiments related to these concepts and hand in a lab report. The most useful thing about these computer experiments is that they give concrete reality to abstract and difficult concepts. It is often the case that physics students are excellent at symbolic manipulation without understanding the meaning of what they are doing. The simulation is a confrontation for them, as they must match their abstract symbols with the data they see on the screen in front of them. Typical experiments have included

Landau damping, parametric instabilities, behavior of test charges, and plasma echoes.

More recently, graduate students and post doctoral researchers who are not specialists in simulation, have begun to use these user friendly codes in their own research. Experimentalists have performed simulations to give them guidance in interpreting experimental results or to give them clues to what kinds of phenomena they ought to be looking for. Theorists have performed simulations to verify their own theories or to observe new phenomena which they may try to model. Experimentalists have adapted to such user friendly simulations more readily than theorists. Perhaps it is because they are not so concerned with whether to "believe" the simulation. They believe only their own experiments anyway. The simulations just give them insight along the way.

One of the criticisms of making simulation codes readily available as "black boxes" is that the codes would be abused by non-experts and applied in regimes where they would produce nonsense. The answer to this is that the interface would have to prevent this. Currently each potential user of a simulation code must be aware of the algorithms and their limitations in order to use a code properly, and this can be a substantial barrier. But the author of the code is the one who should know this best and his code itself should prevent, or at least warn the user when the limitations are being approached. This requires that a certain amount of intelligence must be added to these interfaces.

The simple user friendly interface I have just described is just the beginning. Other examples exist and more are coming. Diglio Simoni from JPL recently generalized the above interface program to be usable by any FORTRAN program in a UNIX environment. The author of any FORTRAN program merely provides a description file of the variables to be included in the input preprocessor, adds an include statement and a read statement to his FORTRAN program, and suddenly it is user friendly.

In another example, Tom Gladd wrote an interface to the MAGIC particle code[2], which uses a mouse to specify the location of conductors, without having to specify numbers. The ability to see the conductor makes incorrect entries much less likely. Clearly this general idea of visually displaying input data is an important way user friendly interfaces can minimize erroneous entries.

Another important feature which should be part of user friendly interfaces is the ability to translate between units. Most codes are written in normalized computer units, which the user must learn for each code. This should not be necessary. The interface should be able to automatically translate from common physical units to the computer units so the user does not need to know about the computer units. Notice that the translation is done by the interface. The original program does not have to be changed.

In summary, user friendly interfaces can be written to make simulation programs accessible to all scientists, not just the computer professional, and this will lead to more widespread use of these codes and better science.

So far I have been talking about computer interfaces which are in the nature of a preprocessor to an existing simulation program, with an accompanying command file to hide the operating system from the user. A closely related trend is that of making the entire code interactive. The best known example of interactive particle simulation programs is the IBM PC based codes developed by Ned Birdsall's group at Berkeley[3]. We have been inspired by Ned to attempt to do similar things at UCLA with our mainframe based codes.

Interactive codes have great potential in improving scientific productivity. The user sees the result immediately and can make new choices quickly. It is usually the case that when working on a new problem, it takes many trial and error runs before a useful and interesting parameter regime is found. Running interactively greatly speeds up this exploratory phase. Furthermore, the easier it is to run, the more likely the scientist will explore parameters more thoroughly and the less likely he or she is to give up on some potentially interesting cases because they would just take too long or are too complicated.

Interactive programs must send output to some graphical terminal, so a graphical interface such as that popularized by the Apple Macintosh, is quite natural, if not always simple to implement. This means that commands do not have to be remembered. Instead choices can be made from a menu by means of devices such as a mouse. Interactive programs require the user to change the simulation while running it. Often this amounts to just turning diagnostics on and off, or a display on and off. But sometimes it can mean turning certain simulations parameters on and off, such as power to an antenna. A well written interactive program would allow simulations to be run that the author of the simulation may not have anticipated.

I should mention as an aside, that students often by genius or accident run cases no one has run before. There is nothing more awe inspiring than to see someone like John Dawson stand in front of a classroom of students running interactive simulations, and watch him figure out the important physics in the most bizarre cases.

Of course, it is not always advantageous to run scientific codes interactively. In particular, once a useful parameter regime has been identified, it may be more productive to do a large series of runs in batch mode. But for many scientists, such as John Dawson, the exploratory phase, where new ideas and concepts are tested, is the most important one. Unfortunately, machines such as the Crays are not well suited to interactive simulations, but personal computers and workstations are improving all the time, and these are well suited.

In summary, interactive simulation will have increasing impact in speeding up the exploratory phase of research and in making it better done.

Animation is the next feature I want to discuss. It may or may not be part of an interactive simulation. But whether it is done interactively or independently, it is a powerful new tool. The reason is that it leads to new insights and discoveries that would most likely not have occurred otherwise. In my own research I have examples of piles of printed graphical output, where I could not understand the underlying physics. But within 5 minutes of animating the same output, it became quite clear. In other cases, I have obtained new insights into physical processes I thought I had already understood. Much of physics involves dynamics in time and our eye is such a marvelous tool for recognizing patterns in time. Animation is also a good tool for conveying complex information to others. I can display the results of 10 CPU hours of calculation in one 30 second animation, and it will be better understood than a long seminar with lots of still pictures. It is clear to me that routine use of animation will lead to much better physics.

Another development in graphics is the increasing ability to display 3D data. This is still an area under intense development, undergoing rapid changes. It is still mostly the domain of experts, so I will not have much to say about it, except to say that the ability to display and comprehend data in 3 or more dimensions will require special training, but should result in a deeper understanding of complex phenomena.

It is probably quite clear by now that writing user friendly interfaces and interactive simulations is not an entirely trivial task. I believe that it will require enough effort that a few codes will begin to dominate common usage. This tendency is already quite clear in other disciplines, such as computational chemistry, where a few codes maintained by a small number of groups are in common use by the general community. The skills needed to write such codes are sufficiently specialized that a new discipline is arising, computational science, as distinct from computer science or physics. Several universities are already offering Ph.D. degrees in computational science. The days when every physicist wrote his own code are fading.

The final topic I want to discuss is parallel processing. Parallel processing is based on the idea that using many slow but cheap processors working together on a problem will be more cost effective than one very fast and expensive processor. Whether it is really more cost effective may depend on whose time you are trying to minimize, yours or the computer's. It also depends on the architecture of the parallel computer. Closely related to parallel processing is distributed processing, which I will not discuss here.

One common architecture uses distributed memory. This is like connecting a collection of IBM PCs together with wires and getting them to work together on some problem. The word hypercube describes one type of distributed memory parallel computer that has a particular kind of wiring topology. The first hypercube at Cal Tech was in fact just such a collection of IBM PCs. Since there is no common memory, the user must partition his problem into parts, in such a way as to minimize the need to move data from one processor to another. This generally means the simulation must be reorganized at this highest level, although lever level subroutines often do not have to be modified.

Paulette Liewer and Robert Ferraro from JPL and myself have implemented particle simulations in such architectures[4], and we have found that the particle pushing part of the simulation can achieve high efficiencies (> 90% for 128 CPUs), although the field solver was less efficient. Experience with other scientific problems on hypercubes has shown that physical problems are usually quite regular and work efficiently in such parallel architectures.

The other common approach in parallel processing is to use a shared, common memory that several processors can access. With shared memory it is possible for a compiler to perform automatic parallelization on a loop level. This has the advantage that algorithms written for vector machines can be used on a parallel machine. However, as the number of processors increases memory conflicts become greater, and eventually the user will mostly likely have to partition his problem into parts to avoid this, as in the distributed memory case.

On the Mark IIIfp Hypercube at JPL, we have already achieved about twice the performance of a single processor Cray 2 running the same problem, although at considerable expense in human time.

Let me tell you about another development which will soon impact parallel processing, namely the development of cheap but extremely fast RISC processors. I recently had an opportunity to benchmark the new IBM RS/6000 workstation with my particle simulation codes. It turned out that this workstation achieved about 1/2 of the Cray X-MP performance on scalar code, and about 1/4 when the Cray is running vectorized code. This workstation costs about $13,000! Imagine what will happen when one puts 128 such processors together in a parallel computer. One would have 30 times the Cray performance for one million dollars or less!

Finally, if one couples this great computational power with user friendly interfaces, then one can see that imaginative scientists like John Dawson will have the opportunity to be very creative. And even the less talented will be able to do extraordinary things. In summary, then, the future seems very bright and exciting to me.

References:

1. V.K. Decyk and J.E. Slottow, "Supercomputers in the Classroom", *Computers in Physics*, vol. 3, No. 2, p. 50, 1989.

2. N.T. Gladd, "NTRFACE: An Object-Oriented Interface System for PIC Codes", *Proc. of the 13th Conf. on Numerical Simulation of Plasmas*, paper PT7, Santa Fe, NM, 1989.

3. John P. Verboncoeur and Vahid Vahedi, "WinGraphics: An Optimized Windowing Environment for Interactive Real-Time Simulations", *Proc. of the 13th Conf. on Numerical Simulation of Plasmas* paper PT6, Santa Fe, NM, 1989.

4. P.C. Liewer and V.K. Decyk, "A General Concurrent Algorithm for Plasma Particle-in-Cell Codes", *J. Computational Phys.*, vol. 85, p. 302 (1989).

A.T. Lin
Physics Department
University of California, Los Angeles

Radiation Generation From Electron Beams

Abstract

This paper describes the subject of radiation generation from electron beams including free electron lasers (FEL) and electron cyclotron masers (ECM), with an emphasis on John Dawson's contributions. He started the investigation of radiation generation mechanisms quite early (1973) and I was fortunate to work under his guidance at that time. The paper considers plasma lasers, free electron lasers, trapped particle instabilities, and the interaction of absolute and convective instabilities in electron cyclotron maser amplifiers.

Introduction

Searching for efficient schemes to convert electron beam kinetic energy into coherent electromagnetic waves has allured many scientists for more than half a century.

For the conversion mechanisms to be efficient, it is essential that the electrons remain in resonance with the wave for a long time. Since electrons could not move at velocities above the velocity of the electromagnetic wave in free space, one must either slow down the electromagnetic wave by means of a dielectric medium or slow wave structure (slow wave device), or speed up the beam modes by using an externally applied magnetic field to undulate the electron motion (fast wave device). The former approach is limited to relatively long wavelengths and low energy beams because the electron-wave coupling decays exponentially from the structure. The latter approach, which is capable of generating coherent radiation covering a wide expanse of the electromagnetic spectrum from the microwave through the ultraviolet regions, places no limitation on the beam energy and the wave frequency and has received a great deal of attention for the last two decades.

In fast wave devices, the electron beam is subjected to a periodic magnetic field (FEL) or a longitudinal uniform magnetic field (ECM) such ad to add an additional momentum ($k_0 = 2\pi/\ell$) to the beam mode ($\omega = kv_0$, v_0 is the electron drifting velocity). Here "ℓ" is the magnetic field periodicity in FEL and is the pitch of the electron gyration in ECM. To achieve strong electron-wave interactions, the electron beam must be bunched. In a FEL the ponderomotive force produced by beating of wave magnetic fields and the electron oscillatory velocity in the wiggler field exerts a periodic longitudinal perturbation on the electron beam, such that the accelerated electrons catch up with the decelerated electrons. If $\omega < (k+k_o)v_o$, the bunched electrons are in the decelerating phase of the wave and will give up their kinetic energy to the electromagnetic wave. On the other hand, the bunching of an electron cyclotron maser relies on the electron mass dependence on its energy (relativistic effect). In a uniform magnetic field, an electron gyrates faster if it loses energy so that the decelerated electrons catch up with the accelerated electrons. The azimuthally-bunched electrons convert their kinetic energies to the wave if $\omega > (k + k_o)v_o$. This paper will concern only fast wave devices.

Plasma Lasers

Consider a plasma with finite-amplitude short-wavelength ion-density fluctuations[1] of wave number k_i and frequency $\omega_i \simeq 0$. These fluctuations are capable of playing the same role as the magnetostatic wiggler, and lead to a strong coupling of a high-frequency positive-energy wave (either electrostatic or electromagnetic) to a slow negative-energy space charge wave associated with the electron beam. In this situation, both disturbances can grow at the expense of the beam drifting energy. The theoretical analysis is most simply handled in a reference frame moving with the electrons. In this frame and with the electrostatic and non-relativistic approximations, the ion density fluctuation is given by

$$n_i = n_0 \left[1 + \epsilon \cos k_i(x + v_0 t)\right] \ . \tag{1}$$

We employ Lagrangian coordinates[2] for the electrons and let $X(x_0)$ be the displacement of an electron from its equilibrium position x_0. The equation of motion for an electron is

$$\ddot{X}(x_0) = -\frac{e}{m_e}E(x_0) , \tag{2}$$

and the electron density fluctuation can be written as

$$n = n_0 \Big/ \left(1 + \frac{\partial X}{\partial x_0}\right) . \tag{3}$$

By Gauss's law, $E(x_0)$ is 4π times the excess charge, which passes from the right of electron x_0 to its left. Thus, we have

$$\begin{aligned} E(x_0) &\simeq 4\pi e n_0 \{1 + \epsilon \cos k_i(x_0 + v_0 t)\} X(x_0) \\ &+ \frac{4\pi e n_0 \epsilon}{k_i} \sin k_i(x_0 + v_0 t) . \end{aligned} \tag{4}$$

The equation of motion thus becomes

$$\begin{aligned} \ddot{X}(x_0) = &-\omega_p^2 \left[1 + \epsilon \cos k_i(x_0 + v_0 t)\right] X(x_0) \\ &+ \omega_p^2 \frac{\epsilon}{k_i} \sin k_i(x_0 + v_0 t) . \end{aligned} \tag{5}$$

The last term gives the steady driving of the electrons due to the passing ion fluctuations, but does not give rise to instability, and so it may be neglected in the stability analysis. Equation (5) can be cast into the form of a standard Mathieu equation[3] by introducing $2\zeta = k_i(x_o + v_o t) - \pi$

$$\frac{\partial^2 X}{\partial^2 \zeta} + \frac{4\omega_p^2}{k_i^2 v_0^2}(1 - \epsilon \cos 2\zeta) X = 0 . \tag{6}$$

For $\epsilon \ll 1$, it exhibits instabilities when the coefficient $2\omega_p/k_i v_o = P$ is close to an integer. The growth rate is peaked at $P \simeq 1$ and is given by

$$\frac{\gamma}{\omega_p} = \frac{1}{2}\left[\frac{\epsilon^2}{4} - (P-1)^2\right]^{1/2} \tag{7}$$

For large ϵ, strong off-resonant growth can also occur. For $P \simeq 2, 3, \ldots$, etc., the growth rate is progressively smaller.

An electrostatic particle code was used to illustrate the plasma laser instability described above. The ions were taken to be infinitely massive with a prescribed spatial density distribution to avoid the two-stream instability. The following parameters were used: $v_o = 12.5 v_T$, $k_i \lambda_D = 0.16$ (v_T is the thermal velocity and λ_D is the Debye length); the ion fluctuation amplitude ϵ was varied. The time evolution of the electric field energy of the most unstable mode is shown in Fig. 1a. The linear growth rate (Fig. 1b), evaluated from simulations (solid dots), is in good agreement with the theoretical prediction from Eq. (7) (solid line). The mechanism for saturation is caused by the trapping of beam electrons by the long-wavelength modes with phase velocities slower then v_0.

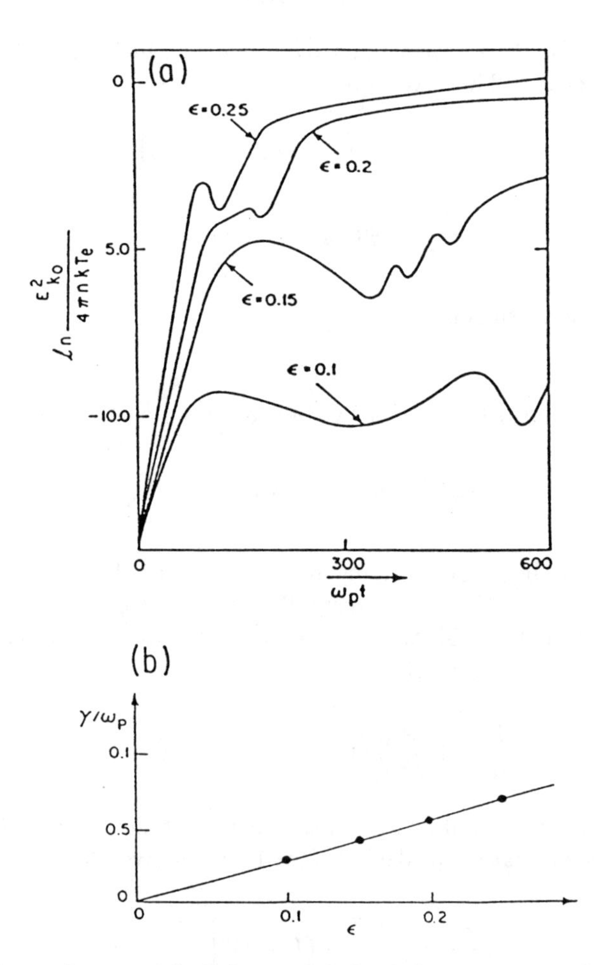

FIGURE 1 Plasma lasers with $k_0 \lambda_D = 0.012$, $k_i \lambda_D = 0.16$, and $v_0 = 12.5 v_T$. **(a):** The time evolution of the averaged wave energy for the most unstable mode. **(b):** Growth rate versus ion density fluctuations (solid dots = simulation, solid line = theory)

The plasma laser instability was also considered in the solar wind environment as the source of emitting non-thermal radiation[4,5]. Recently[6], employing a relativistic electromagnetic particle code and propagating the electron beam obliquely to the ion density fluctuations, a very high frequency ($\omega = 2\gamma^2 k_i v_o$, γ is the beam relativistic factor) electromagnetic wave was efficiently generated based on the plasma laser mechanism.

Free Electron Lasers

In 1973 I spent one year at the Princeton Plasma Laboratory as a research associate. During that time period we developed a non-relativistic electromagnetic particle code. Dawson, who is always ahead of the crowd in pursuing research, suggested the simulaton of free electron lasers. As a result of that investigation, the dispersion relation and simulation results of a free electron laser were published in 1974[7]. After we moved to UCLA, the EM code was modified to include relativistic effects, and the topic of free electron laser simulations was picked up again by a graduate student.

In a free electron laser, a relativistic electron beam passes through a static helical magnetic field. The rippled magnetic field couples the positive energy electromagnetic wave and the negative energy slow space charge beam wave. In the beam frame, the interaction is very similar to the parametric process, which was extensively studied in laser fusion. The resulting dispersion relation[8] is given by

$$\left\{[\omega - (k + k_0)v_0]^2 - \frac{\omega_p^2}{\gamma_0^3}\right\}\left(\omega^2 - k^2c^2 - \frac{\omega_p^2}{\gamma_0}\right) = \frac{\omega_{ce}^2(k + k_0)^2}{2\gamma_0^5 k_0^2}\,, \tag{8}$$

where $\omega_{ce} = eB_w/m_e c$ and B_w is the ripple magnetic strength. To derive the output frequency scaling of a free electron laser, the uncoupled slow space charge beam mode in a ripple magnetic field can be used ($\omega = (k + k_0)v_0$, $\omega_p/\gamma_0^{3/2} \ll k_0 v_0$).

$$\omega = \frac{k_0 v_0}{1 - \frac{v_0}{c}} \approx 2\gamma_0^2 \frac{k_0 v_0}{\left(1 + \frac{\omega_{ce}^2}{c^2 k_0^2}\right)}\,. \tag{9}$$

This scaling illustrates that, employing relativistic electron beams with different energy, free electron lasers are capable of encompassing a large fraction of the entire electromagnetic spectrum.

Of considerable interest is the saturation mechanism, which determines the potential efficiency of a free electron laser. This can be approximately estimated by assuming that trapping of the electrons in the longitudinal ponderomotive potential wave gives rise to the saturation of the instability. For large γ_0, the efficiency can be expressed by

$$\eta = \frac{\gamma_0 - \gamma_{ph}}{\gamma_0 - 1} \approx \frac{1}{2k_0 c \gamma_0^{3/2}}, \tag{10}$$

where γ_{ph} is the γ corresponding to the phase velocity of the wave.

The simulation results of using $k_0 c = 2.2\omega_p$, and $\gamma_0 = 2$ are shown in Figures 2 to 4. The unstable spectrum at $\omega_p t = 40$ is displayed in Fig. 2. Both the wavenumber and frequency matching conditions are satisfied between the unstable modes in the electromagnetic and the electrostatic spectrum. The linear bandwidth of a free electron laser with $\gamma_0 = 1.8$ and $\omega_{ce} = 0.7\omega_p$ is plotted in Fig. 3. The solid curve represents the theoretical prediction (Eq. 8) and solid dots are calculated from simulation results. They agree quite well. The efficiency as a function of γ_0 is given in Fig. 4 for three different ω_{ce}. The results are in reasonable agreement with Eq. (10), which shows that the efficiency decreases as γ_0 increases. For $\gamma_0 = 2.5$, the conversion efficiency is about 10%. To increase the efficiency, Dawson suggested in 1979[9] that the ripple field strength can be increased right before the trapping process takes place. In doing so, a deeper ponderomotive potential well is created, which forces the electrons to give up more energy to the wave. Simulations were carried out to demonstrate this concept and a factor of two efficiency enhancement was achieved.

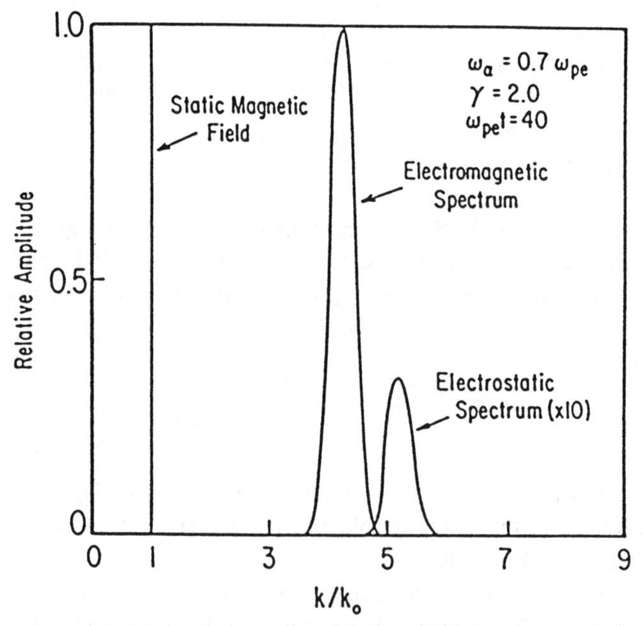

FIGURE 2 The rippled magnetic field, the unstable electromagnetic, and the electrostatic spectrum satisfy the wavenumber matching condition.

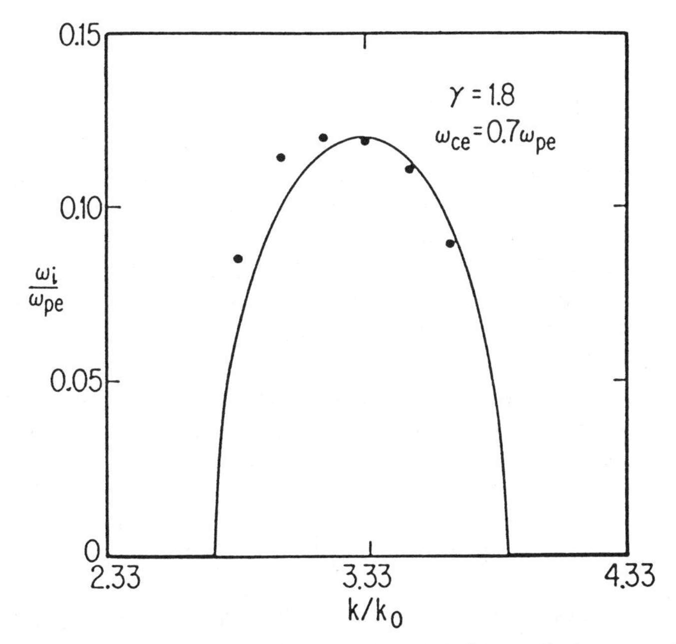

FIGURE 3 Theoretical growth rate versus wavenumber and the growth ratio, obtained from simulation for the case with $\gamma = 1.8$ and $\omega_{ce} = 0.7\,\omega_{pe}$.

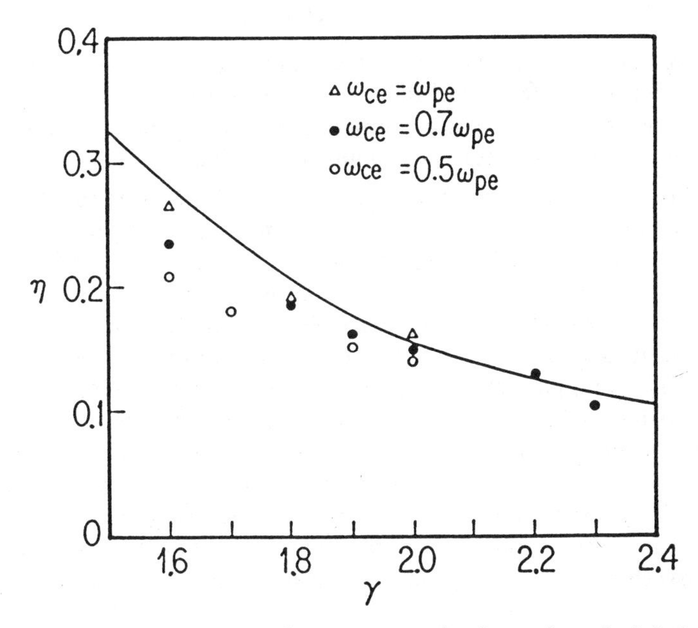

FIGURE 4 Theoretical efficiency of energy transfer from the relativistic electron beam to electromagnetic radiation versus γ, and the results obtained from simulations.

Trapped Particle Instability

The detuning between the electrons and waves caused by the electron energy loss in a free electron laser interaction can be compensated by re-accelerating the electrons[10] or decreasing the wiggler period[11] (Eq. 9). All these efficiency enhancement schemes have to be implemented when the FEL saturation is about to occur. The trapping of electrons in the longitudinal ponderomotive potential was observed to give rise to the saturation of a FEL (previous section). Therefore, it is important to analyze the stability of electron bouncing motion in the wave. As far back as 1969, Dawson[12] already investigated the trapped particle instability in a large amplitude plasma wave. Following closely the procedure of Reference 12 and replacing the plasma wave by the FEL ponderomotive potential wave, the dispersion relation retained the lowest order coupling can be obtained as

$$\left(\Omega^2 - \omega_b^2\right)\left(\omega^2 - c^2k^2\right) = \frac{\omega_t^2\omega_{ce}^2(k_0+k)^2}{2k_0^2\gamma_0^5}\,, \tag{11}$$

where $\Omega = \omega - (k_0 + k)v_{ph}$, and ω_t is the plasma frequency for the trapped electrons. The electron bouncing frequency is approximately given by

$$\omega_b \simeq (k_0 + k_s)\left(\frac{e\phi_0}{m_e\gamma^3}\right)^{1/2}\omega_p\,, \tag{12}$$

where k_s is the wavenumber of the carrier wave and ϕ_0 is the longitudinal potential. In comparing Eq. (8) with Eq. (11), we notice that ω_p^2/γ_0^3 is replaced by ω_b^2. This is because the electrons are trapped and their responses to the perturbation are no longer plasma oscillation but bouncing motion.

The simulations of FEL trapped particle instabilities were carried out using the following parameters: $\gamma_0 = 3, ck_0 = 1.52\omega_p$, and $\omega_{ce} = 0.535\omega_p$. The time evolution of the unstable waves is shown in Fig. 5a. The growth rate of the carrier wave estimated from simulation is 0.08 ω_p, which is in agreement with the theoretical prediction of Eq. (8). The trapping of the electrons by the ponderomotive potential wave causes the saturation to occur at $\omega_p t \simeq 150$ and the efficiency of this case is about 8%. This trapping process becomes very evident in Fig. 5a, which exhibits oscillatory behavior in the time evolution of the carrier wave energy. At the onset of trapping, a broad spectrum of parasitic waves with frequencies lower than that of the carrier wave becomes unstable. These waves grow with a very large growth rate $(0.2\omega_p)$, which is substantially larger than the original carrier wave growth rate. Substituting the appropriate parameters from simulation results into Eq. (11) yields a growth rate of 0.15 ω_p, which is close to the simulation results.

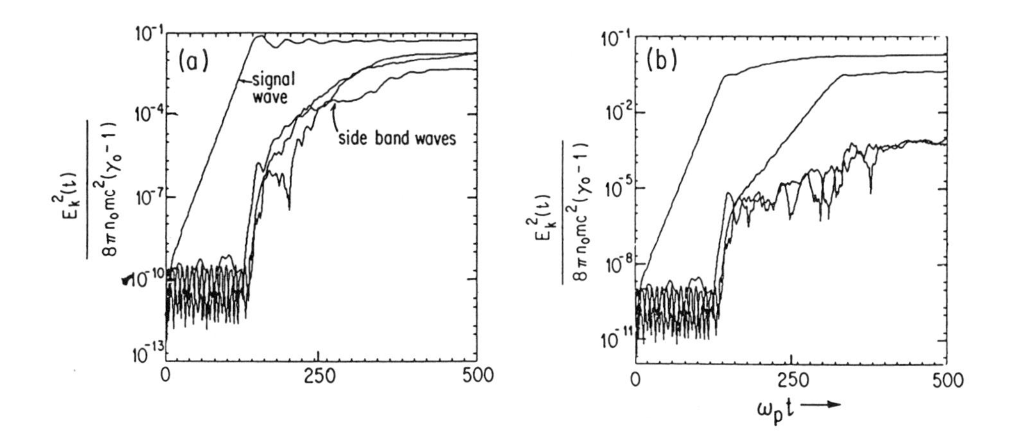

FIGURE 5 Time evolutions of the unstable electromagnetic wave energies: **(a)** without imposing any dc electric field, **(b)** imposing a dc electric field at $\omega_p t = 150$.

When a dc electric field with amplitude $E_0 \simeq (k_0 + k_s)\phi_0$ is applied at the saturation time $(\omega_p t = 150)$, the combined action of the dc and radiation fields causes some of the beam electrons to become runaways, while others remain clamped at the decelerating phase of the radiation field (Fig. 6a). The clamped electrons transfer nearly all the dc electric field energy that went into them to the electromagnetic radiation. The escaped electrons are lost to the interaction. The gain in the output power can, in principle, be extended indefinitely if the side band waves can be prevented from growing. The simulation results show that the unstable side band spectrum is substantially narrowed upon imposing the dc electric field (Fig. 5b). This is due to the distortion of the potential well and phase shift caused by the dc electric field. In fact, only the mode with $ck = 12.2\omega_p$ remains unstable with a reduced growth rate of 0.03 ω_p, and the carrier wave energy is enhanced to six times its original saturation energy. The enhancement process is eventually terminated due to the detrapping of resonant electrons by the ponderomotive wave with $c(k_0+k) = 13.7\omega_p$, whose phase velocity is in the proximity of the original ponderomotive wave (Fig. 6c). At the same time the side band instability also generates a long wavelength ponderomotive wave with wavenumber $c(k_s - k) = 3\omega_p$ (Fig. 6b), which does not give rise to any detrapping. Up to now, only the variable wiggler scheme was successfully implemented in the experiment[13] which produced a factor of five efficiency enhancement.

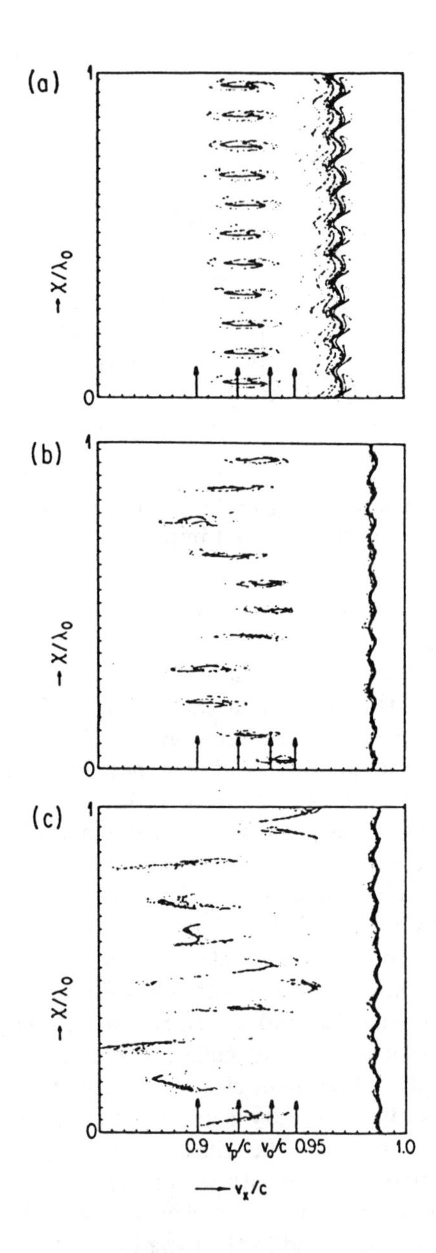

FIGURE 6 Time evolution of the phase space for the case imposing a dc electric field at $\omega_p t = 150$ with amplitude $E_0 \simeq E_{s\ell}$, **(a)**: $\omega_p t = 210$, **(b)**: $\omega_p t = 300$, **(c)**: $\omega_p t = 325$.

Electron Cyclotron Maser Amplifiers

In relativistic microwave devices, a metallic wall waveguide is commonly utilized to confine the propagation of electromagnetic waves generated by electron beams. To simulate the boundary effects, the relativistic electromagnetic particle code was modified to become capable of handling these realistic boundary conditions and both the single[14] and multi[15] transverse mode versions were developed. In this section, the multi mode code will be employed to study the interaction between the convective and absolute unstable modes in electron cyclotron maser amplifiers.

In a fast wave ($v_{ph} > c$) electron cyclotron maser, a relativistic electron beam with finite transverse dc momentum propagates through a waveguide immersed in a longitudinal uniform magnetic field (B_0). The interaction between the transverse electric waveguide mode and the beam cyclotron mode excites various instabilities. These modes are respectively the following:

$$\omega^2 = \omega_{mn}^2 + c^2k^2 \, , \tag{13}$$

where ω_{mn} is the waveguide cutoff frequency and is given by $\omega_{mn} = k_{mn}c/r_w$, where r_w is the waveguide radius and k_{mn} is the nth zero of Bessel's function $J'_m(k_{mn}) = 0$; and

$$\omega = kv_0 + s\Omega \tag{14}$$

where s is the harmonic number and $\Omega = eB_0/m_e\gamma_0$. The uncoupled dispersion curves of Eqs. (13) and (14) are shown in Fig. 7a with the following parameters: $r_w = 0.2654$ cm, $B_0 = 12.38$ kG, $I_b = 1.25$A, $V_b = 87.75$ kV, and $\alpha = v_\perp/v_\parallel = 1$. These parameters were used in a recent experiment[16] to demonstrate the convective amplification of an injected signal based on the electron cyclotron maser mechanism. The experimental observations revealed that the excitation of the absolute unstable mode (TE_{21} and $s = 2$) degrade the performance of the convective amplification (TE_{11} and $s = 1$). The experimental results also illustrated that the TE_{21} mode could be suppressed by increasing the injected signal power (Fig. 8a).

According to Figure 7a, the s = 1 beam line grazes the TE_{11} mode at the location marked "3". The injected signal will be at this frequency and the amplification will reach the saturation stage within the system length L = 15 cm. On the other hand, the s = 2 beam line intersects with the TE_{11} and TE_{21} modes at four different locations. The instabilities at "4" and "5" are convective in nature. They will amplify the noise fluctuation and convect out of the system before reaching significant amplitudes to affect the behavior of the desired mode ("3"). The instabilities at "1" and "2" are absolute in nature because their group velocities are opposite to the electron drift direction which provides an internal feedback mechanism. These modes will grow from the noise fluctuation and will be trapped in the system until they become saturated. Depending on their saturation level, the amplification of the desired mode will either be totally prohibited or somewhat degraded.

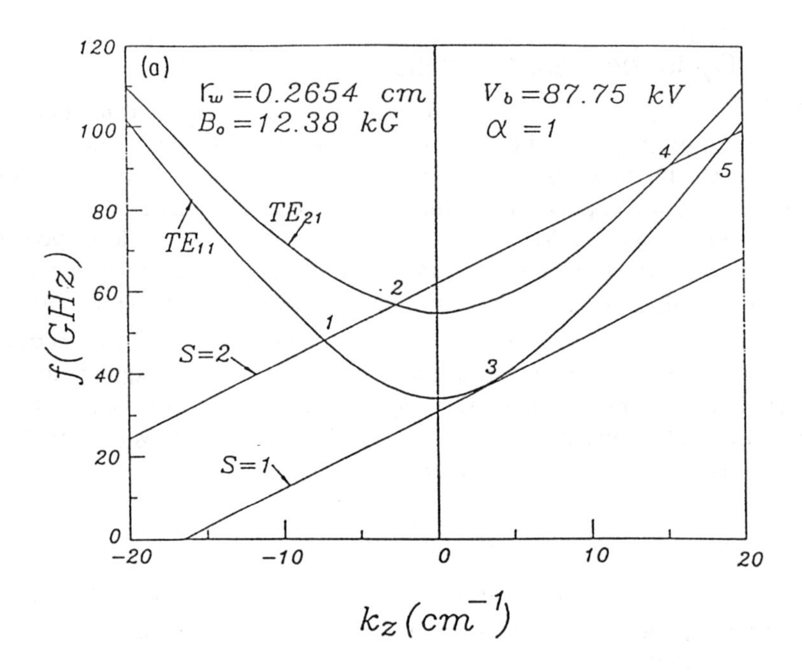

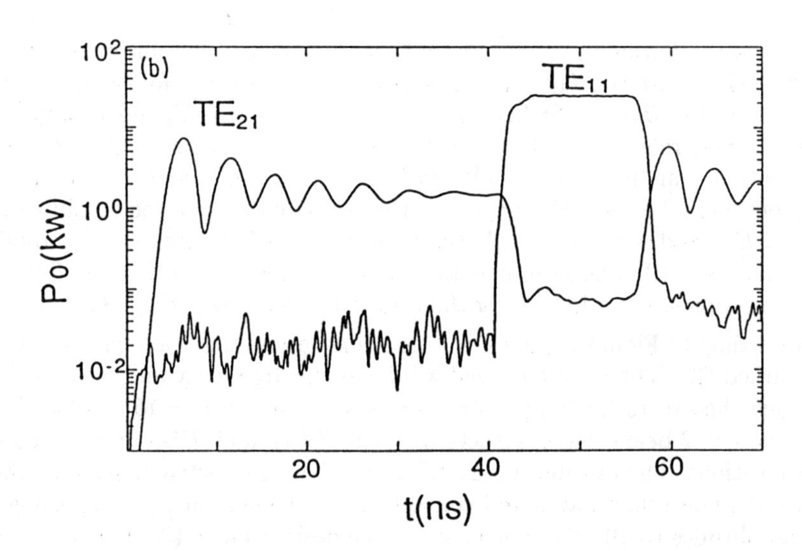

FIGURE 7 Electron cyclotron maser: **(a)** uncoupled dispersion curves, **(b)** simulation results of the TE_{11} and TE_{21} mode power.

A simulation run was carried out to elucidate their mutual influences. The electron beam was sent into the system (t = 0) and the desired mode was injected later (t = 40 nsec.) (Fig. 7b). At the earlier time, the absolute unstable mode (TE_{21}) grew and saturated. After the amplification of the desired mode (TE_{11}), the TE_{21} is suppressed. When the injected signal is turned off, the TE_{21} reappears. In the presence of the TE_{21} mode oscillation, the electron beam acquires a 10% effective thermal spread which has a deleterious effect on the desired amplification. On the other hand, the presence of the amplified TE_{11} mode induces an effective 20% thermal spread (near the saturation regime) and through the Doppler effect (kv), it tends to stabilize the absolute unstable mode. The effects of varying the injected power on the desired mode output (TE_{11}) and the oscillation (TE_{21}) power are shown in Fig. 8a (experiments) and Fig. 8b (simulations). They agree quite well, which ensures us that the particle code can be employed not only to study the basic physics but also to predict experimental outcome of microwave devices.

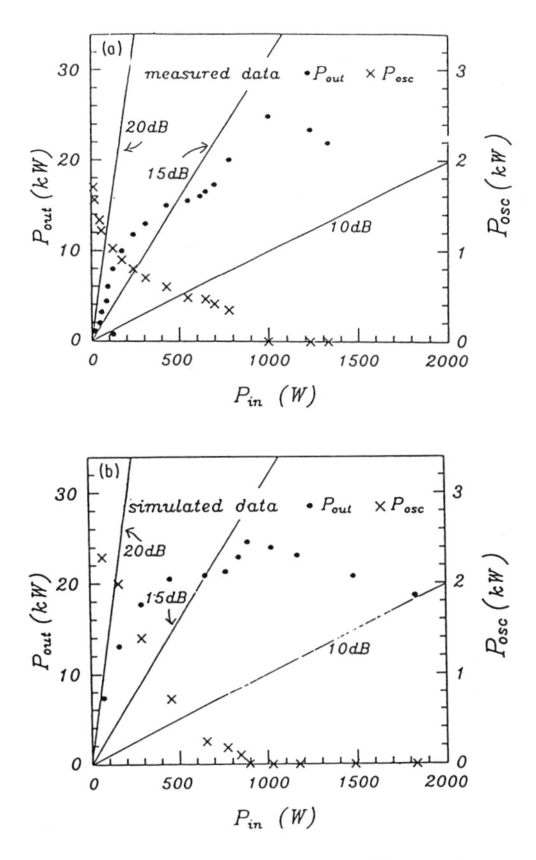

FIGURE 8 The desired mode and the absolute unstable mode output power versus the input power, **(a)** measured data, **(b)** simulated data.

Acknowledgments

This work has been supported by the Air Force Office of Scientific Research Contract No. 91-0006.

References

1. A.T. Lin, P.K. Kaw, and J.M. Dawson, *Physical Review* A, 8, 2618 (1973).
2. J.M. Dawson, *Physical Review*, 118, 381 (1960).
3. N.W. Mclachlan, "Theory and Application of Mathieu Functions," (Dover, New York, Chap. IV), 1964.
4. M.V. Goldman and D.F. DuBois, *Phys. Fluids*, 25, 1062 (1982).
5. D.B. Melrose, *Aust. J. Phys.*, 35, 67 (1982).
6. K.R. Chen, private communication.
7. A.T. Lin, J.M. Dawson, and H. Okuda, *Phys. Fluids*, 17, 1995 (1974).
8. T. Kwan, J.M. Dawson, and A.T. Lin, *Phys. Fluids*, 20, 581 (1977).
9. A.T. Lin and J.M. Dawson, *Phys. Rev. Lett.*, 42, 1670 (1979).
10. A.T. Lin, *Phys. Rev. Lett.*, 46, 1515 (1981).
11. N.M. Kroll, P.L. Morton, and M.N. Rosenbluth, *IEEE J. of Quantum Electronics*, EQ-17, 1436 (1981).
12. W.L. Kruer, J.M. Dawson, and R.D. Sudan, *Phys. Rev. Lett.*, 23, 838 (1969).
13. T.J. Orzechowski et al., *Phys. Rev. Lett.*, 57, 2172 (1986).
14. M. Caplan, A.T. Lin, and K.R. Chu, *Int. J. Elect.*, 53, 659 (1982).
15. A.T. Lin and C-C Lin, "Nuclear Instruments and Methods in Physics Research", A250, 373 (1986).
16. L.R. Barnett, L.H. Chang, H.Y. Chen, K.R. Chu, W.K. Lau, and C.C. Tu, *Phys. Rev. Lett.*, 63, 1062 (1989).

Andrew Sessler
Lawrence Berkeley Laboratories

Accelerators and Light Sources: Opening Remarks to Session VIII

We shall, in this session, first have two talks on plasma accelerators, and then one talk on frequency shifting by means of plasmas. Both topics have been pioneered by John Dawson.

It was in the paper with Toshi Tajima in 1978 that the subject of plasma accelerators using a laser was first seriously proposed. Of course – especially over glasses of beer – accelerator physicists had talked of the use of lasers, but no one had a specific method in mind. That was changed by Tajima and Dawson. The subject of novel accelerator methods "took off": many conferences [at Los Alamos (1982), Malibu (1985), Madison (1986), and Lake Arrowhead (1989); and in Europe at Oxford (1982), Frascati (1984), Orsay (1987), and one planned for East Germany (1991)], hundreds of research papers, and major programs (both theoretical and experimental) at a good number of institutions. I think it is correct to say that the impetus for making the field "respectable", i.e., worthy of effort because not only are the goals important, but it is possible to make progress towards them, was primarily due to John.

Our first paper is only the tip of the iceberg; it serves to remind us of the very large world out there ...

The subject of frequency modification is similar, but time displaced. The basic paper is by Scott Wilks, Warren Mori, and John Dawson in 1988. True, the subject of frequency modification had been known for some years, but primarily as a "bad effect". The work of Dawson et al. points out that the phenomena can be controlled and, furthermore, with modern lasers can be very large and useful. So far, there have not been very many papers on this subject, and only significant experimental work at UCLA, but I think we will hear much more about this topic in the future.

Our second paper introduces us to this very new topic. It is one of the most recent things that John has worked upon and therefore most appropriate for the last paper at this conference ...

R. Bingham, U. de Angelis, and T.W. Johnston*
Rutherford Appleton Laboratory, Chilton, Didcot, Oxon OX11 0QX
*INRS-ENERGIE, University du Quebec, Varennes, Canada

Plasma Accelerators

Recently attention has focused on charged particle acceleration in a plasma by a fast, large amplitude, longitudinal electron plasma wave. The plasma beat wave and plasma wakefield accelerators are two efficient ways of producing ultra-high accelerating gradients. Starting with the plasma beat wave accelerator (PBWA) and laser wakefield accelerator (LWFA) schemes proposed by Tajima and Dawson[1] in 1979 and the plasma wakefield accelerator (PWFA) also proposed by Dawson[2] and co-workers in 1985, steady progress has been made in theory, simulations and experiments.

In the beat wave accelerator, two laser beams of nearly equal frequencies resonantly beat in a plasma in such a way that their frequency and wavenumber differences correspond to the plasma wave frequency and wavenumber. For co-propagating laser beams the plasma wave phase velocity $v_{ph} = (\omega_1 - \omega_2)/(k_1 - k_2) \simeq \omega_{pe}/k_p$ is equal to the group velocity of the laser beams which is slightly less than c, the velocity of light, if the laser frequencies are much greater than the plasma frequency. A very similar mechanism is involved in the stimulated Raman scattering process. The group velocity of the plasma wave is essentially zero for a cold plasma, so that the excited plasma wave is left behind as the laser pulse propagates through the plasma. The laser pulse therefore always propagates into regions of undisturbed plasma leaving a large amplitude plasma wave behind which will eventually evolve into plasma turbulence.

Electrons which are injected with velocities close to the phase velocity of the plasma wave can be trapped and accelerated to higher energies. A significant experiment on the plasma beat wave accelerator was carried out by Joshi's group at UCLA[3] where an accelerating field of 1 GeV/m was demonstrated. Similar fields have since been reported by the Imperial College and RAL group,[4] using a neodymium glass laser and similar plasma densities to that used by the UCLA group.

In the LWFA, a short pulse of laser light, whose frequency is much greater than the plasma frequency, excites a wake of plasma oscillations due to the ponderomotive force. Since the plasma wave is not resonantly driven as in the beat wave accelerator the plasma density does not have to be of a high uniformity to achieve large amplitude waves. With the rapid development of high brightness lasers such as the 10J, 1ps (Tabletop Ten Terawatt) system developed by the LLNL group[5], which could eventually be the front end of a 1kJ, 1ps (Petawatt) system, the LWFA is a promising source of energetic particles of the order of 10-100GeV in distances of the order of a meter. The focal intensities of such lasers will be $\geq 10^{19} W/cm^2$ with $v_{osc}/c \geq 1$, where v_{osc} is the electron quiver velocity in the laser field, which is the strong nonlinear relativistic regime of the LWFA. Any analysis must therefore be in the strong relativistic regime.

The PWFA on the other hand does not require the use of lasers, only a high current but low energy relativistic electron beam, which is injected into a cold plasma. As in the two-stream instability, the streaming electrons lose energy to the background plasma by exciting a plasma wave. The phase velocity of the plasma wave is tied to the velocity of the injected electrons, which is close to c. The concept of the PWFA has been known since the late 50's when theory and experiments were carried out by the Soviets.[6,7] Because of the need for ultra-relativistic beams no detailed studies, however, were carried out until recently by Dawson[2] and his group at UCLA, with experiments being carried out at Argonne National Laboratory.[8,9] The results of these experiments have directly verified the existence of electron acceleration in plasma wakefields by the injection of a test bunch of electrons into a beam-driven plasma wave. The experiments validated relevant predictions of linear wakefield theory such as the excitation and structure in both the longitudinal and transverse directions of the excited plasma wave.

A number of drawbacks associated with the plasma beatwave scheme have to be overcome: (a) the necessity for very uniform plasma, (b) plasma wave saturation due to relativistic frequency detuning, and (c) the resonance condition requiring fine laser tuning. Plasma waves excited by either short pulse lasers (LWFA) or by charged particle beams (PWFA) appear to be easier and more efficient that the beat wave excitation. In the wakefield scheme since the excitation is due to a single pump all the problems raised by the resonance condition are absent and the plasma wave can grow to larger amplitudes.

In the spirit of research pioneered by John Dawson we present in the next two sections two different types of computations developed for the study of the LWFA and the PBWA; these complement the particle in cell methods so long the hallmark of John's research.

Numerical Solution of Model Equations Describing Laser Wakefield Excitation

Using a model based on one fluid, cold relativistic hydrodynamics and Maxwell equations together with a "quasi-static" approximation, a set of two coupled non-linear equations describing the self-consistent evolution in 1-D of the laser pulse vector potential envelope and the scalar potential of the excited wakefield is obtained. Starting from the equation for electron momentum

$$\frac{\partial \mathbf{p}}{\partial t} + v_z \frac{\partial \mathbf{p}}{\partial z} = -\left(e\mathbf{E} + \frac{1}{c}\mathbf{v} \times \mathbf{B}\right) \tag{1}$$

where

$$\mathbf{p} = m_0 \gamma \mathbf{v}, \; \gamma = \left(1 + p^2/m_0^2 c^2\right)^{1/2},$$

m_0 and $\mathbf{v}$ being the electron rest mass and velocity.

In eq. (1) we have assumed that all quantities only depend on z and t, z being the direction of propagation of the (external) pump and

$$\mathbf{E} = -\frac{1}{c}\frac{\partial \mathbf{A}_\perp}{\partial t} - \hat{z}\frac{\partial \phi}{\partial z} \; ; \; \mathbf{B} = \nabla \times \mathbf{A}_\perp ; \quad \mathbf{A}_\perp = \hat{x}A_x + \hat{y}A_y \; , \tag{2}$$

where $\mathbf{A}_\perp$ is the vector potential of the electromagnetic pulse and ϕ the ambipolar potential due to charge separation in the plasma.

Using eqs. (1) and (2), the perpendicular component of the electron momentum is found to be

$$\frac{p_\perp}{m_0 c} = \frac{e}{m_0 c^2}\mathbf{A}_\perp \equiv \mathbf{a}(z, t) \tag{3}$$

and we can write

$$\gamma = \left[1 + \left(\frac{p_\perp}{m_0 c}\right)^2 + \left(\frac{p_z}{m_0 c}\right)^2\right]^{1/2} \equiv \gamma_a \gamma_\parallel \tag{4}$$

where

$$\gamma_a = \left(1 + \mathbf{a}^2\right)^{1/2}; \quad \gamma_\parallel = \left(1 - \beta^2\right)^{-1/2} \tag{5}$$

where $\beta = v_z/c$

The equations derived from this model are now the longitudinal component of eq. (1), the equation of continuity, Poisson's equation and the wave equation for $\mathbf{a}(z,t)$, which are given by

$$\frac{1}{c}\frac{\partial}{\partial t}\left(\gamma_a\sqrt{\gamma_\parallel^2 - 1}\right) + \frac{\partial}{\partial z}\left(\gamma_a\gamma_\parallel\right) = \frac{\partial\phi}{\partial z}; \ \varphi \equiv \frac{e\phi}{m_0c^2} \tag{6}$$

$$\frac{1}{c}\frac{\partial n}{\partial t} + \frac{\partial}{\partial z}\left(n\frac{\sqrt{\gamma_\parallel^2 - 1}}{\gamma_\parallel}\right) = 0 \tag{7}$$

$$\frac{\partial^2\varphi}{\partial z^2} = \frac{\omega_{p0}^2}{c^2}\left(\frac{n}{n_0} - 1\right) \tag{8}$$

$$c^2\frac{\partial^2\mathbf{a}}{\partial z^2} - \frac{\partial^2\mathbf{a}}{\partial t^2} = \omega_{p0}^2\frac{n}{n_0}\frac{\mathbf{a}}{\gamma_a\gamma_\parallel} \ . \tag{9}$$

Assuming a driving pulse of the form

$$\mathbf{a}(z,t) = \frac{1}{2}\mathbf{a}_0(\xi,\tau)e^{-i\theta} + c.c. \ , \tag{10}$$

where $\theta = \omega_0 t - k_0 z, \omega_0$ and k_0 being the central frequency and wave-number, $\xi = z - v_g t$, $v_g = \partial\omega_0/\partial k_0$ is the group velocity and τ is a slow time-scale such that

$$a_0^{-1}\frac{\partial^2 a_0}{\partial\tau^2} \ll \omega_0^2$$

Accounting for changes in the pump due to the plasma reaction, the wave equation becomes

$$\left[2\frac{\partial}{\partial\tau}\left(i\omega_0 a_0 + v_g\frac{\partial a_0}{\partial\xi}\right) + c^2\left(1 - v_g^2/c^2\right)\frac{\partial^2 a_0}{\partial\xi^2} + 2i\omega_0\left(\frac{c^2k_0}{\omega_0} - v_g\right)\frac{\partial a_0}{\partial\xi}\right]e^{-i\theta} + c.c.$$

$$= \left[c^2k_0^2 - \omega_0^2 + \frac{n}{n_0}\frac{\omega_{p0}^2}{\gamma_a\gamma_\parallel}\right]a_0e^{-i\theta} + c.c. \tag{11}$$

where ω_{p0} is the plasma frequency of the unperturbed plasma. Equations (6), (7), (8) and (11) form the basic set for this problem in the "envelope approximation".

A simplified set of equations for the weak pump, weakly-relativistic regime, i.e., $|a_0|^2 < 1, \beta^2 \ll 1$, was derived and solved by Gorbunov et al[10] and Sprangle et al.[11] The solution has the structure of a wakefield growing inside the e.m. pulse and oscillating behind the pulse with the maximum amplitude being reached inside the pulse. Using the quasi-static approximation the time derivative can be neglected in the electron fluid equations (6) and (7) yielding the following constants

$$\gamma_a\left(\gamma_\parallel - \beta_0\sqrt{\gamma_\parallel^2 - 1}\right) - \varphi = 1 \tag{12}$$

$$n\left(\beta_0\gamma_\parallel - \sqrt{\gamma_\parallel^2 - 1}\right) = n_0\beta_0\gamma_\parallel \tag{13}$$

where $\beta_0 = v_g/c$. The constants of integration have been chosen in such a way that

$$n = n_0, \gamma_\parallel = 1, \varphi = 0$$

for

$$\gamma_a = 1\left(|a_0|^2 = 0\right) \tag{14}$$

Using equation (12) and (13), the general system (6)-(9) can be written as two coupled equations describing the evolution of the laser pulse envelope $\mathbf{a}_0$ and the scalar potential φ

$$\frac{\partial^2\varphi}{\partial\xi^2} = \frac{\omega_{p0}^2}{c^2}G \tag{15}$$

$$2i\omega_0\frac{\partial a_0}{\partial\tau} + 2c\beta_0\frac{\partial^2 a_0}{\partial\tau\partial\xi} + \frac{c^2\omega_{p0}^2}{\omega_0^2}\frac{\partial^2 a_0}{\partial\xi^2} = -\omega_{p0}^2 H a_0 \tag{16}$$

where

$$G = \frac{\sqrt{\gamma_\parallel^2 - 1}}{\beta_0\gamma_\parallel - \sqrt{\gamma_\parallel^2 - 1}}\,, \quad H = 1 - \frac{\beta_0}{\gamma_a\left(\beta_0\gamma_\parallel - \sqrt{\gamma_\parallel^2 - 1}\right)}\,.$$

The present set of nonlinear equations (15) and (16) are obtained using a quasi-static approximation, which yields two integrals of the motion given by equations (12) and (13). The model is valid for electromagnetic pulses of arbitrary polarization and intensities $|a_0|^2 \geq 1$.

A simplified version of equation (15) for $\beta_0 = 1$ and $a_0 =$ constant has been solved by Tsintsadze[12] while the coupled system for $\beta_0 = 1$ has been solved by Bulanov et al[13]. Sprangle et al[14] considered the solution of equation (15) together with an equation describing the full wave equation for $\mathbf{a}$ without the envelope approximation but still for the limiting case of $\beta_0 = 1$.

The set of coupled nonlinear equations (15) and (16) are solved numerically in the stationary frame of the pulse. Eq. (15), Poisson's equation for the wakefield, is solved with the initial conditions $\varphi = 0, \frac{\partial\varphi}{\partial\xi} = 0$ by a simple predictor-corrector method. The envelope equation (16) describing the evolution of the laser pulse is written as two coupled equations for the real and imaginary parts of a_0 and solved implicitly.

The quasi-static approximation requires that the quantity $\gamma_c = \gamma_a\left(\gamma_\parallel^2 - 1\right)^{\frac{1}{2}}$ be constant. This quantity is checked frequently during the calculation and is found to eventually change after a number of plasma periods; the greater a_0, the sooner γ_c changes. Numerical solutions of equations 15 and 16 showing the evolution of the excited plasma wakefield potential φ and electric field E_w as well as the envelope of the laser pulse is shown in Figs. 1 and 2. In all the figures we have plotted the fields as normalized quantities $\varphi = e\phi/m_0c^2$, $E_w = e|E|/m_0c\omega_{p0}$ and $|a_0| = \frac{eA_0}{m_0c^2}$, defined previously. The horizontal scale in all cases is the position $\xi = z - v_gt$ normalized by the plasma wavelength λ_p, with the pulse propagating from left to right. Time is in terms of the plasma period, $T_p = 2\pi/\omega_{p0}$, $|a_0^{in}|$ signifies the peak amplitude of the normalized initial vector potential of the laser pulse, and finally σ_r and σ_f represent the Gaussian pulse rise and fall coefficients, respectively (for a symmetric pulse $\sigma_r = \sigma_f$). The present study considers different pulse shapes. Figure 1 shows the evolution of $|a_0|, \varphi$ and the electric field E_w at time $t = 2T_p$ for parameters similar to those considered by Tsintsadze[12]. It is obvious that the most efficient pulse shape for wakefield generation is the one with a steep leading edge and a long trailing edge. Figure 2 shows the evolution of $|a_0|, \varphi$ and E_w at two different times; at late times there is significant distortion of the trailing edge of the laser pulse resulting in photon spikes. This distortion occurs where the wake potential has a minimum and the density has a maximum. The spike arises as a result of the photons interacting with the plasma density inhomogeneity with some photons being accelerated (decelerated) as they propagate down (up) the density gradient. This effect was predicted again by John Dawson and his group and is called the photon accelerator[2]. The distortion of the trailing edge increases with increasing ω_{p0}/ω_0. For a square wave pulse, pulse distortion is found to occur at the leading edge, as shown by Bulanov et al[13] and also by our present simulations (not shown). From Figure 2 we find that the longitudinal potential $\frac{e\phi}{mc^2} > 1$ or $\frac{eE_z}{m_e\omega_{p0}c} > 1$, is significantly greater than fields obtained in the PBWA, which are limited by relativistic de-tuning, no such saturation exists in the LWFA.

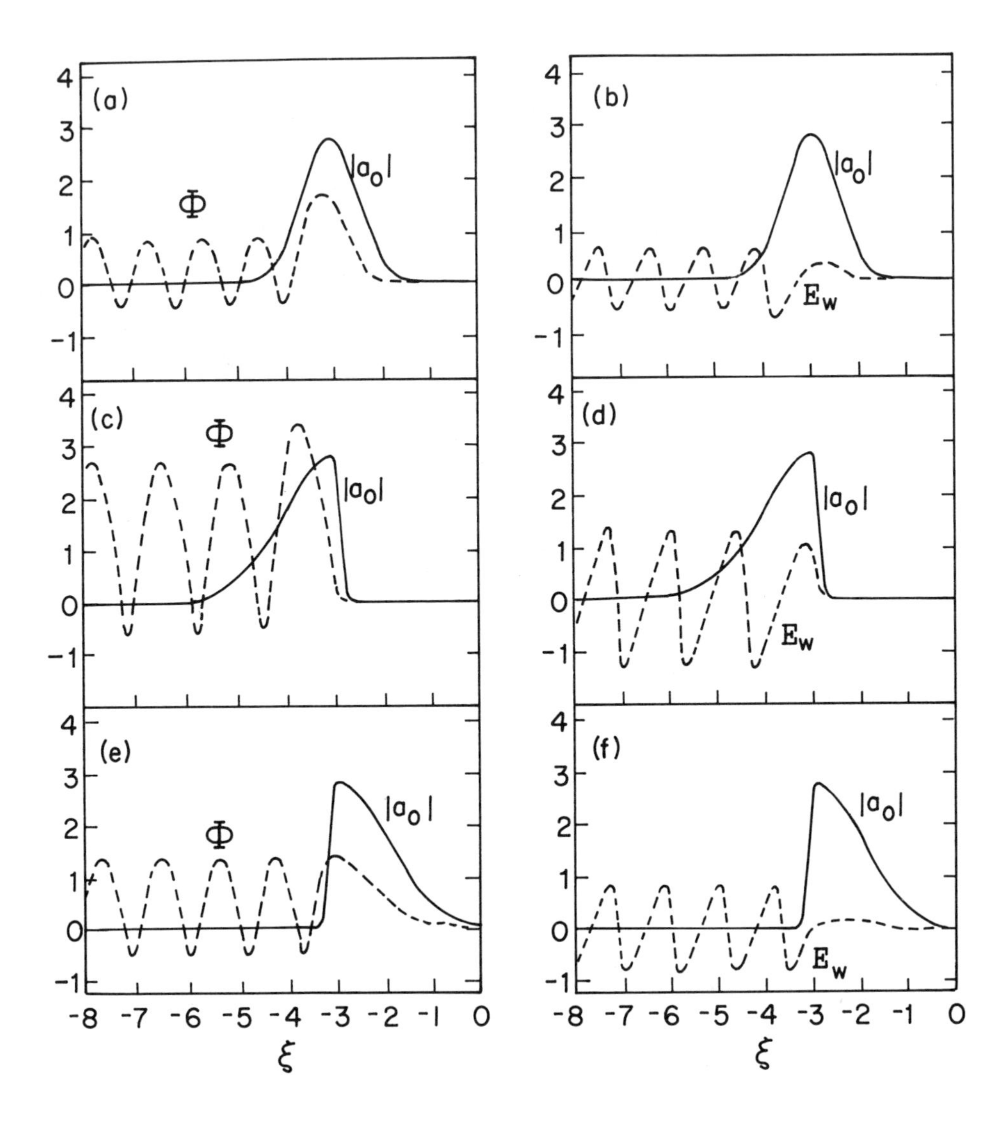

FIGURE 1 The values of the magnitude of the normalized vector potential $|a_0|$ (solid curves) and scalar potential Φ and/or wake-electric field E_w (dashed curves) with position $\xi = z - v_g t$; $|a_0^{in}| = \sqrt{8}$, $\omega_{p0}/\omega_0 = 0.01$, $t = 2T_p$. (a) and (b) Gaussian rise and fall $\sigma_r = \sigma_f = 5c/\omega_{p0} = 0.795\lambda_p$; (c) and (d) Gaussian rise $\sigma_r = c/\omega_{p0} = 0.159\lambda_p$, Gaussian fall $\sigma_f = 9c/\omega_{p0} = 1.433\lambda_p$; (e) and (f) Gaussian rise $\sigma_r = 9c/\omega_{p0} = 1.433\lambda_p$, Gaussian fall $\sigma_f = 9c/\omega_{p0} = 0.159\lambda_p$.

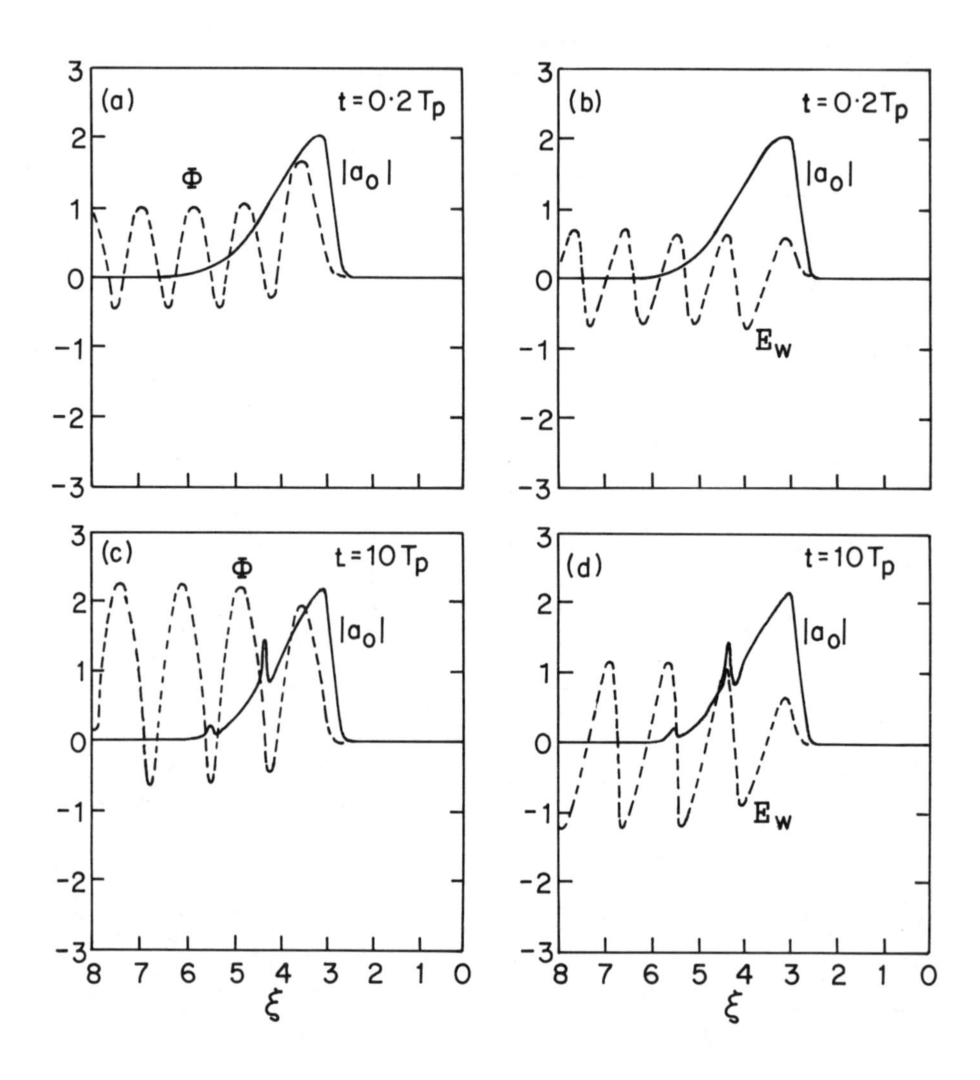

FIGURE 2 The values of the magnitude of the normalized vector potential $|a_0|$ (solid curves) and scalar potential Φ and/or wake-electric field E_w (dashed curves) with position $\xi = z - v_g t$, $|a_0^{in}| = 2$, ω_{p0}/ω_0 =0.1 Gaussian rise $\sigma_r = 0.25$ λ_p, Gaussian fall $\sigma_f = 1.5\lambda_p$; curves (a) and (b) are at time $t = 0.2T_p$; (c) and (d) are at $t = 10T_p$.

Euler-Vlasov Code for the Numerical Simulations of Plasma Beat Waves

The recent results in beatwave interactions using an Euler-Vlasov fluid code[17] allow elucidation of significant phenomena. (For reasons of computational economy these simulations use the previous parameter values[17]: $\omega_{\text{pump}}/\omega_{pe} = 2.6, \omega_{\text{idler}}/\omega_{pe} = 1.562.$); ω_{pump} and ω_{idler} here refer to the two laser beams, $\omega_{\text{pump}} \geq \omega_{\text{idler}}$.

These phenomena are as follows:

1. In spatially periodic simulations of initial-value problems:
   - i) action-sum behavior
   - ii) particle trapping/untrapping in phase space
2. In causal space-time simulations:
   - iii) constant beat frequency
   - iv) chirped beat frequency
   - v) comparison with mode coupling.

In a periodic simulation evolving from an initial state, the behavior can be compared with the classic parametrically coupled 3-oscillator model. As shown in Fig. 3, while the behaviors are generally similar, there are important differences,

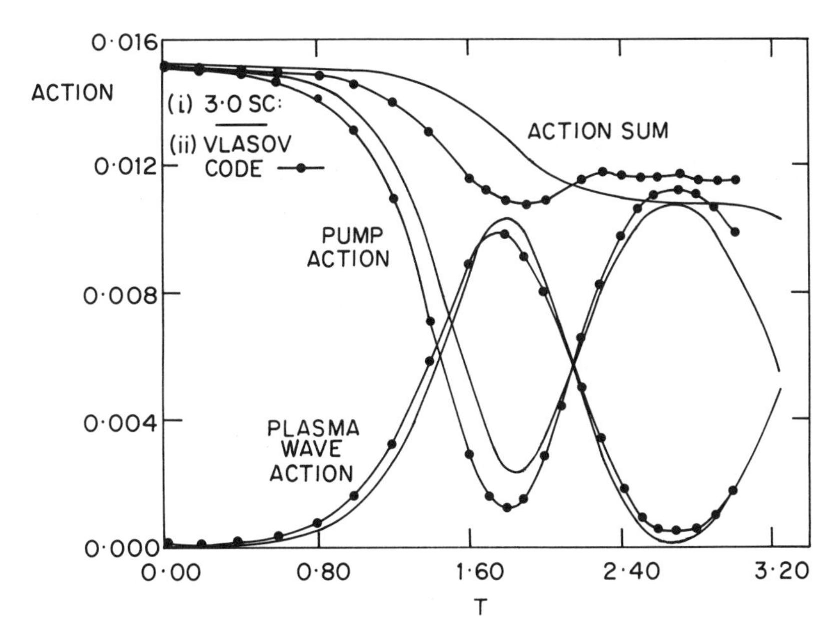

FIGURE 3 i) Typical parametrically coupled 3-oscillator result. ii) Pump + plasma wave action-sum

which can be traced to the nonlinear interaction between the trapped electrons and the plasma wave. Since the plasma wave interacts nonlinearly with both the pump and idler (which are undamped) as well as with the trapped and nearly trapped electrons, the way to isolate the interaction between the plasma wave interaction with the trapped electrons is to consider the action sum of pump + plasma wave. Instead of being linearly proportional to the plasma wave amplitude (the behavior for constant damping), the action sum decay rate has a more complicated behavior due to the details of electron trapping, even reversing as the plasma wave is decreasing while the pump wave is growing from its minimum value. By using the trapping separatrix boundary (assuming wave growth is negligible) one can account for the energy in trapped electrons. Dividing by the plasma oscillation frequency gives the action transfer involved, as shown by good overall conservation of the pump + plasma action sum plus our trapped electron "action." This behavior also makes it clear that a simple coupled-mode theory cannot be adequate if energy transfer to trapped electrons is significant and that electron trapping effects must then be correctly included.

Also noteworthy is the complex electron phase space behavior as shown in Fig. 4. As the plasma wave rises the formation of clear phase space vortices is seen and, as the plasma wave field falls, phase space density folding becomes evident as the previously trapped electrons escape by riding up over the upper separatrix.

Causal (X-T) Simulations

Causal simulations for constant beat frequency have already been published[17] which display phase space features similar to those shown in the periodic cases. An additional feature is the creation of strong slowly-moving spatial modulation in the plasma wave envelope, as seen in Fig. 5 of reference [17], so the phase space folding is now related to particle travel through spatial plasma modulation (rather than to temporal modulations as in the periodic case). (Selected phase space frames making this point were shown in Fig. 7 of reference [17].)

Calculations[18] using the Lagrangian oscillator model of Rosenbluth and Liu[19] indicated that one could increase the plasma wave amplitude to the wavebreaking limit by time variation ("chirp") of the frequency difference (below some maximum value for frequency sweep) to stay in resonance with the plasma oscillation and that the amplitude and frequency relation was reasonably well represented (see Fig. 5) by the Rosenbluth and Liu dispersion relation.

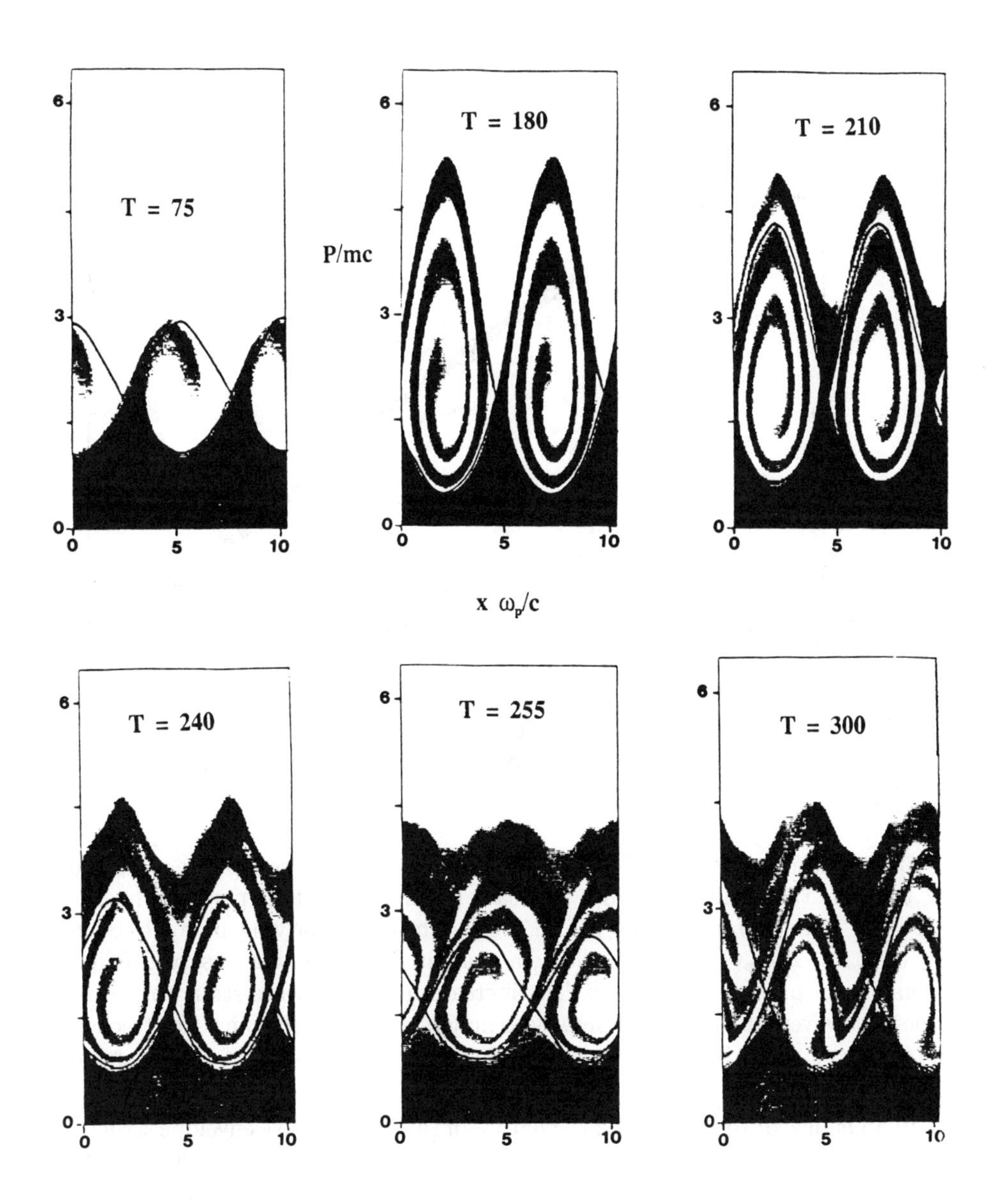

FIGURE 4 Some electron phase space frames for $\omega_p t = 75$, 180, 210, 240, 255, 300.

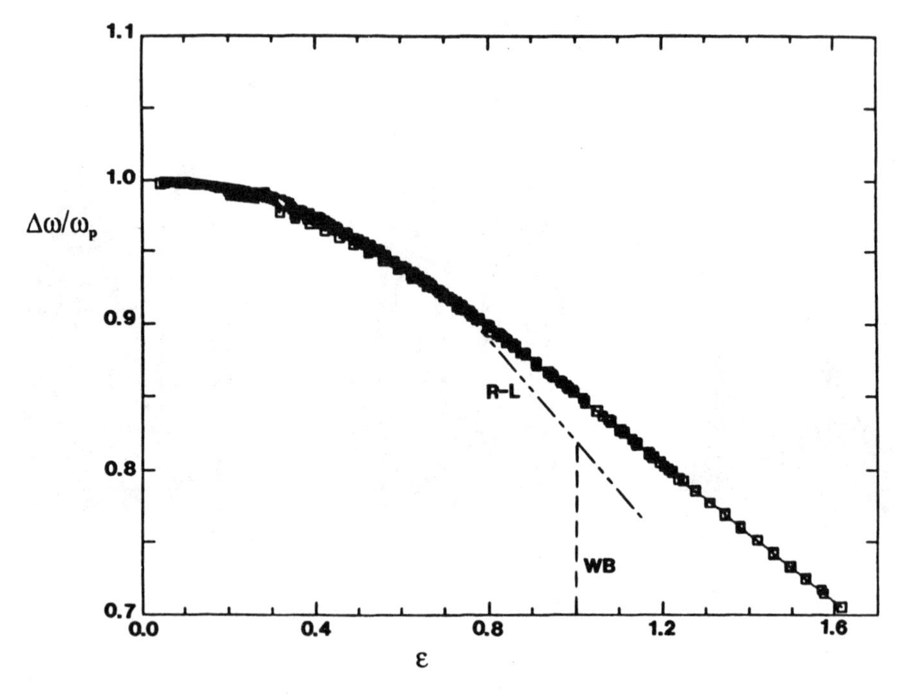

FIGURE 5 Nonlinear frequency-amplitude relationship from the chirped Rosenbluth-Liu oscillator. Their dispersion relation is shown for comparison.

Causal simulations showed the actual maximum effective sweep rate to be about 1/2 that for the oscillator model (at least for the simulation parameters used). Using the same parameters as before, but with a time-varying pump frequency now of $\omega_o = 2.6\omega_p(1 - (\omega_p t/19500))$, a dramatic increase was observed (Fig. 6) for maximum particle momentum. The increase in plasma wave strength was only modest, however, and work is underway to understand how "chirping" the frequency led to such a striking increase in particle energy for such a modest increase in plasma wave field.

In an endeavor to understand the nature of the plasma wave envelope structure, a space-time coupled-mode code[20] (with constant damping) was used and, in order to compare with the (unchirped) Vlasov result, the Vlasov transverse electric field was spatially Fourier transformed, selected by wavevector into pump, idler (Stokes) and anti-Stokes groups and then each transformed back into x-space with the results shown in Fig. 7. It is clear that the general features are well reproduced. The major part of the structure formed is seen to be due to pump depletion/plasma wave buildup cycles, but particle trapping details will have to be included to do better. This encouraging result from the simple coupled mode equations suggest the interesting possibility of handling higher-frequency cases with complex amplitude equations for the pump, idler (Stokes) and anti-Stokes waves and only using the full

Vlasov (or PIC) code for the plasma wave. The beat force for the Vlasov code would be computed from the complex amplitude equations and the real Vlasov plasma output would be converted to complex amplitude form for the mode equations via the Hilbert transform. If this method works, it could provide a saving of the order $(\omega_{\text{pump}}/\omega_p)^2$ (i.e., time x space) in computer time, as compared with the direct Vlasov method with the highest space-time resolution required.

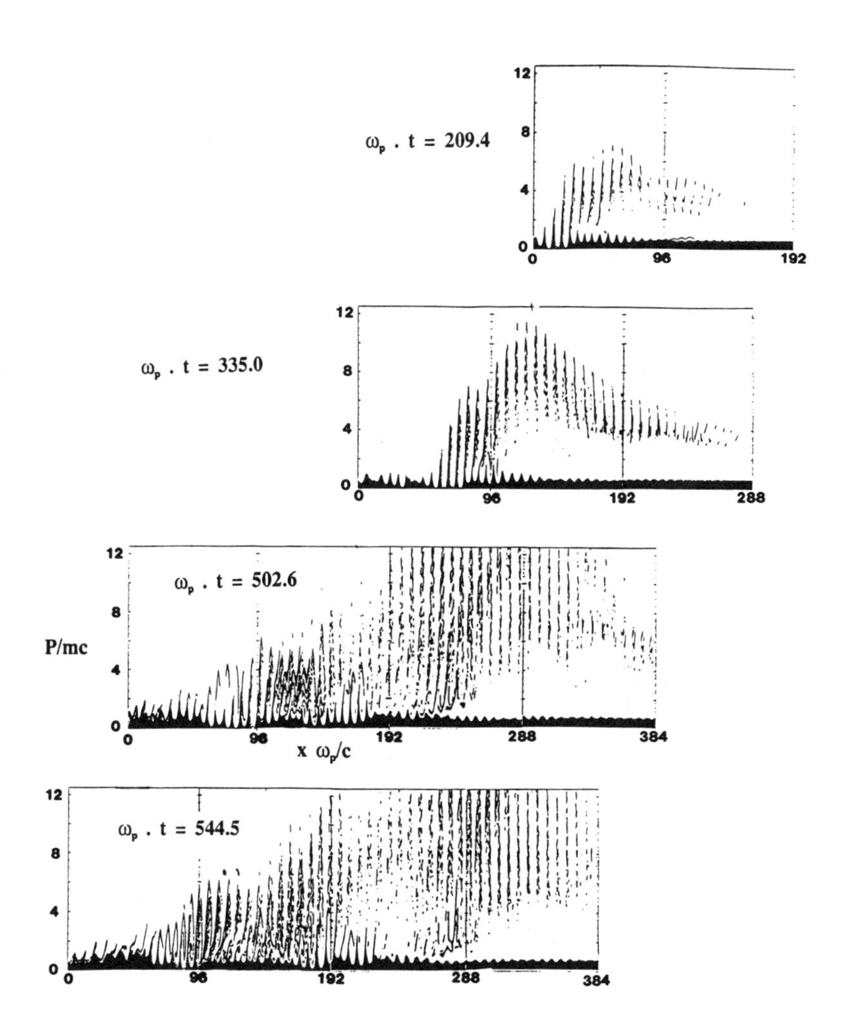

FIGURE 6 Electron phase space frames for the chirped pump. Note the maximum momentum is more than twice that of the unchirped case (Figs. 6 and 7 of reference 17).

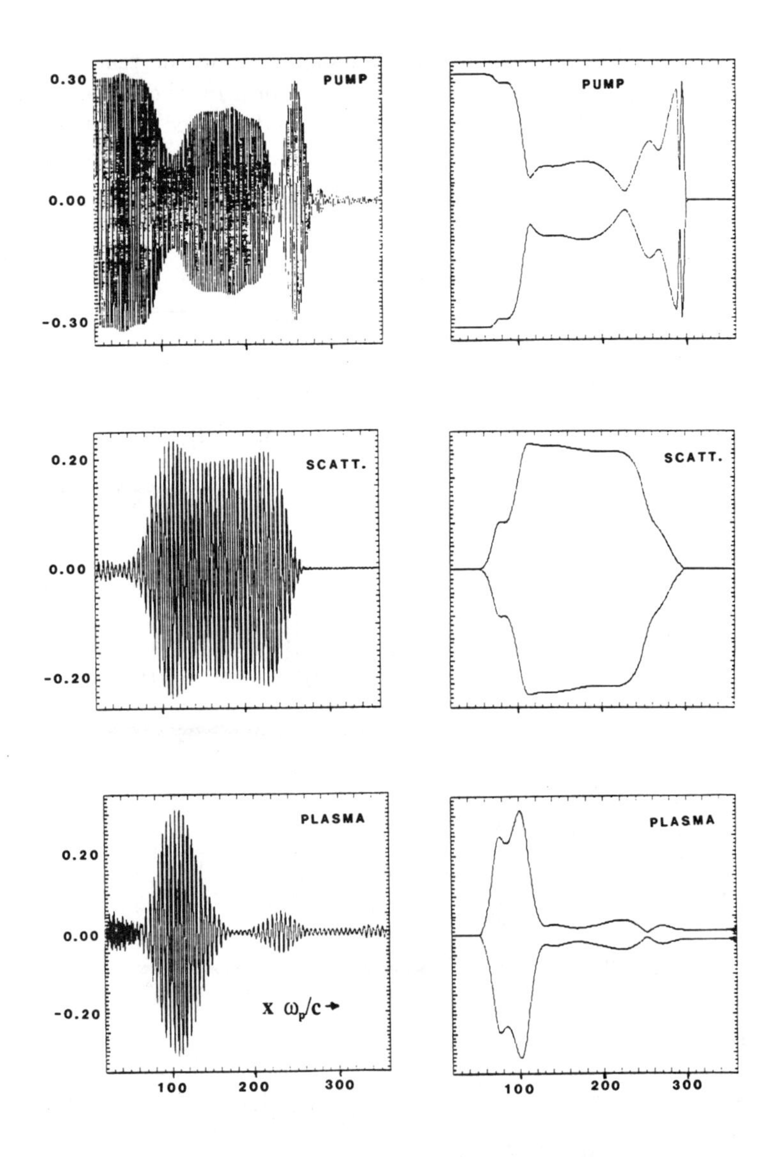

FIGURE 7 Correspondence between Vlasov waves (left) and coupled mode results (right).

While this work is still in progress, some conclusions can be drawn already. It is abundantly clear that it will require careful simulation to interpret any experimental results on electron acceleration using beatwave-induced plasma waves.

Judicious chirping of the beatwave driver should be looked at. An interesting possibility exists to extend detailed modeling to frequency ratios greater than the current practical maximum of 10 or so for Vlasov codes, by using mode equations for the electromagnetic waves rather than the full Vlasov model.

Acknowledgements

This paper is dedicated to Professor John Dawson on the occasion of his 60th birthday. He is now entering his second life cycle and we look forward to more of his stimulating original research and physical intuition.

Many happy returns to a very dear friend.

References

1. T. Tajima, J.M. Dawson, *Phys. Rev. Lett.,* **43**, 267 (1979)
2. P. Chen, J.M. Dawson, R.W. Huff and T. Katsouleas, *Phys. Rev. Lett.,* **54**, 693 (1985)
3. C.E. Clayton, C. Joshi, C. Darrow and D. Umstadter, *Phys. Rev. Lett.,* **54**, 2343 (1985)
4. A.E. Dangor, A.K.L. Dymoke-Bradshaw and A.E. Dyson, *Physica Scripta,* **T30**, 267 (1990)
5. C. Darrow, M.D. Perry, F. Patterson, E.M. Campbell, T. Katsouleas and W.B. Mori, *AIP Conf. Proc. 193 on Advanced Accelerator Concepts* C. Joshi, ed.
6. Ya.B. Fainberg, *Proceedings of CERN Symposium on High Energy Accelerators and Pion Physics, Geneva,* (1956)
7. F. Kharchenko, Ya.B. Fainberg, R.M. Nikoleav, E.A. Kornilov, E.A. Lutsenko and N.S. Pedenko, *Sov. Phys.* J.E.T.P., **11**, 493 (1960)
8. J.B. Rosenzweig, D.B. Cline, B. Cole, H. Figueroa, W. Gai, R. Konecny, J. Norem, P. Schoessow and J. Simpson, *Phys. Rev. Lett.,* **61**, 98 (1988)
9. J.B. Rosezweig, P. Schoessow, B. Cole, W. Gai, R. Konecny, J. Norem and J. Simpson, *Phys. Rev. A.,* **39**, (1989)
10. L. Gorbunov and V. Kirsanov, *Sov Phys* JETP, **66**, 290, (1987)
11. P. Sprangle, E. Esarey, E. Ting and G. Joyce, *App. Phys. Lett.*, **53**, 2146, (1988)

12. N.L. Tsintsadze, *Physica Scripta,* **T30**, 41, (1990)

13. S.V. Bulanov, V.I. Kirsanov and A.S. Sakharov, *Physica Scripta,* **T30**, 208 (1990)

14. P. Sprangle, E. Esarey and A. Ting, *Phys. Rev. Lett.*, **64**, 2011, (1990)

15. U. de Angelis, *Physica Scripta,* **T30**, 210 (1990)

16. S.C. Wilks, J.M. Dawson, W.B. Mori, T. Katsouleas and M.E. Jones, *Phys. Rev. Lett.* **62**, 2600, (1989)

17. P. Bertrand, A. Ghizzo, T.W. Johnston, M. Shoucri, E. Fijalkow and M.R. Feix, *Phys. Fluids*, **B2(5)**, 1028-1037, (1990)

18. T.W. Johnston and J.P. Matte, *Bull. Amer. Phys. Soc.*, **34(9)**, 2024, (1989) abstract 4T24

19. M.N. Rosenbluth and C.S. Liu, *Phys. Rev. Lett.*, **29(11)**, 701-705, (1972)

20. T.W. Johnston, G. Picard, J.P. Matte, V. Fuchs and M. Shoucri, *Plasma Physics and Contr. Fusion*, **24(4)**, 473-492, (1985).

C. Joshi and W.B. Mori
Electrical Engineering Department
University of California, Los Angeles, CA 90024-1594

Frequency Upconversion of Electromagnetic Radiation Using Plasmas

Abstract

In this paper we propose a new class of coherent electromagnetic wave generation devices that, in principle, can cover the range of frequencies from microwaves to the visible range of the electro-magnetic spectrum. In this method the frequency of the source e.m. wave is upshifted by suddenly lowering the refractive index of the medium in time by creating a plasma. This technique of frequency upconversion using plasmas has numerous potential applications, particularly where time dependent frequency shifts on a period by period basis are required[1].

[1]Work supported by DOE contract DE-AS03-83-ER40120 and DOE grant DE-FG03-87-ER13752

Introduction

When an electromagnetic wave impinges upon a medium whose dielectric properties do not vary in space but do vary in time, then the wavenumber of the radiation remains constant while the frequency changes. On the other hand, if the dielectric properties of the medium vary in space only, then the wavenumber will change while the frequency remains constant. One way to understand this conjugate relationship between time and frequency and space and wavenumber is that ω and t, and similarly k and x, are related by Fourier transforms. With the advent of short intense laser pulses it is now possible to produce plasmas with "rapid" density variations. By "rapid" we mean on a time scale of a few cycles of electromagnetic radiation in the few gigahertz frequency range. The refractive index of plasma is simply given by $\left(1-\omega_p^2/\omega^2\right)^{\frac{1}{2}}$ where ω_p is the plasma frequency and ω is the frequency of the electromagnetic wave inside the plasma. Since the refractive index of a plasma can vary from 1 (no plasma) to 0 ($\omega_p=\omega$), it is in principle possible to upshift the frequency of the electromagnetic wave over a large frequency range. Also, if the electromagnetic wave is already immersed in the plasma whose density can be continuously increased in time, it is possible to obtain "chirped" radiation. Such radiation sources will have an impact in many areas of physics and technologies, particularly where time dependent frequency shifts on a period-by-period basis are desirable and/or where other inexpensive sources are as yet unavailable.

Flash Ionization

It has been recognized for many years that the frequency of pre-existing radiation will be increased by instantaneously creating plasma[1]. This has recently[2] received attention because significant advances in laser technology now make it possible to suddenly ionize a gas to create a plasma. The resulting frequencies and modes generated by uniform instantaneous ionization can be derived from the following. Consider a right going pulse of electromagnetic radiation which is embedded in a volume of gas. If the gas has a negligible effect, then the radiation dispersion relation is $\omega_o=ck_o$. At some instant the entire volume of gas is instantly ionized. After the plasma has formed, the dispersion relation for the radiation is $\omega^2=\omega_p^2+c^2k^2$. In order to determine the amplitude of the radiation in the plasma it is necessary to solve an initial value problem subject to the continuity of $\vec{E}$, $\vec{B}$, and $\vec{j}$ at the time of ionization. These are three continuity conditions. Hence, there must be three modes present in the plasma after ionization. These are right and left going electromagnetic waves which satisfy the dispersion relation of light in plasma, and a time independent (static) magnetic field. The fields before ionization ($t\leq 0$) are

$$E_y = E_o \sin(k_o x - \omega_o t) \tag{1a}$$
$$B_z = E_o \sin(k_o x - \omega_o t) \tag{1b}$$
$$\frac{4\pi}{c} J_y = 0 \tag{1c}$$

After ionization the fields are

$$E_y = E_+ \sin(k_o x + \omega_o t) + E_- \sin(k_o x - \omega_o t) \tag{2a}$$

$$B_z = -E_+ \frac{ck_o}{\omega} \sin(k_o x + \omega_o t) + E_- \frac{ck_o}{\omega} \sin(k_o x - \omega_o t) + B_s \sin(k_o x) \tag{2b}$$

$$\frac{4\pi}{c} J_y = -E_+ \frac{\omega_p^2}{c\omega} \cos(k_o x + \omega_o t) + E_- \frac{\omega_p^2}{c\omega} \cos(k_o x - \omega_o t) + B_s k_o \sin(k_o x) \tag{2c}$$

where $\omega^2 = \omega_p^2 + c^2 k_o^2 = \omega_p^2 + \omega_o^2$. The wavenumber does not change during the ionization because the fields cannot change instantaneously. Mathematically, this follows from imposing the continuity conditions, which give

$$E_\pm = \frac{E_o}{2}\left(1 \pm \frac{\omega_o}{\omega}\right) \tag{3a}$$
$$B_s = E_o \frac{\omega_p^2}{\omega^2} \tag{3b}$$

This analysis indicates that it is possible to upshift a significant fraction of the source radiation from ω_o to $\sqrt{\omega_o^2 + \omega_p^2}$. Also, since the source wave is initially inside the medium being ionized, ω_p^2 can be much greater than ω_o^2. This is not the case if the source wave has to enter the plasma from the outside, in which case the electromagnetic wave cannot propagate in a plasma with densities greater than the critical density, or $\omega_p > \omega_o$. In the flash ionization case when $\omega_p \gg \omega_o$, the upshifted frequency, $\approx \omega_p\left(1 + \frac{1}{2}\omega_o^2/\omega_p^2\right)$, continuously approaches the cut-off and although the wavenumber remains fixed by the initial condition, the group velocity tends to zero, leading to the upshifted wave leaking out of the ionized region at an ever-slowing rate as one goes to higher and higher plasma densities.

Non-Instantaneous Ionization

The above analysis is valid when the ionization time is shorter than ω_o^{-1}. In reality it may not be possible to ionize a volume of gas that fast. In order to describe the evolution of the radiation field for longer ionizations times we resort to the differential equation for $\vec{E}$. For simplicity we assume one-dimensional plane waves. The equation for the electric field is

$$\left(\frac{\partial^2}{\partial t^2} - c^2 \frac{\partial^2}{\partial x^2} \right) \vec{E} = 4\pi \frac{\partial \vec{j}}{\partial t} \tag{4}$$

The ionization process influences the functional form for $\vec{j}$. A formula for $\vec{j}$ in terms of the electric field is most easily derived by assuming the plasma consists of an infinite set of species. Each species is labeled by the time at which it is created. The total current density can then be written as

$$\vec{j} = -e \sum n_i v_i$$

where the velocity of the plasma created at t_i is

$$v_i = \frac{-e}{m} \int_{t_i}^{t} dt' E(t')$$

and the density added at t_i is

$$\Delta n_i = dt \, \frac{\partial n}{\partial t} (t_i)$$

We have assumed that new electrons are born with zero velocity. This in general is a good assumption. When each electron is born, it sees a different vector potential of the source electromagnetic wave. As a result it acquires a drift energy[2]. The drift velocity has a periodicity which is the same as the source wave wavelength. The drifting electrons represent a current in the plasma which in turn implies that a static but periodic magnetic field must exist in the plasma. Now, it follows that the functional form of $\vec{j}$ is

$$\vec{j} = -\frac{e^2}{m} \int_0^t dt' \, \frac{\partial n}{\partial t'} \int_{t'}^{t} dt'' \vec{E}$$

In Eq. (4) only $\frac{\partial \vec{j}}{\partial t}$ is needed and it is

$$\begin{aligned} \frac{\partial \vec{j}}{\partial t} &= -\frac{e^2}{m} \left\{ \int_0^t dt' \, \frac{\partial n}{\partial t} \, \vec{E}(t) + \frac{\partial n}{\partial t} \int_t^t dt'' \vec{E} \right\} \\ &= -\frac{e^2}{m} \, n(t) \vec{E} \qquad . \end{aligned}$$

Therefore, the equation which describes the evolution of the electric field is simply

$$\left(\frac{\partial^2}{\partial t^2} - c^2\nabla^2 + \omega_p^2(t)\right)\vec{E} = 0 \tag{5}$$

The reason that $\frac{\partial \vec{j}}{\partial t}$ does not depend on $\frac{\partial n}{\partial t}$ is that at the instant when new plasma is created it cannot have any velocity. This partial differential equation can be Fourier transformed in space, since the coefficients only vary in time, to give

$$\left(\frac{d^2}{dt^2} + 2\nu\frac{d}{dt} + \omega_o^2 + \omega_p^2(t)\right) E_k = 0 \tag{6}$$

where $\vec{E} = \vec{E}_k(t)e^{ik_o x}$ and ν is a phenomenological damping rate. The WKB solution to Eq. (6) for $\nu = 0$ is

$$E_k = E_o\left(\frac{\omega_o^2}{\omega_o^2 + \omega_p^2(t)}\right)^{\frac{1}{4}} e^{-\int dt\sqrt{\omega_o^2+\omega_p^2(t)}} \quad . \tag{7}$$

Therefore in the WKB limit the radiation has a frequency, $\omega(t) = \sqrt{\omega_o^2 + \omega_p^2(t)}$, which increases with time. In addition, the electric field amplitude decreases in time as $\sqrt{\frac{\omega_o}{\omega(t)}}$. If $\nu(t) \ll \omega(t)$, then the effect of damping is to multiply the WKB solution by the factor $e^{-\int dt\nu(t)}$.

In order to obtain significant frequency upshift, it is necessary that $\omega_p \gg \omega_o$. Under these conditions $\omega(t) \cong \omega_p(t)$, which means that the radiation in the plasma is near its cut-off frequency. As a result, much of the radiation is reflected at the plasma-vacuum interface. This needs to be considered when estimating the power conversion efficiency. The amount of power initially present before ionization is

$$P_o = \frac{cE_o^2}{4\pi} \ .$$

The amount of power leaving the plasma-vacuum interface is

$$P_o = \left|\frac{2\sqrt{\varepsilon}}{\sqrt{\varepsilon}+1}\right|^2 \frac{c|E|^2}{4\pi}$$

where $\varepsilon = \left(1 - \frac{\omega_p^2}{\omega^2}\right) = \frac{\omega_o^2}{\omega_2}$ is the dielectric constant and E is the electric field inside the plasma. If the amount of energy leaving the plasma can be neglected, then $|E| = |E_o|\sqrt{\frac{\omega_o}{\omega}}$. To lowest order the power efficiency is therefore

$$\frac{P}{P_o} = \frac{4\left(\frac{\omega_o^2}{\omega^2}\right)}{\left(1 + \frac{\omega_o}{\omega}\right)^2}\frac{\omega_o}{\omega} \ . \tag{8}$$

In the limit that $\omega_o/\omega \ll 1$ Eq. (8) reduces to

$$\frac{P}{P_o} = 4\frac{\omega_o^3}{\omega^3} .$$

If damping is included, the efficiency is considerably reduced for frequencies which are generated at times for which $\int dt\nu(t) \gg 1$.

Simulations

The physical principles underlying flash ionization have been verified by PIC simulations as well as by preliminary experiments. The simulation results are summarized in Fig. 1, where the power spectrum and the electric field leaving the plasma-vacuum interface are plotted for three different ionization rates. In Fig. 1a the ionization was instantaneous, while in Fig. 1b and Fig. 1c the plasma density was ramped linearly in time over $\frac{2\pi}{\omega_o}$ and $10\frac{2\pi}{\omega_o}$, respectively. The newly born simulation particles were always initialized without any velocity. The results are in basic agreement with theory. The maximum final frequency in all three cases was $\sqrt{3}\omega_o$ because the maximum density corresponded to $\omega_p^2 = 2\omega_o^2$. In addition, when the ionization occurs over $10\frac{2\pi}{\omega_o}$, the electric field gradually falls off over time and the power spectrum is chirped, as predicted by the WKB analysis[3].

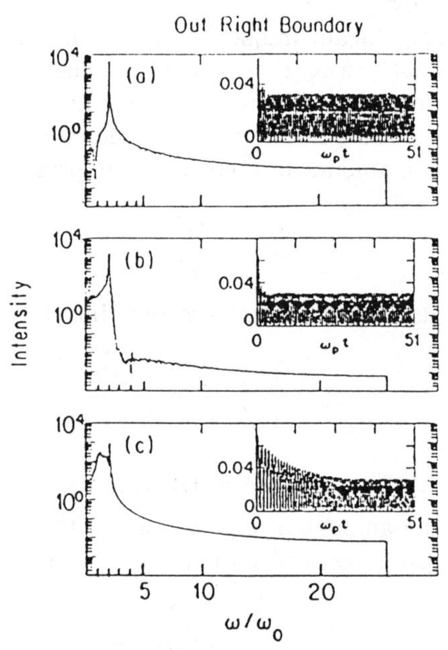

FIGURE 1 Power spectrum for radiation leaving the right-hand boundary of the plasma for increasing ionization times, Δt.

Experimental Verification

Now we describe an experiment which has demonstrated frequency upshifting of microwave radiation by rapid plasma creation[4]. In this experiment, a 33.3 GHz source electromagnetic wave contained in a cylindrical waveguide resonator is shifted up in frequency to 77 GHz by laser ionization of the vapor inside the resonator. The cut-off density for microwave source radiation is approximately 1.3×10^{13} cm^{-3}. Since densities of this order can be rather easily produced by photoionization of organic vapors using rather modest power lasers, we expect to see significant frequency upshifts.

In the experiment, a resonant circular waveguide operating close to its cut-off is filled with microwave radiation. The group velocity of the source radiation is therefore quite small, $\sim$ 0.4 c. The cavity is filled with azulene vapor (C_8H_{10}), which can be easily ionized using a frequency quadrupled Nd:YAG laser. A few nanoseconds long, 0.26μm laser pulse containing 40 mJ of energy is sent through the cavity to ionize the azulene and create a plasma. The transit time of the laser through the cavity is 1 ns. Apart from the effect of the transit time of the laser, the plasma density inside the cavity builds up fairly uniformly as a function of time for azulene pressures below 20 mT. Above this pressure pump depletion of the laser can present a problem, leading to spatial as well as temporal variations of plasma density inside the cavity. Since the plasma frequency is a function of time, the upshifted frequency $\omega^2(t) = \omega_c^2 + k^2c^2 + \omega_p^2(t)$ is also a function of time.

Fig. 2 shows time histories of signals at the source frequency, and frequencies exceeding 36 GHz and 40 GHz. One can see that, coincident with the firing of the laser, the leakage radiation at the source frequency is terminated. Simultaneously, the 36 GHz detector shows the onset of a signal that lasts typically 5ns. The 40 GHz signal peaks 2ns or so after the 36 GHz signal, as might be expected from a chirped signal.

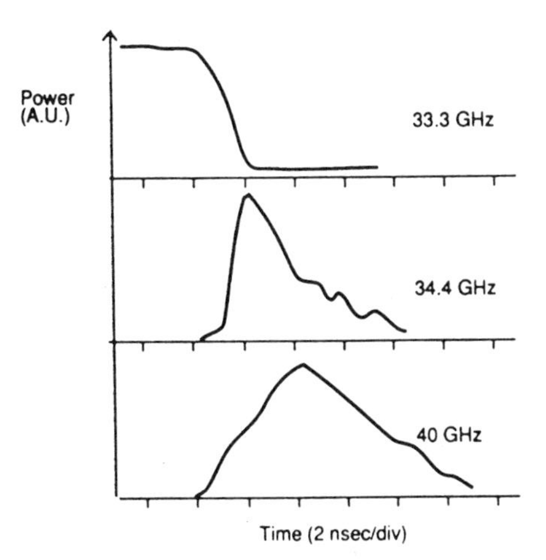

FIGURE 2 Temporal histories of signals on different cut-off frequency detectors

Fig. 3 shows the complete frequency spectrum of the upshifted radiation, after deconvolution of the experimental data. The data can be fitted quite well by a Lorentzian at the high frequency end (50 to 77 GHz). These higher frequency components are thought to arise simply from Fourier transformation of a rapidly but exponentially falling source signal and thus *do not* represent evidence for upshifted frequencies due to the plasma. The data up to 40 GHz, however, do show the expected characteristics of frequency upshift due to time varying plasma density.

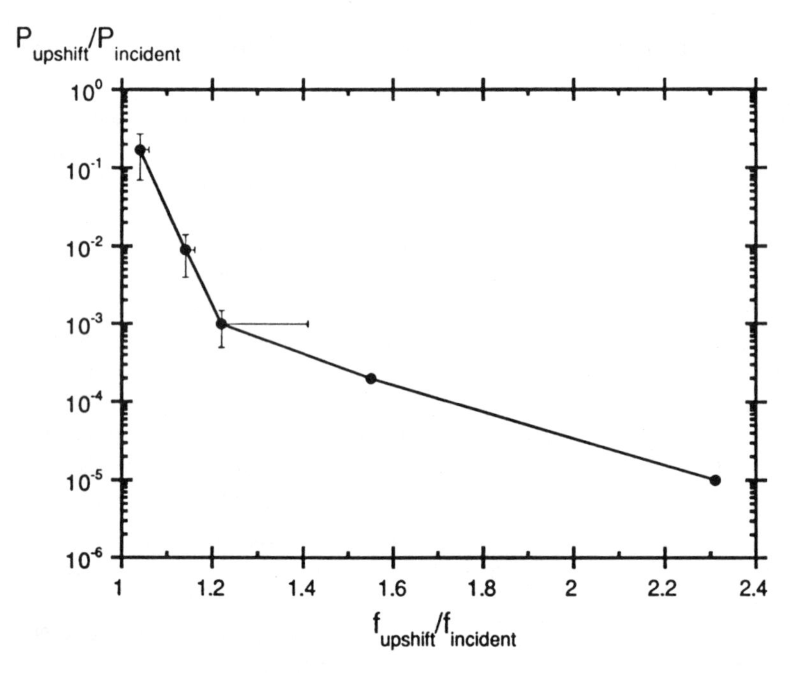

FIGURE 3 Power conversion efficiency vs. frequencies in the plasma frequency upshift experiment (ref. 4).

By using gases that can be more efficiently ionized by ultraviolet laser light, we believe that it is possible to increase the plasma density up to 10^{14} cm^{-3} in this experiment and therefore get a tunable frequency source from 33 GHz all the way up to 100 GHz.

There are some critical questions regarding the efficiency and mode structure of the upshifted radiation. As ω_p becomes much larger that ω_o, the upshifted radiation approaches its own cut-off. Even though the radiation intensity may be below any instability threshold (parametric decay, simulated Brillouin and filamentation in particular), strong damping of this electromagnetic wave can take place due to electron-ion and electron-neutral collisions. This could rapidly reduce the efficiency of the frequency upshift process. Also, for $\omega \gg \omega_o$ the radiation is now in an overmoded guide; consequently, many modes can radiate into free space. Further work is necessary to resolve these important issues.

Future Work

Ionization Fronts

Up to now we have discussed the effect that lowering the dielectric constant in a spatially uniform fashion as a function of time has on the frequency of an embedded electromagnetic wave. An interesting situation arises if the source wave is not embedded inside the medium but rather is directed at a spatial dielectric discontinuity that is propagating at close to the speed of light. A particular case of importance is when a source wave propagating in neutral gas is incident upon a plasma whose boundary is moving towards the wave at nearly the speed of light. Such a relativistically moving interface could be created by a short laser pulse moving through a gas, thus creating a so-called ionization front.

At first, it may appear that a moving ionization front is like a moving mirror. This not so because a moving mirror has energy, while an ionization front does not. Whereas, in the case of a moving mirror, the frequency shift of the wave is a consequence of energy exchange with the mirror (the photon number being conserved), in the case of an ionization front any frequency change must be at the expense of the number of photons, and the total energy of the daughter waves must be the same as that of the original electromagnetic wave.

One procedure for calculating the upshifted frequency is as follows[6]: One first calculates the frequency of the source wave in the rest frame of the front. For the counter-propagating front and the source wave, this is $\omega_f = \gamma(1+\beta)\omega_o$, where γ is the Lorentz factor associated with the velocity of propagation of the front, and β is the normalized velocity of the front. Since the plasma frequency ω_p is Lorentz invariant, the frequency of the transmitted wave in the rest frame of the front is simply given by the dispersion relation for the light wave in the plasma, $\omega_f^2 = \omega_p^2 + c^2 k_f^2$. As expected, the wavenumber changes as the source wave enters the plasma. Upon transforming back to the laboratory frame, the final frequency of the transmitted wave becomes $\gamma(\omega_f - v_o k_f)$, which is approximately $\omega_i(1 + \frac{1}{4}\omega_p^2/\omega_i^2)$.

There are several important points about this simple result. First, even though the front is "underdense" in the front's frame, we have a significant frequency upshift of the source wave. Second, when $\omega_p^2/\omega_i^2 \equiv n/n_c \gg 1$, the incident wave can still penetrate the front. Third, the frequency upshift scales simply linearly with the plasma density rather than as $\sqrt{n}$, as in the flash ionization case. Fourth, although not shown here, in certain limits the transmitted wave is actually propagating in the "reflected" direction with a significant frequency upshift. And finally, the upshifted pulse is compressed in duration by a factor ω/ω_o, i.e., the ratio of the final to incident frequencies, as it must to conserve the number of cycles of oscillation.

We note that, as in the case of the flash ionization, in the ionization front's case there are two additional modes: a reflected wave and a free streaming mode (static magnetic field). The reflected wave, as in the case of a reflection from a relativistically moving mirror, has a frequency shift of $4\gamma_o^2\omega_i$ but a negligible amount of power, unless the front is overdense, $\omega_p > \gamma(1+\beta)\omega_i$. This latter possibility was considered by Lampe et al[7]. some years ago.

The reflection and transmission coefficients for the underdense fronts have been calculated by Mori. Some outstanding theoretical problems are the two dimensional effects, ripples, and finite length fronts. Fortunately, with the advent of very short laser pulses it is now possible to generate a relativistically propagating ionization front that, in the laboratory frame at least, is typically 1 wavelength or less long for microwave source frequencies.

Experimentally, how might one realize the potential of an underdense ionization front to upshift the frequency of e.m. waves? One could use exactly the same setup discussed previously, but now accomplish the ionization using a few picosecond long u.v. laser pulses. The expressions for the upshifted frequency are now somewhat different, because we are now in a waveguide. Nevertheless, for plasma densities on the order 10^{15} cm^{-3}, it should be possible to upshift a 30 GHz source wave to the terrahertz frequency range and obtain pulses that are sub-nanosecond in duration. The use of ionization fronts to upshift the frequency could thus lead to a new class of tunable and coherent light sources with unprecedented capabilities.

Applications

Fig. 4 shows the capability of existing power sources in terms of peak and average power as a function of frequency. To see how plasma-based sources discussed in this paper compare with these existing sources, we have assumed the following set of parameters for a millimeter wave source, based on the plasma upshift technique (Table 1).

TABLE 1

Frequency Range	30 - 600 GHz
Conversion Efficiency	20% at 60 GHz
(30 GHz Source)	0.5% at 600 GHz
Peak Power	100 kw at 30 GHz
	500 w at 600 GHz
Rep Rate	10 KHz

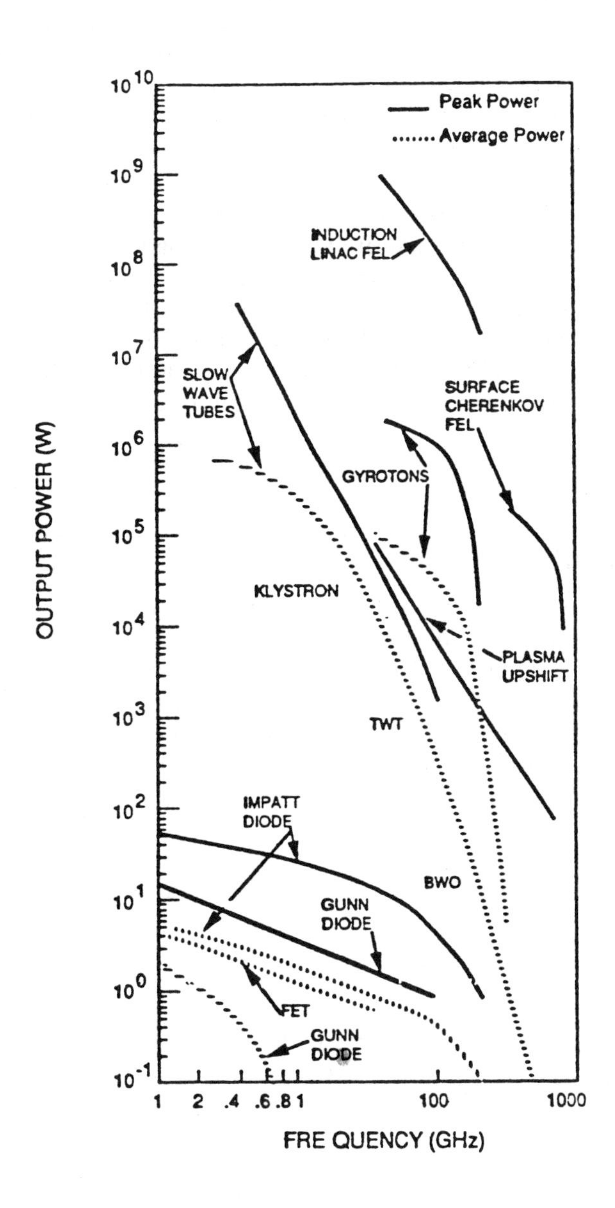

FIGURE 4 Microwave-millimeter wave source power capability summary.

The existing sources fall into the following broad categories: solid state (e.g. IMPATT and GUNN), vacuum electronic (gyrotrons, klystrons, magnetrons), and quantum (lasers). The solid state devices tend to be generally low power, whereas slow wave tubes dominate at higher powers. It is our view that plasma upshift devices will fall in between the two, but will offer a much broader tunability and chirpability in the millimeter wave and, perhaps, in the far infrared region of the electromagnetic spectrum. The dominant uses for such sources are in the areas of sources for advanced acceleration, plasma diagnostics, spectroscopy, radar, remote sensing, radiometry, and communications. Many of these applications require narrow band sources, but for many, availability of tunable and/or broad bandwidth sources having moderate power would represent a breakthrough. For instance, a source with tunability over a wide frequency range would allow measurement of large plasma density variations. Similarly, availability of short and/or chirped pulses would be a real boost in the development of impulse or crafted wave radar.

Acknowledgements

It is indeed a great pleasure to acknowledge Professor John Dawson's interest and contributions to this work. We are also grateful to Dr. Scott Wilks for his simulations, and Dr. Andrew Sessler of LBL for many stimulating interchanges on the topic of frequency upshift using plasmas. J. Slater of Spectra Technology has pointed out many potential applications of this technology. We thank C. Clayton, K. Marsh, and R. Savage Jr. for help with the experimental work.

References

1. F.R. Morgenthaler, *IRE Trans. Microwave Theory Tech.*, vol. MTT-6, p. 167, 1958.

   V. Goteti and D. Kalluri, *IEEE Trans. Plasma Sc.*, vol. 17, p. 828, 1987.

   W. Manheimer and B. Ripin, *Phys. Fluids* **29**(7), 2283 (1986).

   S.C. Wilks, J.M. Dawson and W.B. Mori, *Phys. Rev. Lett.*, vol. 61, p. 337, 1989.

2. P.B. Corkum, W.H. Burnett and F. Brunel, *Phys. Rev. Lett.* **62**, 1259 (1989).
3. S.C. Wilks, Ph.D. Thesis, UCLA (1988).
4. C. Joshi, C.E. Clayton, K. Marsh, D.B. Hopkins and A. Sessler, *IEEE Trans. Plasma Sc.*, vol. 18, no. 5, 814 (1990).
5. W.B. Mori, *to be published.*
6. M. Lampe and E. Orr, *Phys. Fluids* **21**, 42 (1978).

James Maniscalco
TRW Systems Group

A Tribute to John Dawson (Remarks at the Banquet Session of the International "Dawson" Symposium)

Good evening. I was asked by Tom to say a few words about how John came to TRW and what he has meant to us over the years, and I am delighted to do it because I was one of the many scientists and engineers at TRW who benefited from John's contributions to plasma physics.

John was introduced to TRW by Burt Fried, and he began consulting with us in the mid-'70s. This was an exciting time for plasma technology development at TRW, and I think all over the country. In the aftermath of the Arab oil embargo, energy research was enjoying unprecedented government support, and TRW formed an energy group under Johnny Foster, which was there to focus the company's focus on energy research.

John's first – and by far his largest – contribution to our plasma technology development, was the development of a plasma isotope separation process, which is given his name, "Dawson Separation Process". TRW spent $15,000 at that time, (yes, a whole $15,000!), to provide and perform a group of demonstrations on the

Taken from a recording of Jim Maniscalco's comments at the Banquet Session on Monday evening, September 24, 1990.

process which John defined. These would sufficiently convince the Department of Energy to pick up the development of the Dawson Process for uranium enrichment, and over the next 10 years the Department of Energy and TRW spent about $170 million for development.

The process was taken to a point where its feasibility for full-scale uranium enrichment was proven, and it is available should it ever be needed. The Dawson Process was also shown to be more economical than the existing calutron process for separating isotopes, and it can be used for almost any element in the periodic table. In my opinion, one of the most important successes of the Dawson Process has been the separation of enriched palladium.

This enriched palladium is irradiated and the radiation, of course, produces a radioactive palladium, which is used in the treatment of prostate cancer. Prostate cancer, for those of you who do not know, was the 3rd largest killer of men over the age of 45. Rather than performing risky surgery, tiny seeds of this enriched palladium are planted in the prostate tumor, and the radioactivity from the palladium kills the cancer cells. The Food and Drug Administration gave Theragram Corporation permission to commercially distribute the radioactive material, which they call "Therazize" in January of 1978, and that company determined that the Dawson Process module that is at TRW was the only source that could produce a large enough quantity of this at a cost that could make this treatment affordable.

In just four short months, TRW produced and delivered enough enriched palladium for about 20 to 50 thousand treatments of this. We just heard from Theragram Corporation that they are now doing something like 20 to 30 treatments a week, and the number is growing.

I know that John has always been motivated by the humanistic applications of plasma physics, and so I know that he is extremely proud of the Dawson Process that he invented and of the role it is playing in the treatment of this disease.

John, congratulations for something that has come together and to full fruition long before age 60.

I just want to finish by saying that all of us who have had the pleasure of working with you at TRW say "Thank you, John", and we all want to wish you a happy 60th birthday, with many more fruitful years to follow.

John and Nancy Dawson

Some Personal Comments (Remarks After the Banquet Honoring Professor Dawson)

On the evening of Monday, September 24, 1990, a banquet was held in honor of Professor John Dawson. Tributes to John were made by Dr. Jim Maniscalco on behalf of TRW (text contained in this volume) and by Professor Roberto Peccei on behalf of the UCLA Physics Department. Following a number of toasts, presentations, and reminiscences by members of the audience, John Dawson and his wife, Nancy, added their personal remarks on this occasion.

Nancy Dawson:

I'm not one to stand up in front of people, but I feel there is an irony here that I wish to point out. John came up with the Dawson Separation Process, which was used to separate Palladium 102 for treating prostate cancer after his illness. It was to prostate cancer that we thought we would lose him. But thanks to John's work, it kept him going and we have him here today.

John Dawson:

In thinking about this meeting, it reminded me of a drowning man and how his life passes in front of his face. Except that it is sort of like after you drowned you end up in heaven, and only the good things are there. All I can say is thank you, everybody, for being so kind and so generous.

A career in physics turned out differently than what I had expected when I was a young man. First of all, I had never thought I would be a world traveler, but I must say that that part has enriched me greatly. The time I spent in Japan was partially described by Professor Husimi this afternoon. He left out a few very important things: His hospitality; the extremely good way they treated me; the skiing trip at New Year's which is a long story that I will not describe. This led to a second skiing trip which is even more interesting, which I will also not describe except to say that I found out that my name was worth a 30% discount at a hotel. I think that was due to some exaggeration of my stature by some Japanese graduate students.

Frank Chen's pictures brought back some interesting memories of Japan. One was of a meeting in Japan. Al Simon was there, as perhaps you recall from the pictures. The morning after the dinner with the Geishas, Al was complaining about how his legs ached and were stiff. He thought he had picked up some horrible oriental disease. However, it was just from sitting on the floor for several hours during the party which he was not used to.

Chan Joshi mentioned my "photographic memory". My photographic memory goes back to this story. As Tom said, I grew up in Maryland and I got all my degrees there. I actually lived with my folks during that time and sometimes I drove an old station wagon that we had down to the university. Sometimes I would hitchhike and catch a ride. One day I went to the university and hitchhiked home. I looked over and the station wagon was gone, and I asked "What happened to the station wagon?" "Oh, you drove that to school!" So I guess the fact that I cannot remember my telephone number is not a sign of senility. At least, I hope not.

Well, this has been a particularly privileged occasion for me. First of all, my wife Nancy is here, and as everyone has said, I could not have done all of these things without the support that she has given me over all these years. She has put up with all of my idiosyncracies and all of the travel that I have done, and things like that. It is also a privilege to have my children and their spouses here, but it is a real privilege to have my parents here. To have your parents be able to attend an occasion like this, I think, is a rare privilege, and I really appreciate it.

Thank you to all who have inspired me, sparked me along, and contributed so much to me. But I think that my thanks really goes to all of my students who have come to work with me; perhaps I have also enriched them a little bit.

John Dawson
Department of Physics
University of California at Los Angeles 90024-1547

Resume

Born in Champaign, Illinois; September 30, 1930

Education

University of Maryland, B.S. in Physics, 1952
University of Maryland, M.S. in Physics, 1954
University of Maryland, Ph.D. in Physics, 1957

Positions Held Since PH.D.

1989-Present: Associate Director, Institute for Plasma and Fusion Research, University of California at Los Angeles

1973-Present: Professor of Physics, University of California at Los Angeles

1987-1989: Member of Institute for Plasma and Fusion Research, University of California at Los Angeles

1976-1987: Director, Center for Plasma Physics and Fusion Engineering, University of California at Los Angeles

1969-1971: Scientific Advisor to the Director of Research, Grade 18, Naval Research Laboratories

1966-1973: Head Theoretical Group, Plasma Physics Laboratory, Visiting Lecturer with Rank of Professor, Princeton University

1962-1973: Senior Research Physicist, Princeton University

1956-1962: Research Physicist, Plasma Physics Laboratory, Princeton University

Publication List

1954

1. Michels, A., Van Straaten, W., and Dawson, J.M., "Isotherms and Thermodynamical Functions of Ethane at Temperatures Between 0! C and 150! C and Pressures up to 200 ATM," *Physica*, **20**, 17 (1954).

1955

2. Jansen, L. and Dawson, J.M., "On Intermolecular Forces and Crystal Structures of Rare Gases," *J. Chem. Phys.*, **22**, 1619 (1954) and **23**, 482 (1955).

1956

3. Castle, B.J., Dawson, J.M., and Jansen, L., "Effect of Orientational Forces on Second Virial Coefficients," *J. Chem. Phys.*, **23**, 1733 (1955) and **24**, 1078 (1956).
4. Michels, A., Wassinaar, T., Wolkers, G.J., and Dawson, J.M., "Thermodynamic Properties of Xenon as a Function of Density Up to 520 Amagat and as a Function of Pressure up to 2800 Atmospheres, at Temperatures Between 0! C and 150! C," *Physica*, **22**, 17 (1956).

1958

5. Berger, J.M., Newcomb, W.A., Dawson, J.M., Frieman, E.A., Kulsrud, R.M., and Lenard, A., "Heating of a Confined Plasma by Oscillating Electromagnetic Fields," *Phys. Fluids*, **1**, 301 (1958).

1959

6. Dawson, J.M., "Nonlinear Electron Oscillations in a Cold Plasma," *Phys. Rev.*, **113**, 383 (1959).
7. Dawson, J.M. and Oberman, C., "Oscillations of a Finite Cold Plasma in a Strong Magnetic Field," *Phys. Fluids*, **2**, 103 (1959).

1960

8. Dawson, J.M., "Plasma Oscillations of a Large Number of Electron Beams," *Phys. Rev.*, **118**, 381 (1960).
9. Dawson, J.M., Hsi, C.G., and Shanny, R., "Some Investigations of Nonlinear Plasma Behavior in One-Dimensional Plasma Models," *Plasma Physics and Controlled Nuclear Fusion Research* (International Atomic Energy Agency, Vienna) **1**, 735 (1960).

1961

10. Dawson, J.M., "On Landau Damping," *Phys. Fluids*, **4**, 869 (1961).

1962

11. Dawson, J.M., "A One-Dimensional Plasma Model," *Phys. Fluids*, **5**, 445 (1962).
12. Dawson, J.M., "Investigation of the Double-Stream Instability," *Nuc. Fusion*, **3**, 1033 (1962).
13. Dawson, J.M. and Oberman, C., "High-Frequency Conductivity and the Emission and Absorption Coefficients of a Fully Ionized Plasma," *Phys. Fluids*, **5**, 517 (1962).
14. Oberman, C., Ron, A., and Dawson, J.M., "High-Frequency Conductivity of a Fully Ionized Plasma," *Phys. Fluids*, **5**, 1514 (1962).

1963

15. Dawson, J.M. and Oberman, C., "Effect of Ion Correlations on High-Frequency Plasma Conductivity," *Phys. Fluids*, **6**, 394 (1963).
16. Ron, A., Dawson, J.M., and Oberman, C., "Influence of Collisions on Scattering of Electromagnetic Waves by Plasma Fluctuations," *Phys. Rev.*, **132**, 497 (1963).
17. Dawson, J.M., Oberman, C., and Kulsrud, R.M., "Breakdown of Helium in a Stellarator by Large Amplitude Magnetic Pumping," *Nuc. Fusion*, **3**, 265 (1963).

1964

18. Oberman, C. and Dawson, J.M., "Plasma Stability Criteria from Conservation Laws," *Phys. Fluids*, **7**, 773 (1964).
19. Dawson, J.M., "Thermal Relaxation in a One-Species, One-Dimensional Plasma," *Phys. Fluids*, **7**, 419 (1964).
20. Dawson, J.M., "On the Production of Plasma by Giant Pulse Lasers," *Phys. Fluids*, **7**, 981 (1964).
21. Guerney, R., Oberman, C., Dawson, J.M., and Ron, A., "Comments on 'High-Frequency Conductivity of a Fully Ionized Plasma'," *Phys. Fluids*, **7**, 921 (1964).

1965

22. Balazs, N.L. and Dawson, J.M., "On Thermodynamic Equilibrium in a Gravitational Field," *Physica*, **31**, 222 (1965).
23. Birmingham, T., Dawson, J.M., and Oberman, C., "Radiation Processes in Plasmas," *Phys. Fluids*, **8**, 297 (1965).
24. Kuckes, A.F. and Dawson, J.M., "Electron Cyclotron Harmonic Radiation from a Plasma," *Phys. Fluids*, **8**, 1007 (1965).
25. Dawson, J.M. and Uman, M.F., "Heating a Plasma by Means of Magnetic Pumping," *Nuc. Fusion*, **5**, 242 (1965).

1966

26. Dawson, J.M. and Nakayama, T., "Kinetic Structure of a Plasma," *Phys. Fluids*, **9**, 252 (1966).

27. Tidman, D.A., Birmingham, T.J., Dawson, J.M. and Nakayama, T., "Comments on Kinetic Structure of a Plasma," *Phys. Fluids*, **9**, 1881 (1966).

28. Birmingham, T.J., Dawson, J.M., and Kulsrud, R.M., "Contribution of Electron-Electron Collisions to the Emission of Bremsstrahlung by a Plasma," *Phys. Fluids*, **9**, 2014 (1966).

29. Kofoid, M.J. and Dawson, J.M., "Anomalous Skin Depth in a Gaseous Plasma," *Phys. Rev. Lett.*, **17**, 1086 (1966).

1967

30. Nakayama, T. and Dawson, J.M., "Derivations of Hierarchies for N-Particle Systems and Vlasov Systems by Means of the Functional Calculus," *J. Math. Phys.*, **8**, 553 (1967).

31. Shanny, R., Dawson, J.M., and Greene, J.M., "One-Dimensional Model of a Lorentz Plasma," *Phys. Fluids*, **10**, 1281 (1967).

32. Dawson, J.M., "Electrical Conductivity of a Multi-Ion Species Plasma," *Phys. Fluids*, **10**, 2280 (1967).

33. Dawson, J.M., "Investigations of Nonlinear Behavior in One-Dimensional Plasma Model," *Proceedings of the Symposium on Computer Simulation of Plasma and Many-Body Problems*, Williamsburg, Virginia, April, 1967, pp. 25-30 (1967).

34. Langdon, B. and J.M. Dawson, "Investigation of a Sheet Model for a Bounded Plasma with Magnetic Field and Radiation," *Proceedings of the Symposium on Computer Simulation of Plasma and Many-Body Problems*, Williamsburg, Virginia, April, 1967, pp. 39-40 (1967).

1968

35. Dawson, J.M. and Shanny, R., "Some Investigation of Nonlinear Behavior in a One-Dimensional Plasma," *Phys. Fluids*, **11**, 1506 (1968).

36. Winsor, N., Johnson, J.L., and Dawson, J.M., "Geodesic Acoustic Waves in Hydromagnetic Systems," *Phys. Fluids*, **11**, 2448 (11968).

37. Dawson, J.M., Shanny, R., and Kruer, W.L., "Investigations of Electrostatic Oscillations of a Non-Maxwellian Plasma", *Proceedings of the APS Topical*

Conference on Numerical Simulation of Plasma, Los Alamos, New Mexico, September 18-20, paper A-2 (1968).

38. Dawson, J.M., "Radiation from Plasmas," *Advances in Plasma Physics*, A. Simon and W.B. Thompson, Eds. (Interscience Publishers, New York) (1968).

1969

39. Dawson, J.M., Shanny, R., and Birmingham, T.J., "Collisional Absorption and Emission of Longitudinal Waves in a One-Dimensional Plasma," *Phys. Fluids*, **12**, 687 (1969).

40. Dawson, J.M., Kaw, P., and Green, B., "Optical Absorption and Expansion of Laser-Produced Plasmas," *Phys. Fluids*, **12**, 875 (1969).

41. Kaw, P.K. and Dawson, J.M., "Laser-Induced Anomalous Heating of a Plasma," *Phys. Fluids*, 12, 2586 (1969).

42. Kulsrud, R.M., Oberman, C., Dawson, J.M., and Rosenbluth, M.N., "Comments on 'Enhanced Bremsstrahlung from Supraluminous and Subluminous Waves in a Isotropic, Homogeneous Plasma'," *Phys. Fluids*, **12**, 1957 (1969).

43. Shanny, R., Dawson, J.M., and Greene, J.M., "Comments on Collisional Damping of Plasma Oscillations," *Phys. Fluids*, **12**, 2227 (1969).

44. Kruer, W.L., Dawson, J.M., and Sudan, R.M., "Trapped-Particle Instability," *Phys. Rev. Lett.*, **23**, 838 (1969).

45. Dawson, J.M., Hsi, C.G., and Shanny, R., "Some Investigations of Nonlinear Plasma Behavior in One-Dimensional Plasma Models," *Third Conference on Plasma Physics and Controlled Nuclear Fusion Research*, Novosibersk, U.S.S.R., August, 1968, in Plasma Physics and Controlled Nuclear Fusion Research, IAEA, Vienna, 1, 735 (1969).

1970

46. Kaw, P.K. and Dawson, J.M., "Relativistic Nonlinear Propagation on Laser Beams in Cold Overdense Plasmas," *Phys. Fluids*, **13**, 472 (1970).

47. Ramanatham, G.V., Dawson, J.M., and Kruskal, M.D., "Entropy Principle for the Derivation of a New Cluster Expansion," *J. Math. Phys.*, **11**, 339 (1970).

48. Boris, J.P., Dawson, J.M., Orens, J.H., and Roberts, K.V., "Computation on Anomalous Resistance," *Phys. Rev. Lett.*, **25**, 706 (1970).

49. Kruer, W.L., Kaw, P.K., Dawson, J.M., and Oberman, C., "Anomalous High-Frequency Resistivity and Heating of a Plasma," *Phys. Rev. Lett.*, **24**, 987 (1970).

50. Winsor, N.K., Johnson, J.L., and Dawson, J.M., "A Numerical Model for Toroidal Plasma Containment with Flow," *J. Comp. Phys.*, **6**, 430 (1970).

51. Kruer, W.L. and Dawson, J.M., "Sideband Instability," *Phys. Fluids*, **13**, 2747 (1970).

52. Dawson, J.M., Papadopoulos, K., and Shanny, R., "Comments on Collisionless Electrostatic Shocks," *Phys. Fluids*, **13**, 1650 (1970).

53. Kaw, P., Valeo, E., and Dawson, J.M., "Interpretation of an Experiment on the Anomalous Absorption of an Electromagnetic Wave in a Plasma," *Phys. Rev. Lett.*, **25**, 430 (1970).

54. Kruer, W.L. and Dawson, J.M., "Anomalous Damping of Large-Amplitude Electron Plasma Oscillations," *Phys. Rev. Lett.*, **25**, 174 (1970).

55. Haber, I., Wagner, C.E., Boris, J.P., and Dawson, J.M., "A Self-Consistent Electromagnetic Particle Code," *Proceedings of the Fourth Conference on Numerical Simulation of Plasmas*, edited by J.P. Boris and R. Shanny (Office of Naval Research, Department of the Navy), Arlington, Virginia (1970), p. 126.

56. Dawson, J.M., Kaw, P., Kruer, W.L., Oberman, C., and Valeo, E., *Proceedings of the International Conference on Laser-Produced Plasmas*, Moscow, U.S.S.R., November (1970).

57. Dawson, J.M., Winsor, N.K., Bowers, E.C., and Johnson, J.L., "Bulk Viscosity, Magnetic Field Corrugation and Containment in Toroidal Configurations," *IV European Conference on Controlled Fusion and Plasma Physics*, Rome, Italy, August 31-September 4, p. 9 (1970).

58. Rosen, B., Kruer, W.L., and Dawson, J.M., "A New Version of the Dipole Expansion Scheme," *Proceedings of the Fourth Conference on Numerical Simulation of Plasmas*, J.P. Boris and R. Shanny, Eds. (Office of Naval Research, Dept. of the Navy, Arlington, Virginia) 561 (1970).

59. Dawson, J.M., "The Electrostatic Sheet Model for a Plasma and Its Modification to Finite Size Particles," *Method in Computational Physics*, B. Alder, S. Fernbach and M. Rotenberg, Eds. (Academic Press, New York) Vol. 9, p. 1 (1970).

1971

60. Kaw, P.K. and Dawson, J.M., "External Control of Ion Waves in a Plasma by High-Frequency Fields," *Phys. Fluids*, **14**, 792 (1971).

61. Kruer, W.L. and Dawson, J.M., "Anomalous Heating of Plasma Electrons Driven by a Large Transverse Field at wpe," *Phys. Fluids*, **14**, 1003 (1971).

62. Dawson, J.M., Furth, H.P., and Tenney, F.H., "Production of Thermonuclear Power by Non-Maxwellian Ions in a Closed Magnetic Field Configuration," *Phys. Rev. Lett.*, **26**, 1156 (1971).

63. Papadopoulos, K., Davidson, R.C., Dawson, J.M., Haber, I., Hammer, D.A., Krall, N.A., and Shanny, R., "Heating of Counterstreaming Ion Beams in an External Magnetic Field," *Phys. Fluids*, **14**, 849 (1971).

64. Dawson, J.M., Okuda, H., and Carlile, R.N., "Numerical Simulation of Plasma Diffusion Across a Magnetic Field in Two Dimensions," *Phys. Rev. Lett.*, **27**, 491 (1971).

65. Dawson, J.M., "Computer Simulation of Plasmas," *Astrophys. and Space Sci.*, **13**, 446 (1971).

66. Kaw, P.K., Dawson, J.M., Kruer, W.L., Oberman, C., and Valeo, E., "Anomalous Heating of Plasma by Laser Irradiation," *Kwantovnaya Elektronika*, **3**, 3 (1971) (in Russian).

67. Stamper, J.A., Papadopoulos, K., Sudan, R.N., Dean, S.O., McLean, E.A., and Dawson, J.M., "Spontaneous Magnetic Fields in Laser-Produced Plasmas," *Phys. Rev. Lett.*, **26**, 1012 (1971).

68. Dawson, J.M., Hertzberg, A., Kidder, R.E., Vlases, G.C., Ahlstrom, H.G., and Steinhauer, L.C., "Long-Wavelength, High-Powered Lasers for Controlled Thermonuclear Fusion," *Plasma Physics and Controlled Nuclear Fusion Research* (International Atomic Energy Agency, Vienna) **1**, 673 (1971).

69. 69. Winsor, N. K., Bowers, E. C., Hellberg, M. A., and J.M. Dawson, "Rotating Toroidal Equilibria and Shocks", Plasma Physics and Controlled Nuclear Fusion Research (International Atomic Energy Agency, Vienna) 11, 393 (1971).

70. Dawson, J.M., Kruer, W.L., Hertzberg, A., Vlases, G.C., Ahlstrom, H.G., Steinhauer, L.C., and Kidder, R.E., "Controlled Fusion Using Long-Wavelength Laser Heating with Magnetic Confinement," *Fundamental and Applied Laser Physics, Proc. of the Esfahan Symposium*, August 29-September 5 (1971).

71. Birdsall, C.K. and Dawson, J.M., "Plasma Physics," *Computers and Their Role in the Physical Science*, Chapter 13, S. Fernbach and A. Taub, Eds. (Gordon and Breach, New York) p. 247 (1971).

72. Dawson, J.M., "Thermokinetic Expansion Theory," *Laser Interaction*, H.J. Schwarz and H. Hora, Eds. (Plenum Press, New York) p. 355 (1971).

73. Dawson, J.M., Kruer, W.L., and Rosen, B., "Investigation of Ion Waves," *Dynamics of Ionized Gases*, Tokyo, Japan, September 13-17, 1971, M.J. Lighthill, I. Imai and H. Sato, Eds. (University of Tokyo Press) p. 47.

1972

74. Kruer, W.L. and Dawson, J.M., "Anomalous High-Frequency Resistivity of a Plasma," *Phys. Fluids*, **15**, 446 (1972).

75. Kawabe, T., Kawai, Y., Saka, O., Nakamura, Y., and Dawson, J.M., "Effects of Electron-Neutral Collisions on Propagation of Electron Plasma Waves," *Phys. Rev. Lett.*, **28**, 889 (1972).

76. Hellberg, M.A., Winsor, N.K., and Dawson, J.M., "Effect of Rotational Transform on Toroidal Confinement," *Phys. Rev. Lett.*, **28**, 1022 (1972).

77. Dawson, J.M., "Particle Simulation of Plasmas," *Comments on Plasma Phys. and Cont. Fusion*, **1**, Vol. 51, (1972).

78. Okuda, H. and Dawson, J.M., "Numerical Simulation on Plasma Diffusion in Three Dimensions," *Phys. Rev. Lett.*, **28**, 1625 (1972).

79. Okuda, H., Dawson, J.M., and Hooke, W.M., "Interpretation of an Enhanced Diffusion Observed in a Solid-State Plasma," *Phys. Rev. Lett.*, **29**, 1658, (1972).

80. Kendel, J.M., Okuda, H., and Dawson, J.M., "Parametric Instabilities and Anomalous Heating of Plasmas Near the Lower Hybrid Frequency," *Phys. Rev. Lett.*, **29**, 995 (1972).

81. Chu, T.K., Hendel, H., and Dawson, J.M., "Laboratory Parametric Instability Experiments at a Plasma Frequency," *Comments on Plasma Phys. and Cont. Fusion*, 111, Vol. 4 (1972).

82. Kainer, S., Dawson, J.M., and Coffey, T., "Alternating Current Instability Produced by the Two-Stream Instability," *Phys. Fluids*, **15**, 2419, (1972).

83. Kainer, S., Dawson, J.M., Shanny, R., and Coffey, T., "Interaction of a Highly Energetic Electron Beam with a Dense Plasma," *Phys. Fluids*, **15**, 493 (1972).

84. Dawson, J.M., "Contributions of Computer Simulation to Plasma Theory," *Proceedings of the First European Conference on Computational Physics*, Geneva, Switzerland, April 10-14 (1972); also in *The Impact of Computers on Physics*, G.R. MacLeod, Ed. (North Holland, Amsterdam) p. 79 (1972).

85. Dawson, J.M., "Anomalous Absorption of Radiation by Plasmas," *Proc. of the U.S.-Japan Seminar on Laser Interaction with Matter*, September 24-29,

1972, Kyoto, Japan, C. Yamanaka, Ed. (Japan Society for the Promotion of Science), p. 235.

1973

86. Okuda, H., and Dawson, J.M., "Theory and Numerical Simulation on Plasma Diffusion Across a Magnetic Field," *Phys. Fluids*, **16**, 408 (1973).
87. Kruer, W.L., Dawson, J.M., and Rosen, B., "The Dipole Expansion Method for Plasma Simulation," *J. Comp. Phys.*, **13**, 114 (1973).
88. Okuda, H., and Dawson, J.M., "Interpretation of Plasma Confinement Minima Near the Resonance Surface in a Stellarator," *Phys. Fluids*, **16**, 2336 (1973).
89. Kaw, P.K., Lin, A.T., and Dawson, J.M., "Quasiresonant Mode Coupling of Electron Plasma Waves," *Phys. Fluids*, **16**, 1967 (1973).
90. Okuda, H., and Dawson, J. M., "Effect of Magnetic Shear on Convective Plasma Transport," *Phys. Fluids*, **16**, 1456 (1973).
91. Lin, A.T., Kaw, P.K., and Dawson, J.M., "A Possible Plasma Laser," *Phys. Rev.*, **8**, 2618 (1973).
92. Johnston, T.W. and Dawson, J.M., "Correct Values for High-Frequency Power Absorption by Inverse Bremsstrahlung in Plasmas," *Phys. Fluids*, **16**, 772 (1973).
93. Kainer, S., Dawson, J.M., and Coffey, T., "Erratum: Alternating Current Instability Produced by the Two-Stream Instability," *Phys. Fluids*, **16**, 1382, 1973.

1974

94. Lin, A.T. and Dawson, J.M., "Enhancement of Plasma Direct Currents by Intense Alternating Current Fields," *Phys. Fluids*, **17**, 987 (1974).
95. Wong, A.Y., Dawson, J.M., Gekelman, W., and Lucky, Z., "Production of Negative Ions and Generation of Intense Neutral Beams by Laser Irradiation," *Appl. Phys. Lett.*, **25**, 579 (1974).
96. Vlases, G., Rutkowski, H., Hertzberg, A., Hoffman, A., Steinbauer, L., Dawson, J.M., Cohn, D.R., Halverson, W., Lax, B., Daugherty, J.D., Eninger, J.E., Pugh, E.R., Chu, T.K., Johnson, J.C., and Lovelace, R.V., "Progress on CO2 Laser Gas Fusion," *Fifth Conference on Plasma Physics and Controlled Nuclear Fusion Research*, Tokyo, November (1974).

97. Canosa, J., Krommes, J., Oberman, C., Okuda, H., Tsang, K., Dawson, J.M., and Kamimura, T., "Theory and Numerical Simulation of Collective Transport of Plasma in Magnetic Fields," *Fifth Conference on Plasma Physics and Controlled Nuclear Fusion Research*, Tokyo, November (1974).

98. Lin, A.T., Dawson, J.M., and Okuda, H., "Application of Electromagnetic Particle Simulation to the Generation of Electromagnetic Radiation," *Phys. Fluids*, **17**, 1995 (1974).

1975

99. Lin, A.T. and Dawson, J.M., "Stimulated Compton Scattering of Electromagnetic Waves in Plasma," *Phys. Fluids*, **18**, 201 (1975).

100. Lin, A.T. and Dawson, J.M., "Scattering of Electromagnetic Waves into Plasma Oscillations Via Plasma Particles," *Phys. Fluids*, **18**, 1542 (1975).

101. Wong, A.Y., Nakamura, Y., Quon, B.H., and Dawson, J.M., "Surface Magnetic Confinement," *Phys. Rev. Lett.*, **35**, 156 (1975).

102. Chu, C., Dawson, J.M., and Okuda, H., "Anomalous Electron Transport and Lower-Hybrid Wave Damping," *Phys. Fluids*, **18**, 1762 (1975).

103. Okuda, H., Chu, C., and Dawson, J.M., "Turbulent Damping of the Convective Cells and the Lower Hybrid Waves," *Phys. Fluids*, **18**, 243 (1975).

1976

104. Kamimura, T. and Dawson, J.M., "Effect of Mirroring on Convective Transport in Plasmas," *Phys. Rev. Lett.*, **36**, 313 (1976).

105. Busnardo-Neto, J., Dawson, J.M., Kamimura, T. and Lin, A.T., "Ion-Cyclotron Resonance Heating of Plasma and Associated Longitudinal Cooling," *Phys. Rev. Lett.*, **36**, 28 (1976).

106. Chu, C., Dawson, J.M., and Okuda, H., "Plasma Heating at Frequencies Near the Lower Hybrid," *Phys. Fluids*, **19**, 981 (1976).

107. Dawson, J.M., Okuda, H., and Rosen, B., "Collective Transport in Plasmas," *Adv. Comp. Phys., Methods in Computational Physics*, **16**, 281 (1976).

108. Hershkowitz, N. and Dawson, J.M., "Fusion Reactor with Picket-Fence Walls," *Nuclear Fus.*, **16**, 4, 639 (1976).

109. Dawson, J.M., Kim, H.C., Arnush, D., Fried, B.D., Gould, R.W., Heflinger, L.O., Kennel, C.F., Romesser, T.E., Stenzel, R.L., Wong, A.Y., and Wuerker, R.F., "Isotope Separation in Plasmas by Use of Ion Cyclotron Resonance," *Phys. Rev. Lett.*, **37**, 1547 (1976).

1977

110. Lin, A.T. and Dawson, J.M., "Stimulated Compton Scattering of Electromagnetic Waves Off Plasma Ions," *Phys. Fluids*, **20**, 538 (1977).

111. Chen, F.F., Dawson, J.M., Fried, B.D., Furth, H.P., and Rosenbluth, M.N., "Comments on 'Generalized Criterion for Feasibility of Controlled Fusion and Its Application to Nonideal D-D Systems'," *J. Appl. Phys.*, **48**, 415 (1977).

112. Busnardo-Neto, J., Pritchett, P.L., Lin, A.T., and Dawson, J., "A Self-Consistent Magnetostatic Particle Code for Numerical Simulation of Plasmas," *J. Comp. Phys.*, **23**, 300 (1977).

113. Kwan, T., Dawson, J.M., and Lin, A.T., "The Free Electron Laser," *Phys. Fluids*, **20**, 581 (1977); also in *Plasma Physics, Nonlinear Theory and Experiments*, Hans Wilhelmsson, Ed., (Plenum Press, New York) p. 486 (1977).

114. Tajima, T., Mima, K., and Dawson, J.M., "The Alfven Ion-Cyclotron Instability: Its Physical Mechanism and Observation in Computer Simulation," *Phys. Rev. Lett.*, **39**, 201 (1977).

1978

115. Lin, A.T. and Dawson, J.M., "Computer Simulation of Current Penetration in a Plasma," *Phys. Fluids*, **21**, 109 (1978).

116. Tajima, T., Leboeuf, J.N., and Dawson, J.M., "Double-Layer Forward Shocks in a Magnetohydrodynamic Fluid," *Phys. Rev. Lett.*, **40**, 652 (1978).

117. Okuda, H., Dawson, J.M., Lin, A.T., and Lin, C.C., "Quasi-Neutral Particle Simulation Model with Application to Ion Wave Propagation," *Phys. Fluids*, **21**, 476 (1978).

118. Pritchett, P.L., and Dawson, J.M., "Phase Mixing in the Continuous Spectrum of Alfven Waves," *Phys. Fluids*, **21**, 516 (1978).

119. Kamimura, T., Wagner, T., and Dawson, J.M., "Simulation Study of Bernstein Modes," *Phys. Fluids*, **21**, 1151 (1978).

120. Dawson, J.M., Huff, R.H., and Wu, C.C., "Plasma Simulation on the UCLA CHI Computer System," *AFIPS Conf. Proc., National Computer Conference*, **47**, 395 (1978).

121. Conn, R.W., Arnush, D., Dawson, J.M., Kerst, D.W., Vanek, V., "Aspects of Octopoles as Advanced Fuel Cycle Fusion Reactors," *International Atomic Energy Agency Publication on Fusion Reactor Design Concepts*, IAEA-TC-145/47, 721, Vienna, Austria (1978).

122. Pritchett, P.L., Wu, C.C., and Dawson, J.M., "Interchange Instabilities in a Compressible Plasma," *Phys. Fluids*, **21**, 1543 (1978).

123. Lin, A.T., Dawson, J.M., and Okuda, H., "Thermal Magnetic Fluctuations and Anomalous Electron Diffusion," *Phys. Rev. Lett.*, **41**, 753 (1978).

124. Forrester, A.T. and Dawson, J.M., "Neutral Beam Line Improvements Resulting from a Reduction of Gas Flow," *Plasma Science*, **PS-6**, 574 (1978).

125. Leboeuf, J.N., Tajima, T., Kennel, C.F., and Dawson, J.M., "Global Simulation of the Time-Dependent Magnetosphere," *Geophys. Res. Lett.*, **5**, 609, (1978).

126. Dawson, J.M., Kamimura, T., Naito, H., Huff, R.W., and Wu, C.C., "Computer Simulation of Heat Transport on a 2-1/2-D Model and Investigations of Large-Scale Plasma Simulations on the CHI Computer," *Proc. of the 8th Conference on Numerical Simulation of Plasmas*, 111, 277 (1978).

1979

127. Naitou, H., Kamimura, T., and Dawson, J.M., "Kinetic Effects on the Convective Plasma Diffusion and the Heat Transport," *J. Phys. Soc. Japan*, **46**, 258 (1979).

128. Decyk, V.K., Dawson, J.M., and Morales, G.J., "Excitation of Lower Hybrid Waves in a Finite Plasma," *Phys. Fluids*, **22**, 507 (1979).

129. Decyk, V.K. and Dawson, J.M., "Computer Model for Bounded Plasma," *J. Comp. Phys.*, **30**, 407 (1979).

130. Leboeuf, J.N., Tajima, T., Dawson, J.M., "Enhanced Drag by Radiation for Runaway Electrons," *Phys. Rev. Lett.*, **43**, 1321 (1979).

131. Leboeuf, J.N., Tajima, T., and Dawson, J.M., "A Magnetohydrodynamic Particle Code for Fluid Simulation of Plasmas," *J. Comp. Phys.*, **311**, 379 (1979).

132. Okuda, H., Lin, A.T., Lin, C.C., and Dawson, J.M., "Splines and High Order Interpolations in Plasma Simulations," *Comp. Phys. Comm.*, **17**, 227 (1979).

133. Kwan, T., and Dawson, J.M., "Investigation of the Free Electron Laser with a Guide Magnetic Field," *Phys. Fluids*, **22**, 1089 (1979).

134. Leboeuf, J.N., Tajima, T., Kennel, C.F., and Dawson, J.M., "Global Magnetohydrodynamic Simulation of the Two-Dimensional Magnetosphere," *Geophys. Mono.* **21**, 536, 1979.

135. Lin, A.T. and Dawson, J.M., "High-Efficiency Free-Electron Laser," *Phys. Rev. Lett.*, **42**, 1670 (1979).

136. Tajima, T. and Dawson, J.M., "An Electron Accelerator Using a Laser," *IEEE Trans. Nucl. Sci.*, NS26, 4188 (1979).

137. Tajima, T. and Dawson, J.M., "Laser Electron Accelerator," *Phys. Rev. Lett.*, **43**, 267 (1979).

1980

138. Tajima, T., Leboeuf, J.N., and Dawson, J.M., "Magnetohydrodynamic Particle Code with Force Free Electrons for Fluid Simulations," *J. Comp. Phys.*, **38**, 237 (1980).

139. Leboeuf, J.N., Tajima, T., Dawson, J.M., and Lin, A.T., "Particle Simulations of Time-Varying X-Points," *Proc. Int. Conf. Plas. Phys.*, **1**, 65 (1980).

140. Brunel, F., Tajima, T., Leboeuf, J.N., Dawson, J.M., "Endloss from a High-Beta Plasma Column," *Phys. Rev. Lett.*, **44**, 1494 (1980).

141. Decyk, V.K., Morales, G.J., Dawson, J.M., "Simulation of Lower Hybrid Heating in a Nonuniform Plasma Slab," *Phys. Fluids*, **23**, 826 (1980).

142. Lin, A.T., and Dawson, J.M., "Nonlinear Saturation and Thermal Effects on the Free Electron Laser Using an Electromagnetic Pump," *Phys. Fluids*, **23**, 1224 (1980).

143. Tajima, T. and Dawson, J.M., "Ion Cyclotron Resonance Heating and the Alfven-Ion Cyclotron Instability," *Nucl. Fusion*, **20**, 129 (1980).

144. Wagner, J.S., Tajima, T., Kan, J.R., Leboeuf, J.N., Akasofu, S.-I., and Dawson, J.M., "V-Potential Double Layers and the Formation of Auroral Arcs," *Phys. Rev. Lett.*, **45**, 803 (1980).

145. Dawson, J.M., MacKenzie, K., "Heating of Tokamak Plasmas by Means of Energetic Ion Beams with Z ¿ 1," *Proc. of the 2nd Joint Grenoble-Varenna Int'l Symposium, September 1980*, p. 953, Vol. 1 (1980).

146. Decyk, V.K., Morales, G.J., and Dawson, J.M., "Computer Modelling of RF Heating," *Proc. of the 2nd Joint Grenoble-Varenna Int'l Symposium, September 1980*, p. 365 (1980).

147. Goede, H., Dawson, J.M., and MacKenzie, K.R., "Observation of Anomalous Penetration of Fields in a Collisionless Plasma," *Phys. Rev. Lett.*, **44**, 1066 (1980).

148. Ashour-Abdalla, M., Leboeuf, J.N., Dawson, J.M., and Kennel, C.F., "A Simulation Study of Cold Electron Heating by Loss Cone Instabilities," *Geophys. Res. Lett.*, **7**, 889 (1980).

149. Lin, A.T. and Dawson, J.M., "Particle Simulation of Free Electron Laser," *Phys. Quantum Elec.*, **7**, 555 (1980).

150. Lin, A.T., Dawson, J.M., and Okuda, H., "Cross-Field Electron Transport Due to Thermal Electromagnetic Fluctuations," *Phys. Fluids*, **23**, 1316 (1980).

1981

151. Tajima, T. and Dawson, J.M., "Laser Beat Accelerator," *IEEE Trans. on Nuclear Science*, NS-28, 3416 (1981).

152. Kindel, J.M., Lin, A.T., Dawson, J.M., and Martinez, R.M., "Nonlinear Effects of Ion Cyclotron Heating of Bounded Plasmas," *Phys. Fluids*, **24**, 498 (1981).

153. Tajima, T., Goldman, M.V., Leboeuf, J.N., and Dawson, J.M., "Breakup and Reconstitution of Langmuir Wavepackets," *Phys. Fluids*, **24**, 182 (1981).

154. Ashour-Abdalla, M., Leboeuf, J.N., Tajima, T., Dawson, J.M., and Kennel, C.F., "Ultrarelativistic Electromagnetic Pulses in Plasmas," *Phys. Rev. A.*, **23**, 1906 (1981).

155. Pritchett, P.L., Ashour-Abdalla, M., and Dawson, J.M., "Simulation of the Current-Driven Electrostatic Ion Cyclotron Instability," *Geophys. Res. Lett.*, **8**, 611 (1981).

156. Leboeuf, J.N., Tajima, T., Kennel, C.F., and Dawson, J.M., "Global Simulations of the Three-Dimensional Magnetosphere," *Geophys. Res. Lett.*, **8**, 257 (1981).

157. Liewer, P.C., Lin, A.T., and Dawson, J.M., "Theory of an Absolute Instability of a Finite-Length Free Electron Laser," *Phys. Rev. A.*, **23**, 3, 1251 (1981).

158. Liewer, P.C., Lin, A.T., Dawson, J.M., and Caponi, M., "Particle Simulations of Finite-Length Free Electron Laser," *Phys. Fluids*, **24**, 7, 1364 (1981).

159. Joshi, C., Tajima, T., Dawson, J.M., Baldis, H.A., and Ebrahim, N.A., "Forward Raman Instability and Electron Acceleration," *Phys. Rev. Lett.*, **47**, 1285 (1981).

160. Dawson, J.M., "Advanced Fusion Reactors," *Fusion*, Vol. 1, Part B, p. 453, Academic Press, Inc. (1981); also in *Nagoya Lectures in Plasma Physics and Controlled Fusion*, Tokai University Press, Y. Ichikawa and T. Kamimura, eds., p. 429 (1989).

161. Wagner, J.S., Kan, J.R., Akasofu, S.-I., Tajima T., Leboeuf, J.N., and Dawson, J.M., "A Simulation Study of V-Potential Double Layers Auroral Arc Deformations," *Physics of Auroral Arc Formation*, Geophys. Monograph Series, **25**, 304, (1981).

162.Brunel, F., Leboeuf, J.N., Tajima, T., Dawson, J.M., Makino, M., and Kamimura, T., "Magnetohydrodynamic Particle Code: Lax-Wendroff Algorithm with Finer Grid Interpolations," *J. Comp. Phys.*, **43**, 268 (1981).

163.Wu, C.C., Walker, R.J., and Dawson, J.M., "A Three Dimensional MHD Model of the Earth's Magnetosphere," *Geophys. Res. Lett.*, **8**, 523 (1981).

164.Dawson, J.M., "Simulation of Space Plasma Phenomenan," *Physics of Auroral Arc Formation*, Geophys. Monograph Series, **25**, 270 (1981).

165.Leboeuf, J.N., Tajima, T., and Dawson, J.M., "Magnetic X-Points, Islands Coalescence and Intense Plasma Heating," *Physics of Auroral Arc Formation*, Geophys. Monograph Series, **25**, 337 (1981).

1982

166.Brunel, F., Tajima, T., and Dawson, J.M. "Fast Magnetic Reconnection Processes," *Phys. Rev. Lett.*, **49**, 323 (1982).

167.Menyuk, C.R., Dawson, J.M., Decyk, V.K., Fried, B.D., and Morales, G.J., "Nonlinear Evolution of Obliquely Propagating Langmuir Waves," *Phys. Rev. Lett.*, **48**, 1104 (1982).

168.Humanic, D.G., Goede, H., and Dawson, J.M., "Expulsion of a Plasmoid from a Spatially Nonuniform Magnetic Field," *Phys. Fluids*, **25**, 271, (1982); Erratum, *Phys. Fl.*, **25**, 1292 (1982).

169.Dawson, J.M. and Kaw, P.K., "Current Maintenance in Tokamaks by Use of Sychrotron Radiation," *Phys. Rev. Lett.*, **48**, 1730 (1982).

170.Dawson, J.M., "Waves and Instabilities in a Magnetized Plasma," *Phys. Scr.* (Sweden), Vol. T2, no. 1, p. 20 (1982 International Conference on Plasma Physics, Goteborg, Sweden, 9-15 June 1982).

171.Leboeuf, J.N., Ashour-Abdalla, M., Tajima, T., Kennel, C.F., Coroniti, F.V., and Dawson, J.M., "Ultrarelativistic Waves in Oversense Electron-Positron Plasmas," *Phys. Rev. A.*, **25**, 1023 (1982).

172.Lebeouf, J.N., Tajima, T., and Dawson, J.M., "Dynamic Magnetic X-Points," *Phys. Fluids*, **25**, 784 (1982).

173.Lin, A.T., Lin, C.C., and Dawson, J.M., "Plasma Heating From Upper-Hybrid Mode Conversion in an Inhomogeneous Magnetic Field," *Phys. Fluids*, **25**, 646 (1982).

174.Lebeouf, J.N., Dawson, J.M., Ratliff, S.T., Rhodes, M., and Luhmann, Jr., N.C., "Inductive Ion Acceleration and Heating in Picket Fence Geometry: Theory and Simulations," *Phys. Fluids*, **25**, 2045 (1982).

175. Wang, L.T., Dawson, J.M., Lin, A.T., Menyuk, C.R., and Tajima, T., "Computer Simulation of Synchrotron Radiation in Two Dimensions," *J. Plasma Phys.*, **28**, 133 (1982).

176. Decyk, V.K., Morales, G.J.,and Dawson, J.M., "Simulation of RF Heating and Current Drive with Lower Hybrid Waves," *Proc. of 3rd Joint Varenna-Grenoble Int'l Symposium*, Vol. II, p. 517 (1982).

1983

177. Dawson, J.M., Leboeuf, J.N., Ratliff, S.T., Luhmann, Jr., N.C., and Rhodes, M., "Toroidal Confinement Experiment Featuring Direct Ohmic Heating of the Ions," *Nucl. Instru. and Methods Phys. Res.*, (Netherlands), Vol. 207, no. 1-2, p. 197 (1983).

178. Lembege, B., Ratliff, S.T., Dawson, J.M., and Ohsawa, Y., "Ion Heating and Acceleration by Strong Magnetosonic Waves," *Phys. Rev. Lett.*, **51**, 264 (1983).

179. Katsouleas, T. and Dawson, J.M., "Unlimited Electron Acceleration in Laser-Driven Plasma Waves," *Phys. Rev. Lett.*,**51**, 392 (1983).

180. Ratliff, S.T., Dawson, J.M., Leboeuf, J.N., Rhodes, M., and Luhmann, Jr., N.C., "Computer Simulations of the Toroidal Cusp Experiment," *Nucl. Fusion*, **23**, 987 (1983).

181. Pritchett, P.L. and Dawson, J.M., "Electromagnetic Radiation from Beam-Plasma Instabilities," *Phys. Fluids*, **26**, 1114 (1983).

182. Katsouleas, T., Joshi, C.Joshi, Mori, W., and Dawson, J.M., "Prospects of the Surfatron Laser Plasma Accelerators," *Proc. 12th Int'l Conf. on High Energy Accelerators*, ed. by F.T. Cole and R. Donaldson, p. 460, August 1983.

183. Katsouleas, T. and Dawson, J.M., "A Plasma Wave Accelerator – Surfatron I," *IEEE Trans. on Nucl. Sci.*, NS-30, 3241 (1983).

184. Mori, W., Joshi, C., and Dawson, J.M., "A Plasma Wave Accelerator – Surfatron II," *IEEE Trans. on Nucl. Sci.*, NS-30, 3244 (1983).

185. Dawson, J.M., Decyk, V.K., Huff, R.W., Jechart, I., Katsouleas, T., Leboeuf, J.N., Lembege, B., Martinez, R.M., Ohsawa, Y., and Ratliff, S.T., "Damping of Large-Amplitude Plasma Waves Propagating Perpendicular to the Magnetic Field," *Phys. Rev. Lett.*, **50**, 1455 (1983).

186. Ratliff, S.T., Dawson, J.M., and Leboeuf, J.N., "Instability of Streaming Electrons Confined by Surface Magnetic Fields," *Phys. Rev. Lett.*, **50**, 1990 (1983).

187.Goede, H., Humanic, D.G. and Dawson, J.M., "Simulations of Compressional Stabilization of the Interchange Mode," *Phys. Fluids*, **26**, 1812 (1983).

188.Dawson, J.M., "Particle Simulation of Plasmas," *Rev. of Mod. Phys.*, **55**, 403 (1983); also in *3rd Int'l School for Space Simulation*, Cepadues-Editions, France **189**, 17 (1987).

189.Dawson, J.M., Fried, B.D., Morales, G.J., Lin, A.T., Decyk, V.K., and Caplan, M., "Theory and Simulation of RF Heating and Current Drive," *9th International Conference on Plasma Physics and Controlled Nuclear Fusion Research*, IAEA-CN-41/V-9 (1983).

190.Decyk, V.K., Morales, G.J., and Dawson, J.M., "Velocity Redistribution in Lower Hybrid Current Drive," *Proc. of IAEA Technical Committee Meeting*, Culham Laboratory, April 18-21, 1983; p. 190.

191.Dawson, J.M., "Current Drive by (1) Injection of Moderately Heavy Ions at Mev/Nucleon; (2) Plasma Synchrotron Radiation," *Proc. of IAEA Technical Committee Meeting*, Culham Laboratory, April 18-21, 1983; p. 366.

1984

192.Dawson, J.M. and Lin, A.T., "Chapter 7.1-Particle Simulation," *Handbook of Plasma Physics*, Vol. 2, M.N. Rosenbluth and R.Z. Sagdeev, Eds. (Elsevier Science Publishers, North Holland) p. 555 (1984); also in *Basic Plasma Physics*, selected chapters, A.A. Galeev and R.N. Sudan, Eds. (North Holland Publishers), p. 461 (1989).

193.Rhodes, M., Dawson, J.M., Gao, P., Luhmann, Jr., N.C., Ratliff, S.T., and Leboeuf, J.N., "Hot-Ion Plasma in the Toroidal Cusp Experiment," *Proc. of 10th Int'l Conf. on Plasma Physics and Controlled Nuclear Fusion Research*, London, U.K., September 12-19, 1984.

194.Dawson, J.M., "Acceleration of Particles by Beat Wave Accelerator and Surfatron," *Conf. on Intersections between Particle and Nuclear Physics*, Steamboat Springs, p. 61 (1984).

195.Shukla, P.K. and Dawson, J.M., "Stimulated Compton Scattering of Hydromagnetic Waves in the Interstellar Medium," *Astrophys. J.*, **276**, 149 (1984).

196.Joshi, C., Mori, W.B., Katsouleas, T., Dawson, J.M., Kindel, J.M., and Forslund, D.W., "Ultrahigh Gradient Particle Acceleration by Intense Laser-Driven Plasma Density Waves," *Nature*, **311**, 525 (1984).

197.Leboeuf, J.N., Brunel, F., Tajima, T., Sakai, J., Wu, C.C., and Dawson, J.M., "Computer Modeling of Fast Collisionless Reconnection," *Geophys. Mono.*, **30**, 282 (1984).

198.Ohsawa, Y. and Dawson, J.M., "Simulation Studies on Stability of EBT," *Phys. Fluids*, **27**, 1491 (1984).

199.Ratliff, S.T. and Dawson, J.M., "Ion Skin Depth Scaling and Cusp Distortion in Picket Fence Geometry," *Phys. Fluids*, **27**, 1743 (1984).

200.Ohsawa, Y., and Dawson, J.M., "Simulation of Bumpy Torus with Hot-Ion Rings," *Phys. Fluids*, **27**, 2287 (1984).

201.Lembege, B., and Dawson, J.M., "Plasma Heating and Acceleration by Strong Magnetosonic Waves Propagating Obliquely to a Magnetostatic Field," *Phys. Rev. Lett.*, **53**, 1053 (1984).

202.Leboeuf, J.N., Ratliff, S.T., and Dawson, J.M., "Theoretical Study of Generalized Toroidal Cusp Configurations," *Nucl. Fusion*, **24**, 1269 (1984).

203.Lembege, B., and Dawson, J.M., "Plasma Heating and Acceleration by Strong Magnetosonic Waves Propagating Obliquely to a Magnetostatic Field," *Phys. Rev. Lett.*, **53**, 1053 (1984).

204.Antani, S., Dawson, J.M., Decyk, V.K., Fried, B.D., Leboeuf, J.N., Lin, T., Moralies, G.J., and Ratliff, S.T., "Fusion Theory and Computer Simulations at UCLA," *Proc. of 10th Int'l Conf. on Plasma Physics and Controlled Nuclear Fusion Research*, London, U.K., p. 655, September 12-19, 1984.

205.Katsouleas, T. and Dawson, J.M., "Comment on Unlimited Electron Acceleration in Laser-Driven Plasma Waves," *Phys. Rev. Lett.*, **53**, 1026 (1984).

206.Dawson, J.M., "Beat Wave and Surfatron Accelerator of Particles," *Proc. of Int'l Conf. on Plasma Physics*, Lausanne, Switzerland, p. 837, ed. by M.Q. Tran and R.J. Verbak, June 27-July 3, 1984.

1985

207.Ohsawa, Y. and Dawson, J.M., "Stability to the High-Frequency Hot Electron Interchange Mode," *J. Phys. Soc. of Japan*, **54**, 454 (1985).

208.Forslund, D.W., Kindel, J.M., Mori, W.B., Joshi, C., and Dawson, J.M., "Two-Dimensional Simulations of Single-Frequency and Beat-Wave Laser-Plasma Heating," *Phys. Rev. Lett.*, **54**, 558 (1985).

209.Chen, P., Dawson, J.M., Huff, R.W., and Katsouleas, T., "Acceleration of Electrons by the Interaction of a Bunched Electron Beam with a Plasma," *Phys. Rev. Lett.*, **54**, 693 (1985).

210.Mori, W.B., Joshi, C.,, Dawson, J.M., Lee, K., Forslund, D.W., and Kindel, J.M., "Studies of the Plasma Droplet Accelerator Scheme," *IEEE Trans. on Nucl. Sci.*, NS-32, 3503 (1985).

211. Dawson, J.M., "MHD Hall Term Model," *Proceedings of the US-Japan Workshop on Advanced Plasma Modeling*, Nagoya, Japan, September 24-27, 1985.

212. Ogino, T., Walker, R.J., Ashour-Abdalla, M., and Dawson, J.M., "An MHD Simulation of By-Dependent Magnetospheric Convection and Field-Aligned Currents During Northward IMF," *J. Geophys. Res.*, **90**, 10835 (1985).

213. Katsouleas, T., Joshi, C., Dawson, J.M., Chen, F.F., Clayton, C., Mori, W.B., Darrow, C., and Umstadter, D., "Plasma Accelerators," *Proc. Second Workshop on Laser Acceleration of Particles*, Malibu, CA, p. 63, Jan 7-18, 1985.

214. Chen, P. and Dawson, J.M., "The Plasma Wakefield Accelerator," *Proc. Second Workshop on Laser Acceleration of Particles*, Malibu, CA, p.201, Jan. 7-18, 1985.

215. Katsouleas, T., Dawson, J.M., Sultana, D., and Yan, Y.T., "A Side-Injected-Laser Plasma Accelerator," *IEEE Trans. on Nucl. Sci.*, NS-32, 3554 (1985).

216. Dawson, J.M., "The Future of Space Plasma Simulation," *Space Sci. Rev.*, **42**, 187 (1985).

217. Dawson, J.M., "Highlights of the Working Group on Plasma Accelerators," *Proc. Second Workshop on Laser Acceleration of Particles*, Malibu, CA, p. 55, Jan. 7-18, 1985.

218. Bingham, R., Mori, W., and Dawson, J.M., "Some Nonlinear Processes Relevant to the Beat Wave Accelerator," *Proc. Second Workshop on Laser Acceleration of Particles*, Malibu, CA, p. 138, Jan. 7-18, 1985.

1986

219. Dawson, J.M., "Computer Simulation of Plasma," *Proc. Scientific Advances in the Supercomputer Age Conference*, Cornel University, New York, p. 33, October 1986.

220. Chen, P., Su, J.J., Dawson, J.M., Bane, K.L.F., and Wilson, P.B., "Energy Transfer in the Plasma Wake Field Accelerator," *Phys. Rev. Lett.*, **56**, 1252 (1986).

221. Sentman, D.D., Leboeuf, J.N., Katsouleas, T., Huff, R.W., and Dawson, J.M., "Electrostatic Instabilities of Velocity-Space-Shell Distributions in Magnetized Plasmas," *Phys. Fluids*, **29**, 2569 (1986).

222. Ogino, T., Walker, R.J., Ashour-Abdalla, M., and Dawson, J.M., "An MHD Simulation of the Effects of the Interplanetary Magnetic Field By Component on the Interaction of the Solar Wind With the Earth's Magnetosphere During Southward Interplanetary Magnetic Field," *J. Geophys. Res.*, **91**, 10029 (1986).

223. Yan, Y.T. and Dawson, J.M., "ac Free-Electron Laser," *Phys. Rev. Lett.*, **57**, 1599 (1986).

224. Dawson, J.M., "Predicting the Magnetospheric Plasma of Weather," *Proc. of NASA/OAST Workshop on Space Tech. Plasma Issues in 2001*, p. 93, Sept. 24-26, 1986.

225. Lembege, B. and Dawson, J.M., "Self-Consistent Plasma Heating and Acceleration by Strong Magnetosonic Waves for q = 90°. Part 1: Basic Mechanisms," *Phys. Fluids*, **29**, 821 (1986).

226. Conn, R.W., Dawson, J.M., Logan, B.G., Boozer, A., Gordon, J.D., and Wagner, C., "The Microwave Tokamak," *Proc. of Specialists' Meeting on Tokamak Concept Innovations*, IAEA, Vienna, p. 589, January 13-17, 1986.

227. Decyk, V.K., Morales, G.J., Dawson, J.M., and Abe, H., "Radial Diffusion of Plasma Current Due to Secondary Emission of Electrostatic Waves by Tail Electrons," *13th European Conference on Controlled Fusion and Plasma Heating*, Schliersee, 358, Vol. II (1986).

1987

228. Lembege, B. and Dawson, J.M., "Self-Consistent Study of a Perpendicular Collisionless and Nonresistive Shock," *Phys. Fluids*, **30**, 1767 (1987).

229. Katsouleas, T., Wilks, S., Chen, P., Dawson, J.M., and Su, J.J., "Beam Loading in Plasma Accelerators," *Particle Accel.*, **22**, 81 (1987).

230. Chen, P., Su, J.J., Katsouleas, T., Wilks, S., and Dawson, J.M., "Plasma Focusing for High Energy Beams," *IEEE Trans. on Plasma Science*, PS-15, 218 (1987).

231. Lembege, B., and Dawson, J.M., "Plasma Heating Through a Supercritical Oblique Collisionless Shock," *Phys. Fluids*, **30**, 1110 (1987).

232. Dawson, J.M., and Decyk, V.K., "Particle Modeling of Plasmas on Super-Computers," *International Journal of Supercomputer Applications*, **1**, 24, 1987.

233. Jechart, I., Katsouleas, T., and Dawson, J.M., "Anomalous Thermal Relaxation of a Two–Dimensional Magnetized Plasma," *Phys. Fluids*, **30**, 65 (1987).

234. Dawson, J.M., "A Positron Factory," in *Advanced Accelerator Concepts*, Frederick E. Mills, ed., AIP Conf. Proc. No. 156, New York, 1987, p. 194; also in *Proc. of UCLA Workshop Linear-Collider BB Factory Conceptual Design*, ed. by D.H. Stork, p. 417, January 26-30, 1987.

235. Wilks, S., Dawson, J.M., and Katsouleas, T., "The Focusing of a Relativistic Electron Beam Using a Performed Ion Channel," submitted to *Phys. Rev. A.*, 1987.

236. Ohasawa, Y., Dawson, J.M., and Van Dam, J.W., "Stability of the Low-Frequency Hot-Electron Interchange Mode For a Bumpy Torus Simulation Model," *Phys. Fluids*, **30**, 3237 (1987).

237. Yan, Y.T., McKinstrie, C.J., Katsouleas, T., and Dawson, J.M., "Counter-Streaming Electron-Beam Beat-Wave Accelerator," *Phys. Rev. A.*, A36, 5455 (1987).

238. Su, J.J., Dawson, J.M., Katsouleas, T., Wilks, S., Chen, P., Jones, M., and Keinigs, R., "Stability of the Driving Bunch in the Plasma Wakefield Accelerator," *Proc. of 1987 IEEE Particle Accelerator Conference*, ed. E.R. Lindstrom and L.S. Taylor, Vol. 1, Washington, D.C., p. 127, March 1987.

239. Joshi, C., Katsouleas, T., Dawson, J.M., Yan, Y.T., and Chen, F.F., "Plasma Wave Wigglers for Free Electron Lasers," *Proc. of 1987 IEEE Particle Accelerator Conference*, ed. E.R. Lindstrom and L.S. Taylor, Vol. 1, Washington, D.C., p. 199, March 1987.

240. Wilks, S., Katsouleas, T., Dawson, J.M., and Su, J.J., "Beam Loading Efficiency in Plasma Accelerators," *Proc. of 1987 IEEE Particle Accelerator Conference*, ed. E.R. Lindstrom and L.S. Taylor, Vol. 1, Washington, D.C., p. 100, March 1987.

241. Katsouleas, T., Dawson, J.M., Mori, W.B., Su, J.J., and Wilks, S., "Theoretical Work on Plasma Accelerators at UCLA," *Proc. of New Developments in Particle Acceleration Techniques*, June 29-July 4, 1987, Orsay, France, Vol. II, p. 401.

242. Mori, W.B., Joshi, C., Dawson, J.M., Forslund, D.W., and Kindel, J.M., "Two Dimensional Simulations of Intense Laser Irradiation of Underdense Plasmas," *Laser Interaction and Related Plasma Phenomena*, Vol. 7, ed. H. Hora and G.H. Miley, Plenum Press, New York, 1987.

243. Dawson, J.M. and Kazeminezjad, F., "A Vlasov Ion Zero Mass Electron Model for Plasma Simulations," *Proc. of the US-Japan Workshop on Advanced Plasma Modeling III*, March 23-17, 1987, Nagoya, Japan.

244. Sydora, R.D., Decyk, V.K., and Dawson, J.M., "Modeling of Subtle Kinetic Processes in Plasma Simulation," *Proc. of the US-Japan Workshop on Advanced Plasma Modeling III*, March 23-17, 1987, Nagoya, Japan.

245. Chen, F.F., Clayton, C.E., Darrow, C., Dawson, J.M., Joshi, C., Katsouleas, T., Leemans, W., Marsh, K., Mori, W., Su, J.J., Umstadter, D., and Wilks, S., "Particle Acceleration by Plasma Waves," *Proc. of Int'l Conf. on Plasma Physics*, Kiev, USSR, p. 797, (1987).

1988

246. Liewer, P.C., Decyk, V.K., Dawson, J.M., and Fox, G.C., "Plasma Particle Simulations on the Mark III Hypercube," *Math. and Computer Modeling*, 11, 53 (1988).

247. Mori, W.B., Joshi, C., Dawson, J.M., Forslund, D.W., and Kindel, J.M., "The Evolution of Self-Focusing of Intense Electromagnetic Waves in Plasma," *Phys. Rev. Lett.*, **60**, 1298 (1988).

248. Bingham, R., Bryant, D.A., Hall, D.S., Dawson, J.M., Kazeminejad, F., Su, J.J., and Nairn, C.M.C., "AMPTE Observations and Simulation Results," *Comp. Phys. Comm.*, **49**, 257, 1988.

249. Sydora, R.D., Lee, W.W., Hahn, T.S., Dawson, J.M., Decyk, V.K., Dimits, A., Naitou, H., Kamimura, T., and Abe, Y., "Three Dimensional Gyrokinetic Plasma Simulation Models with Application to Study of Low Frequency Fluctuations in Tokamaks," submitted to *Proc. 12th ICPP*, Nice, France, October 1988.

250. Katsouleas, T., Su, J.J., and Dawson, J.M., "Plasma Lens Work at UCLA," *Proc. of European Particle Accelerator Conference*, Rome, Italy, June 7-11, 1988.

251. Wilks, S.C., Dawson, J.M., and Mori, W.B., "Frequency Up-Conversion of Electromagnetic Radiation with Use of an Overdense Plasma," *Phys. Rev. Lett.*, **61**, 337, 1988.

252. Katsouleas, T., Su, J.J., and Dawson, J.M., "Underdense Plasma Lenses for Focucing Particle Beams," *Proc. 1988 Linear Accelerator Conference*, CEBAF, Williamsburg, VA.

253. Tang, W.M., Bretz, N.L., Hahm, T.S., Lee, W.W., Perkins, F.W., Redi, M.H., Rewoldt, G., Zarnstorff, M.C., Zweben, S.J., Sydora, R.D., Dawson, J.M., Decyk, V.K., Naitou, H., Kamimura, T., and Abe, Y., "Theoretical Studies on Enhanced Confinement Properties in Tokamaks," *Plasma Physics and Controlled Nuclear Fusion Research* (IAEA; Nice, France, 1988)

254. Liewer, P.C., Decyk, V.K., Dawson, J.M., and Fox, G.C., "A Universal Concurrent Algorithm for Plasma Particle-In-Cell Simulation Codes," *Proc. of Third Conf. on Hypercube Concurrent Computers and Applications*, Pasadena, CA, January 1988; ed. Geoffrey Fox Assoc. for Computing Machinery, New York, 1988, vol. II, p. 1101.

1989

255. Lembege, B., and Dawson, J.M., "Formation of Double Layers Within an Oblique Collisionless Shock," *Phys. Rev. Lett.*, **62**, 2683 (1989).

256. Dawson, J.M., "Rosenbluth-Liu Amplitude Limit for the Beat Wave Accelerator," in *From Particles to Plasmas*, J.W. Van Dam, ed., Addison-Wesley Publishing Co, p. 131-140, 1989.

257. Katsouleas, T. and Dawson, J.M., "Plasma Acceleration of Particle Beams," *AIP Conf. Proceedings 184 Physics of Particle Accelerators*, ed. M. Month and M. Dienes, p. 1798.

258. Wilks, S.C., Dawson, J.M., Mori, W.B., Katsouleas, T., and Jones, M.E., "A Photon Accelerator," *Phys. Rev. Lett.*, **62**, 2600, 1989.

259. Liewer, P.C., Zimmerman, B.A., Decyk, V.K., and Dawson, J.M., "Application of Hypercube Computers to Plasma Particle-In-Cell Simulation Codes," *Proc. of 4th International Conference on Supercomputing*, Santa Clara, CA (1989).

260. Dawson, J.M., "Plasma Particle Accelerators," *Scientific American*, **260**, 54, 1989.

261. Katsouleas, T., Su, J.J., Joshi, C., Mori, W.B., Dawson, J.M., and Wilks, S., "A Compact 100 MeV Accelerator Based on Plasma Wakefields," *SPIE Conf. Proc. OE/LASE '89*, Vol. 1061, Los Angeles, CA, p. 428.

262. Lembege, B., and Dawson, J.M., "Dynamics of a Relativistic Electron-Positron Plasma in a Non-Linear Magnetosonic Wave," sub. to *Phys. Rev. Lett.*, March 1989.

263. Lembege, B., and Dawson, J.M., "Relativistic Particle Dynamics in a Steepening Magnetosonic Wave," *Phys. Fluids B*, **1**, 1001, (1989).

264. Dawson, J.M., Book Review of "The Physics of Laser Plasma Interactions" by William L. Kruer, *Physics Today*, August 1989, p. 69-70.

265. McClements, K.G., Bingham, R., Dawson, J.M. Dawson, Spicer, D.S., and Su, J.J., "Simulation Studies of Electron Acceleration by Ion Ring Distributions in Solar Flares," *Proc. of Workshop on Short Duration Radio Emission During Solar Flares*, Brannwald, Switzerland, August 21-25, 1989.

266. Liewer, P.C., Leaver, E.W., Decyk, V.K., and Dawson, J.M., "Concurrent PIC Codes and Dynamic Load Balancing on the JPL/CalTech Mark III Hypercube," *Proc. of 13th Conf. on Numerical Simulation of Plasmas*, IM2, Santa Fe, NM, September 17-20, 1989.

267. Huff, R.W., and Dawson, J.M., "Particle Simulation on the NCUBE Concurrent Multiprocessor," *Proc. of 13th Conf. on Numerical Simulation of Plasmas*, PMA8, Santa Fe, NM, September 17-20, 1989.

268. Katsouleas, T., Su, J.J., Wilks, S., Mori, W.B., and Dawson, J.M., "The Role of Plasmas in Future Accelerators," *Proc. of HEACC-89*, Tsukuba, Japan, August 22-26, 1989, to appear in *Particle Accelerators.*

269. Dawson, J.M., and Sessler, A., "DNA Base Pair Sequencer: Scanning Tunneling Microscope Plus Infrared Radiation," *Proc. of X-Ray Microimaging for the Life Sciences*, Univ. California Berkeley, May 24-26, 1989; p. 108-113.

270. Wilks, S.C., Katsouleas, T., and Dawson, J.M., "The Photon Accelerator: A Novel Method of Frequency Upshifting Sub-Picosecond Laser Pulses," *AIP Conference Proceedings 193, Advanced Accelerator Concepts*, ed. by C. Joshi, p. 448, 1989.

271. Dawson, J.M., "Summary of the Plasma Working Group," *AIP Conference Proceedings 193, Advanced Accelerator Concepts*, ed. by C. Joshi, p. 57, 1989.

1990

272. Whittum, D.H., Sessler, A.M., and Dawson, J.M., "An Ion-Channel Laser," *Phys. Rev. Lett.*, **64**, 2511 (1990).

273. Dawson, J.M., "Unconventional FEL's," *SPIE Conf. Proc. OE/LASE '90*, Los Angeles, CA, January 1990, Vol. 1227, p. 40.

274. Su, J.J., Katsouleas, T., Dawson, J.M., and Fedele, R., "Plasma Lenses for Focusing Particle Beams," *Phys. Rev. A.*, **41**, 3321 (1990).

275. Katsouleas, T., Su, J.J., Mori, W.B., and Dawson, J.M., "Plasma Physics at the Final Focus of High Energy Colliders," *Phys. Fluids B*, **2**, 1384 (1990).

276. Katsouleas, T., Mori, W.B., Dawson, J.M., and Wilks, S., "Physics of Plasmas with Short Pulse Lasers," *Proc. of SPIE OS/LASE T90*, Los Angeles, CA; January 14-19, 1990.

277. Chen, K.R., Katsouleas, T., and Dawson, J.M., "On the Amplification Mechanism of the Ion-Channel Laser," sub. to *Phys. Rev. Lett.*, (1990).

278. Dawson, J.M., Sydora, R.D., and Decyk, V.K., "Physics Modeling of Tokamak Transport: A Grand Challenge for Controlled Fusion," sub. to *Int'l J. Supercomputing Applications*, (1990).

279. Sydora, R.D., Hahm, T.S., Lee, W.W., and Dawson, J.M., "Fluctuations and Transport Due to Ion Temperature Gradient-Driven Instabilities," *Phys. Rev. Lett.*, **64**, 2015 (1990).

280. Liewer, P.C., Leaver, E.W., Decyk, V.K., and Dawson, J.M., "Dynamic Load Balancing in a Concurrent Plasma PIC Code on the JPL/Caltech Mark III Hypercube," *Proc. 5th Distributed Memory Computing Conf.*, Charleston, SC, April 1990.

281. Chen, K.R., Dawson, J.M., Lin, A.T., and Katsouleas, T., "Unified Theory and Comparative Study of CARMs, ICLs, and FELs," sub. to *Phys. Fluids B*, September 1990.

282. LoDestro, L.L., Cohen, B.I., Cohen, R.H., Dimits, A.M., Matsuda, Y., Nevins, W.M., Newcomb, W.A., Williams, T.J., Koniges, A.E., Dannevik, W.P., Crotinger, J.A., Amala, P.A.K., Sydora, R.D., Dawson, J.M., Ma, S., Decyk, V.K., Lee, W.W., Hahm, T.S., Naitou, N., and Kamimura, T., "Comparison of Simulations and Theory of Low-Frequency Plasma Turbulence," *Proc. of IAEA*, Washington, DC, October 1990.

283. Liewer, P.C., Decyk, V.K., Dawson, J.M., and Lembeg, B., "Numerical Studies of Electron Dynamics in Oblique Quasi-Perpendicular Collisionless Shock Waves," sub. to *J. Geophysical Research*, November 1990.

International Steering Committee

B. Bingham	Rutherford Appleton Lab
B. Coppi	MIT
B. Cohen	LBL
P. Drake	LLNL
R. Gould	Cal. Inst. of Technology
W. Horton	Univ. Texas at Austin
T. Kamimura	Japan Inst. Fusion Sci.
T. Katsouleas	UCLA
P. Kaw	Institute for Plasma Research, India
J. Kindel	LANL
W. Kruer	LLNL
R. Kulsrud	Princeton University
A.B. Langdon	LLNL
J.N. Leboeuf	ORNL
J. Lominadze	USSR Acad. of Sciences
C.S. Liu	Univ. of Maryland
J. Nuckolls	LLNL
R.S. Pease	Rutherford Appleton Lab
A. Sessler	LBL
P. Staudhammer	TRW
R. Sudan	Cornell University
H. Wilhelmsson	Chalmers University

Internal Organizing Committee

M. Abdalla
F.F. Chen
R.W. Conn
V.K. Decyk
J. Foster
B.D. Fried
C. Joshi
C.F. Kennel
W.B. Mori
C. Pellegrini
R. Peccei